Solar energy

The physics and engineering of
photovoltaic conversion,
technologies and systems

Solar energy

The physics and engineering of photovoltaic conversion, technologies and systems

Arno HM Smets
Klaus Jäger
Olindo Isabella
René ACMM van Swaaij
Miro Zeman

UIT CAMBRIDGE,
ENGLAND

Published by

UIT Cambridge Ltd
www.uit.co.uk

PO Box 145
Cambridge
CB4 1GQ
England
+44 (0) 1223 302 041

Copyright © 2016 UIT Cambridge Ltd.
All rights reserved.

Subject to statutory exception and to the provisions
of relevant collective licensing agreements, no part
of this book may be reproduced in any manner
without the prior written permission of the publisher.

First published in 2016, in England

Arno HM Smets, Klaus Jäger, Olindo Isabella,
René ACMM van Swaaij and Miro Zeman
have asserted their moral rights under the
Copyright, Designs and Patents Act 1988.

Cover design by Andrew Corbett
Cover images ©jokerpro production/BigStock and vionet/iStock

Typeset by the authors in DejaVu Sans
Condensed and URW Palatino with LaTeX

ISBN: 978 1 906860 32 5 (paperback)

Available as a free ebook:
ISBN: 978 1 906860 73 8 (ePub)
ISBN: 978 1 906860 75 2 (pdf)

Disclaimer: the advice herein is believed to be correct
at the time of printing, but the authors and publisher
accept no liability for actions inspired by this book.

10 9 8 7 6 5 4 3 2 1

Contents

Foreword — xi

Dean's message — xiii

Preface — xv

About this Book — xvii

Nomenclature — xix

I Introduction — 1

1 Energy — 3
- 1.1 Some definitions — 4
- 1.2 Human energy consumption — 5
- 1.3 Methods of energy conversion — 9
- 1.4 Exercises — 11

2 Status and prospects of PV technology — 13

3 The working principle of a solar cell — 21

II PV fundamentals — 25

4 Electrodynamic basics — 27
- 4.1 The electromagnetic theory — 27
- 4.2 Electromagnetic waves — 28
- 4.3 Optics of flat interfaces — 29
- 4.4 Optics in absorptive media — 31
- 4.5 Continuity and Poisson equations — 32
- 4.6 Exercises — 33

5 Solar radiation — 35
- 5.1 The Sun — 35
- 5.2 Radiometric properties — 37
- 5.3 Blackbody radiation — 39

5.4	Wave-particle duality	42
5.5	Solar spectra	42
5.6	Exercises	45

6 Basic semiconductor physics — 47
- 6.1 Introduction — 47
- 6.2 Atomic structure — 48
- 6.3 Doping — 50
- 6.4 Carrier concentrations — 52
- 6.5 Transport properties — 58
- 6.6 Exercises — 62

7 Generation and recombination of electron-hole pairs — 65
- 7.1 Introduction — 65
- 7.2 Bandgap-to-bandgap processes — 67
- 7.3 Shockley–Read–Hall recombination — 72
- 7.4 Auger recombination — 76
- 7.5 Surface recombination — 77
- 7.6 Carrier concentration in non-equilibrium — 79
- 7.7 Exercises — 80

8 Semiconductor junctions — 83
- 8.1 *p-n* homojunctions — 83
- 8.2 Heterojunctions — 97
- 8.3 Metal-semiconductor junctions — 100
- 8.4 Exercises — 105

9 Solar cell parameters and equivalent circuit — 111
- 9.1 External solar cell parameters — 111
- 9.2 The external quantum efficiency — 115
- 9.3 The equivalent circuit — 117
- 9.4 Exercises — 120

10 Losses and efficiency limits — 125
- 10.1 The thermodynamic limit — 125
- 10.2 The Shockley–Queisser limit — 128
- 10.3 Additional losses — 133
- 10.4 Design rules for solar cells — 136
- 10.5 Exercises — 144

III PV technology — 147

11 A short history of solar cells — 149

12 Crystalline silicon solar cells — 153
- 12.1 Crystalline silicon — 153
- 12.2 Production of silicon wafers — 158
- 12.3 Designing c-Si solar cells — 160
- 12.4 Fabricating c-Si solar cells — 166
- 12.5 High-efficiency concepts — 169
- 12.6 Exercises — 174

13 Thin-film solar cells — 177
- 13.1 Transparent conducting oxides — 178
- 13.2 The III-V PV technology — 180
- 13.3 Thin-film silicon technology — 188
- 13.4 Chalcogenide solar cells — 199
- 13.5 Organic photovoltaics — 205
- 13.6 Hybrid organic-inorganic solar cells — 208
- 13.7 Exercises — 212

14 A closer look to some processes — 217
- 14.1 Plasma-enhanced chemical vapour deposition — 217
- 14.2 Physical vapour deposition — 218
- 14.3 Screen printing technology — 220
- 14.4 Electroplating technology — 221

15 PV modules — 223
- 15.1 Series and parallel connections in PV modules — 223
- 15.2 PV module parameters — 226
- 15.3 Bypass diodes — 227
- 15.4 Fabrication of PV modules — 229
- 15.5 PV module lifetime testing — 231
- 15.6 Thin-film modules — 232
- 15.7 Some examples — 234
- 15.8 Concentrator photovoltaics (CPV) — 235
- 15.9 Exercises — 237

16 Third generation concepts — 239
- 16.1 Multi-junction solar cells — 240
- 16.2 Spectral conversion — 240
- 16.3 Multi-exciton generation — 244
- 16.4 Intermediate band solar cells — 246
- 16.5 Hot carrier solar cells — 246
- 16.6 Exercises — 247

IV PV systems — 253

17 Introduction to PV systems — 255
- 17.1 Introduction — 255

17.2 Types of PV systems 255
17.3 Components of a PV system 259
17.4 Exercise 261

18 Location issues   263
18.1 The position of the Sun 263
18.2 Irradiance on a PV module 269
18.3 Direct and diffuse irradiance 271
18.4 Exercise 274

19 Components of PV systems   275
19.1 Maximum power point tracking 275
19.2 Power electronics 281
19.3 Batteries 297
19.4 Charge controllers 306
19.5 Cables 309
19.6 Exercises 311

20 PV system design   317
20.1 A simple approach for designing stand-alone systems . . . 319
20.2 Load profiles 325
20.3 Meteorological effects 325
20.4 Designing grid-connected PV systems 333
20.5 Designing stand alone PV systems 341
20.6 Exercises 347

21 PV system economics and ecology   351
21.1 PV system economics 351
21.2 PV system ecology 357
21.3 Exercises 359

V Alternative solar energy conversion technologies   363

22 Solar thermal energy   365
22.1 Solar thermal basics 365
22.2 Solar thermal heating 368
22.3 Concentrated solar power (CSP) 376
22.4 Exercises 378

23 Solar fuels   381
23.1 Electrolysis of water 383
23.2 Photoelectrochemical (PEC) water splitting 384
23.3 Exercises 389

Appendix   **393**

A   Derivations in electrodynamics   **395**
    A.1   The Maxwell equations 395
    A.2   Derivation of the electromagnetic wave equation 396
    A.3   Properties of electromagnetic waves 397

B   Derivation of homojunction *J-V* curves   **401**
    B.1   The *J-V* characteristic in the dark 401
    B.2   *J-V* characteristic under illumination 405

C   Some aspects of surface recombination   **409**
    C.1   Infinite surface recombination velocity S 409
    C.2   Surface recombination velocity $S = 0$ 410
    C.3   Open circuit voltage for *p-n* junction solar cells 412

D   The morphology of selected TCO samples   **413**
    D.1   Surface parameters 413
    D.2   Examples 415

E   Some aspects on location issues   **417**
    E.1   The position of the Sun 417
    E.2   The equation of time 425
    E.3   Angle between the Sun and a PV module 426
    E.4   Modules mounted on a tilted roof 426
    E.5   Length of the shadow behind a PV module 429

F   Derivations for DC-DC converters   **433**
    F.1   Buck converter 433
    F.2   Boost converter 434
    F.3   Buck-boost converter 435

G   Fluid-dynamic model   **437**
    G.1   Framework of the fluid-dynamic model 437
    G.2   Convective heat transfer coefficients 439
    G.3   Other parameters 442
    G.4   Evaluation of the thermal model 443

Bibliography   **445**

Index   **453**

Foreword

The implementation of a sustainable world-wide energy supply system is one of the most important measures to be taken to prevent further climate change. Solar energy can play an instrumental role in such a system. Solar energy is abundantly available and is a very versatile energy source.

Solar energy has been used for heating for centuries. Since the invention of the crystalline silicon solar cell by Gerald Pearson, Daryl Chapin and Calvin Fuller in 1954, solar cells have become a very important option for the large scale production of solar electricity. In 2015 photovoltaic electricity already contributes 1% to the global electricity production. The 2014 IEA Roadmaps for Solar Photovoltaic and Solar Thermal Electricity envisage a total share of 27% of the global electricity production by 2050.

Solar energy is used already for supplying small amounts of electricity and heat in rural areas, thereby contributing to the economic development of these areas. Millions of small photovoltaic systems are operational, providing energy, for example, for lighting and telecommunications. Solar energy systems can be integrated very well in the built environment and are contributing substantially to the impressive growth of the utilisation of solar energy that we see today. Solar energy can be used for large scale production of electricity in power plants by means of flat plate and concentrator photovoltaic (PV) systems, as well as by thermal concentrated solar power (CSP) systems.

The utilisation of solar energy is growing very fast. In addition, the goals for solar energy as laid down in government policies on national and European level are very ambitious. As a result many newcomers are entering the field, taking up the challenges. In order to do so, adequate training and education is required. The training needs to be focussed on each level, e.g. the academic level, the level of the system engineer, the level of installers, etc.

The solar-energy field and in particular photovoltaics is very broad. The field of photovoltaics ranges from optics, material and device physics for solar-cell development, to module and power electronics required for the design of complete stand alone and grid-connected systems. For the newcomer to the field, but also for the specialist, it is often difficult to obtain a good overview of the whole field. On one hand it is important that cell and module designers have enough basic knowledge of photovoltaic systems and applications. On the other hand system designers should have sufficient knowledge of the various solar-cell technologies to make the right selection, and once the selection is made, how to use the cells for optimum energy yield.

In this book, first a comprehensive and clear treatment of the fundamental aspects of photovoltaic energy conversion is given. Subsequently, both existing and emerging solar-cell technologies are discussed, and a number of new approaches for future efficiency improvements, like spectral conversion, are introduced. Next an introduction is given into both off-grid and grid-connected photovoltaic systems. This book is completed with an overview of other solar technologies such as flat plate solar collectors and CSP technologies, and solar fuels.

This is an excellent book for education and self-teaching at academic level, but it can also be used by those having a technical background. Though the emphasis is on photovoltaics, the book gives a very good overview of other solar-energy technologies as well.

June 2015

Ronald J. Ch. van Zolingen
Emeritus professor
Eindhoven University of Technology

Dean's message

Providing the world with energy in an environmentally sustainable and climate friendly way poses one of the greatest challenges to mankind. Solar energy will play a prominent role in the generation of electrical energy, whose global consumption is growing even faster than the total energy consumption. The rapid growth of the use of electrical energy is driven by the pervasive digitization of our society, rapid urbanization and the growth of public and private electrical transportation. Annually, the global production of energy is close to a Zetta joule (10^{21} J), of which about 20% of this is electrical energy. In turn today about 1% of this electrical energy is generated by solar systems.

By coincidence the world population generates a Zettabyte of information per year. One byte of information per joule consumed. Fortunately, the information generation is growing a lot faster than our energy consumption, which is hopefully indicative of our potential to find smart solutions for the generation, transport and use of energy. For the increased use of solar electricity, such smart solutions are inevitable. Smart grids are essential for the transport and efficient use of electricity generated by distributed and variable sources. Smart integration of solar power generation in urban environments is another enabler of its increased use.

The solar energy research program at Delft University of Technology is one of the leading programs of our faculty. Its researchers hold key positions in the global solar energy community and their innovative solutions are applied in our society, both by large corporations, start-up companies and in the public domain. In addition, the research team plays a significant role in sustainable energy education at our university and large groups of students follow their courses. Sustainable energy attracts many Dutch students as well international students of master programs such as Sustainable Energy Technology.

The pervasive presence of digitization, as mentioned earlier, has had a significant effect on our educational opportunities as well. This is illustrated by TU Delft's leading position in Massive Open and Online Courses (MOOCs) and other forms of online education. One of our first MOOCs was about solar energy. This course, provided through the edX-platform, was hugely successful and attracted close to 60 thousand participants in its first edition. Now, in its third run the total number of enrolments will be well over a hundred thousand.

In addition to the English language version, this MOOC runs in Arabic and will be presented in Chinese too with the support of our partner organisations. The enormous size and enthusiasm of course participants investing seriously in solar energy knowledge and competencies has inspired the teachers to make this (e-)BOOK with the MOOC. The widespread knowledge and enthusiasm about solar electricity production is certainly indicative of the forthcoming rapid increase of the solar energy fraction of our energy consumption. We are very happy to see that our courses aroused the interest of so many learners in the science and application of solar energy. Several of them were inspired to create a solar energy solution for their local community and decided to go for it.

August 2015

Professor Dr Rob Fastenau
Dean of the Faculty Electrical Engineering, Mathematics and Computer Science
Dean of the Extension School
Delft University of Technology

Preface

At Delft University of Technology we believe that the energy system of the future will be completely different from the system we know today. All the energy we use in the future will come from the Sun; directly via solar photovoltaic modules and thermal collectors, or indirectly in the form of wind and biomass. In this future energy system the conversion and utilization of energy will be highly efficient. These two components, renewable energy sources and energy efficiency, are the key components of sustainable energy. The transition towards a sustainable energy system is a major societal challenge needed to preserve Earth for future generations.

This transition means that electricity will gain a more dominant role in energy demand. We expect electricity to become a universal energy carrier and the backbone of energy supply in the future. By writing a book on solar energy with focus on the direct conversion of solar energy into electricity, so-called photovoltaics (PV), we aim to make more people familiar with this fascinating energy conversion technology. We believe that this book is our contribution to facilitating and accelerating the energy transition towards sustainable energy.

We hope that our book Solar Energy Conversion: Fundamentals, Technologies and Systems will be a useful source for readers studying the different topics on solar energy. These topics are discussed in three courses on photovoltaics at Delft University of Technology: PV Basics, PV Technologies, and PV Systems. In addition, this book also covers other aspects of solar energy, in particular solar thermal applications and solar fuels. Hopefully this book inspires students and professionals around the world to contribute to the realization of a sustainable energy infrastructure, for example by building their own PV system. This book is an excellent supplement to the Massive Open Online Course (MOOC) on Solar Energy (DelftX, ET.3034TU) that is presented by Arno Smets on the edX and edraak platforms.

We received a lot of support and help during the preparation of the book. We are very grateful to Ronald van Zolingen, professor at Eindhoven University of Technology, for reviewing this book. He carried out this task very thoroughly and provided us with many comments and suggestions that resulted in a better and more compact work. We are happy that he also wrote the foreword.

We want to express our special thanks to Gireesh Ganesan Nair for supporting us with figures, exercises, and text editing. We thank Mathew Alani and Adwait Apte for text editing. Giorgos Papakonstantinou together with Dimitris Deligiannis are acknowledged for providing some of the text and figures for Chapter 14. We thank Arianna Tozzi for text and visual material regarding the model estimating the effect of wind speed and irradiance on the module temperature that is presented in Section 20.3. We are grateful for the information on real-life PV systems provided by Stephan van Berkel. Ravi Vasudevan and

Do Yun Kim are acknowledged for many discussions on PV systems, Mirjam Theelen for giving feedback on CIGS and CdTe technologies, and Andrea Ingenito for his insights on the fabrication of crystalline silicon solar cells. We want to thank Wilson Smith and Paula Perez Rodriguez for their support on the chapter on solar fuels, Christiane Becker from Helmholtz-Zentrum Berlin for her remarks on thin-film c-Si solar cells, Rowan MacQueen from the University of New South Wales for his review on third-generation concepts, and Nishant Narayan for his assistance in developing exercises and for contributing to the PV Systems part. Further, we thank all students who followed the lectures at Delft University of Technology or the MOOC Solar Energy, and who provided us with feedback and helped us to reduce the number of typing errors and mistakes.

Enjoy the book!

The AUTHORS

Delft, the Netherlands and Berlin, Germany
September 2015

About this book

This book aims to cover all the topics that are relevant for obtaining a broad overview on the different aspects of *Solar Energy*, with a focus on *photovoltaics*, which is the technology that allows energy carried by light to be converted directly into electrical energy.

The organization of this book is roughly linked to the three lecture series on photovoltaics (PV) that are given at the *Faculty for Electrical Engineering, Mathematics and Computer Science* of *Delft University of Technology* throughout the academic year: PV Basics, which roughly covers the topics covered in Part II on PV Fundamentals; PV Technologies, which covers the topics treated in Part III; and PV Systems, which are treated in Part IV.

In total, this book contains *five parts*. In the introductory Part I we provide the reader with some general facts on energy in Chapter 1, summarize the current status of PV in the world in Chapter 2, and give a first short explanation on how solar cells work in Chapter 3.

Part II aims to cover all the physical fundamentals that are required for understanding solar cells in general and the different technologies in particular. After discussing some basics of electrodynamics in Chapter 4 and solar radiation in Chapter 5, we spend several chapters explaining the most important concepts of semiconductor physics. Following the discussion on the basics in Chapter 6, we elaborate on the different generation and recombination mechanisms in Chapter 7 and introduce different types of semiconductor junctions in Chapter 8. After introducing the most important parameters for characterizing solar cells in Chapter 9, we conclude Part II with a discussion on the efficiency limits of photovoltaic devices in Chapter 10, from which we distil some general design rules that are very important in Part III.

The different PV technologies are discussed in Part III. After summarizing the history of solar cells in Chapter 11, we discuss crystalline silicon technology, which is by far the most important PV technology, in Chapter 12. We continue the discussion by taking a look at the different thin-film technologies in Chapter 13. After that, we take a closer look at some processing technologies in Chapter 14 and discuss how to fabricate PV modules from solar cells in Chapter 15. Part III is concluded with a discussion on several third-generation concepts that aim to combine high efficiencies with low cost in Chapter 16.

Part IV is dedicated to the planning of real PV systems. After a short introduction on PV systems in Chapter 17, we discuss the position of the Sun and its implications in great detail in Chapter 18. The different components of a PV system, starting from the modules, but also including all the balance-of-system components, are introduced in Chapter 19. With all this knowledge we elaborate on designing PV systems – for both off-grid and grid-connected situations in Chapter 20. This part concludes with a discussion on the ecological and economical aspects of PV systems in Chapter 21.

In Part V two alternative solar energy conversion technologies are discussed: we introduce different concepts related to solar thermal energy in Chapter 22. In Chapter 23, which is the last chapter of the regular text, we discuss solar fuels, which allow long-term storage of solar energy in the form of chemical energy.

Most chapters contain exercises in the last section, which allow the reader to assess the studied topics. The book is an extensive source of information that allows students to teach themselves all relevant topics on solar energy. The book concludes with an Appendix, where some derivations are shown that are too lengthy for the book.

Nomenclature

Abbreviations

AM	air mass, –
AOI	angle of incidence, –
BOS	balance of system
DoD	depth of discharge, –
DHI	diffuse horizontal irradiance, Wm^{-2}
DNI	direct normal irradiance, Wm^{-2}
EQE	external quantum efficiency, –
GHI	global horizontal irradiance, Wm^{-2}
IQE	internal quantum efficiency, –
ppm	parts per million
SoC	state of charge, –
SRH	Shockley-Read-Hall (recombination)
SVF	sky view factor, –
TCO	transparent conducting oxide

Latin letters

A	absorption profile, $\text{cm}^{-3}\text{s}^{-1}$
A	surface, m^2
a, **a**	acceleration, ms^{-2}
D	diffusion coefficient, m^2s^{-1}
E	energy, J
f	Fermi-Dirac distribution, –
F, **F**	force, N
FF	fill factor, –
G	generation rate, $\text{m}^{-3}\text{s}^{-1}$
g	density of states function, $\text{m}^{-3}\text{J}^{-1}$

G_M	irradiance on a PV module, Wm^{-2}	
I	current, A	
I_e	irradiance, Wm^{-2}	
$J, \mathbf{J}$	current density, Am^{-2}	
k	wave number, m^{-1}	
ℓ_n, ℓ_p	width of space charge region, m	
L	diffusion length, m	
L_e	radiance, $Wm^{-2}sr^{-1}$	
m	mass, kg	
m^*	effective mass, kg	
M_e	radiant emittance, $Wm^{-2}sr^{-1}$	
N	particle density, m^{-3}	
n	electron concentration, m^{-3}	
n	refractive index (real part), –	
P	power, W	
p	hole concentration, m^{-3}	
Q	heat, J	
R	recombination rate, $m^{-3}s^{-1}$	
R	reflectivity, –	
S_r	surface recombination velocity, ms^{-1}	
T	temperature, K	
T	transmittance, –	
U	heat exchange coefficient, WK^{-1}	
V	electric potential, V	
$\mathbf{v}, v$	velocity, ms^{-1}	
W	work, J	
z	height (function), m	

Greek letters

α	absorption coefficient, –
α	albedo, –
γ	angle of incidence, –
ϵ	electric permittivity, –
ϵ	emissivity, –

ζ	magnetic field, Am^{-1}	
η	efficiency, –	
θ	polar angle, generic angle, –	
κ	refractive index (imaginary part), –	
λ	wavelength, m	
μ	mobility, $\text{m}^2\text{V}^{-1}\text{s}^{-1}$	
ν	frequency, s^{-1}	
ξ	electric field, Vm^{-1}	
ρ	charge density, $\text{A}\cdot\text{s}\cdot\text{m}^{-3}$	
σ	capture cross section, m^2	
σ_r	rms roughness, m	
τ	lifetime, relaxation time, s	
Υ	volume, m^3	
φ	azimuth angle, –	
ϕ	work function, V	
Φ_{ph}	photon flux, $\text{m}^{-2}\text{s}^{-1}$	
χ	dielectric susceptibility, –	
χ	electron affinity, V	
Ψ_{ph}	photon flow, s^{-1}	
Ω	solid angle, –	
ω	angular frequency ($\omega = 2\pi\nu$), s^{-1}	

Subscripts

0	*in vacuo*
A	acceptor
C	conduction band
D	donor
d	drift
F	Fermi
G	bandgap
i	intrinsic, incident
λ	spectral property parameterized by wavelength
L	light
mpp	maximum power point

ν	spectral property parameterized by frequency
oc	open circuit
p	plasma
ph	photon
r	reflected
sc	short circuit
t	transmitted
th	thermal
V	valence band

Constants

c_0	speed of light *in vacuo* (299 792 458 ms^{-1})
ϵ_0	vacuum permittivity (8.854 187 × 10^{-12} AsV^{-1}m^{-1})
F	Faraday constant (96 485.3365 As mol^{-1})
h	Planck constant (6.626 069 × 10^{-34} Js)
k_B	Boltzmann constant (1.380 649 × 10^{-23} JK^{-1})
μ_0	vacuum permeability (4π × 10^{-7} VsA^{-1}m^{-1})
q	elementary charge (1.602 × 10^{-19} C)
σ	Stefan-Boltzmann constant (5.670 373 × 10^{-8} Wm^{-2}K^{-4})
Z_0	impedance of free space (367.7 Ω)

Part I

Introduction

1
Energy

As this book is on *solar energy*, it is good to start the discussion with some general thoughts on *energy*. We begin with a quote from *The Feynman Lectures on Physics* [1].

> There is a fact, or if you wish, a *law*, governing all natural phenomena that are known to date. There is no known exception to this law—it is exact so far as we know. The law is called the *conservation of energy*. It states that there is a certain quantity, which we call energy, that does not change in the manifold changes which nature undergoes. That is a most abstract idea, because it is a mathematical principle; it says that there is a numerical quantity which does not change when something happens. It is not a description of a mechanism, or anything concrete; it is just a strange fact that we can calculate some number and when we finish watching nature go through her tricks and calculate the number again, it is the same.
>
> ...
>
> Energy has a large number of *different forms*, and there is a formula for each one. These are: gravitational energy, kinetic energy, heat energy, elastic energy, electrical energy, chemical energy, radiant energy, nuclear energy, mass energy. If we total up the formulas for each of these contributions, it will not change except for energy going in and out.
>
> It is important to realize that in physics today, we have no knowledge of what energy is. We do not have a picture that energy comes in little blobs of a definite amount. It is not that way. However, there are formulas for calculating some numerical quantity, and when we add it all together it gives ... always the same number. It is an abstract thing in that it does not tell us the mechanism or the reasons for the various formulas.

1.1 Some definitions

We will now state some basic physical connections between the three very important physical quantities of *energy, force, and power*. These connections are taken from classical mechanics but are generally valid. We start with the *force F*, which is any influence on an object that changes its motion. According to Newton's *second law*, the force is related to the acceleration a of a body via

$$\mathbf{F} = m\mathbf{a}, \tag{1.1}$$

where m is the mass of the body. The bold characters denote that $\mathbf{F}$ and $\mathbf{a}$ are vectors. The unit of force is *newton* (N), named after Sir Isaac Newton (1642–1727). It is defined as the force required to accelerate the mass of 1 kg at an acceleration rate of $1\,\mathrm{m\,s^{-2}}$, hence $1\,\mathrm{N} = 1\,\mathrm{kg\,m\,s^{-2}}$.

Energy E, *the* central quantity of this book, is given as the product of F times the distance s,

$$E = \int F(s)\,\mathrm{d}s. \tag{1.2}$$

Energy is usually measured in the unit of *joule* (J), named after the English physicist James Prescott Joule (1818–1889). It is defined as the amount of energy required to apply the force of 1 newton through the distance of 1 m, 1 J = 1 Nm.

Another important physical quantity is *power P*, which tells us the rate of doing work, or, which is equivalent, the amount of energy consumed per time unit. It is related to energy via

$$E = \int P(t)\,\mathrm{d}t, \tag{1.3}$$

where t denotes the time. P is usually measured in the unit of *watt* (W), after the Scottish engineer James Watt (1736–1819). 1 W is defined as one joule per second, 1 W = 1 J/s and 1 J = 1 Ws.

As we will see later on, 1 J is a very small amount of energy compared to human energy consumption. Therefore, in the energy markets, such as the electricity market, often the unit *kilowatt hour* (kWh) is used. It is given as

$$1\,\mathrm{kWh} = 1,000\,\mathrm{Wh} \times 3,600\,\frac{\mathrm{s}}{\mathrm{h}} = 3,600,000,\mathrm{Ws}. \tag{1.4}$$

On the other hand, the amounts of energy in solid state physics, the branch of physics that we will use to explain how solar cells work, are very small. Therefore, we will use the unit of *electron volt*, which is the energy a body with a charge of one elementary charge ($q = 1.602 \times 10^{-19}$ C)[1] gains or loses when it is moved across an electric potential difference of 1 volt (V),

$$1\,\mathrm{eV} = q \times 1\,\mathrm{V} = 1.602 \times 10^{-19}\,\mathrm{J}. \tag{1.5}$$

---

[1] Often, the symbol e is used for the elementary charge. However, in order not to confuse the elementary charge with the Euler number, we use q, just as many others in the solar cell and semiconductor device communities.

1.2 Human energy consumption

After these somewhat abstract definitions we will look at the *human energy consumption*. The human body is at a constant temperature of about 37 °C. It therefore contains *thermal energy*. As the body is continuously cooled by its surroundings, thermal energy is lost to the outside. Further, blood is pumped through the blood vessels. As it travels through the vessels, its *kinetic energy* is reduced because of internal friction and friction at the walls of the blood vessels, i.e. the kinetic energy is converted into heat. To keep the blood moving, the heart consumes energy. Also, if we want our body to move, this consumes energy. Further, the human brain consumes a lot of energy. All of this energy has to be supplied to the body from the outside in the form of food. An average body of a human adult male requires about 10,000 kJ every day.[2] We can easily show that this consumption corresponds to an average power of the human body of 115.7 W. We will come back to this value later.

In modern society, humans not only require energy to keep their body running but in fact consume energy for many different purposes. We use energy for heating the water in our houses and for heating our houses. If water is heated, its thermal energy increases, and this energy must be supplied from the outside. Further, we use a lot of energy for transportation of people and products, by cars, trains, trucks and planes. We use energy to produce our goods and also to produce food. At the moment, you are consuming energy when you are reading this book on a computer or tablet. But also if you are reading it in a printed version, you implicitly consume the energy that was required to print it and to transport it to your place.

As mentioned above, energy is never produced but always converted from one form to another. The form of energy may change in time, but the total amount does not change. If we want to utilize energy to work for us, we usually convert it from one form to another more useable form. An example is the electric motor, in which we convert electrical energy to mechanical energy.

To measure the amount of energy humankind consumes, we refer to two concepts: first, *primary energy*, which 'is the energy embodied in natural resources prior to undergoing any human-made conversions or transformations. Examples of primary energy resources include coal, crude oil, sunlight, wind, running rivers, vegetation[3], and uranium [2]. Humans do not directly use carriers of primary energy, but converted forms of energy, which are called *secondary energy* or *final energy*. Examples of secondary energy carriers are electricity, refined fuels such as gasoline or diesel, and heat which is transported to consumers via district heating.

Modern society is very much based on the capability of humankind to convert energy from one form to another. The most prosperous and technologically developed nations are also the ones which have access to and are consuming the most energy per inhabitant. Table 1.1 shows the primary energy consumption per capita and the average power consumed per capita for several countries. We see that the average US citizen uses an average power of 9,319 W, which is about 80 times what his body needs. In contrast, an average citizen from India only uses about 800 W, which is less then a tenth of the US consumption.

Many people believe that tackling the *energy problem* is among the biggest challenges

---

[2]The energy content of food usually is given in the old-fashioned unit of kilocalories (kcal). The conversion factor is 1 kcal = 4.184 kJ. An average adult male human requires about 2500 kcal a day.

[3]Or biomass *authors note*.

Table 1.1: Total primary energy consumption per capita and average power used per capita of some countries in 2011 [3].

Country	Energy consumption (kWh/capita)	Average power use (W/capita)
USA	81,642	9,319
Netherlands	53,963	6,160
China	23,608	2,695
Colombia	7,792	890
India	6,987	797
Kenya	5,582	637

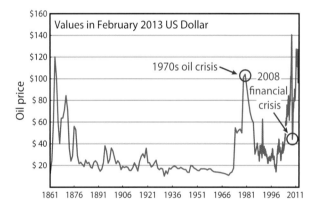

Figure 1.1: The history of the oil price per barrel normalized to the February 2013 value of the US dollar [4].

for humankind in the 21st century. This challenge consists of several problems: First, humankind is facing a supply–demand problem. The demand is continuously growing as the world population is rapidly increasing – some studies predict a world population of 9 billion around 2040, in contrast to the 7 billion people living on the planet in 2014. All these people will need energy, which increases the global energy demand. Further, in many countries the living standard is rapidly increasing; like China and India, where approximately 2.5 billion people are living, which represents more than a third of the world's population. Also the increasing living standards lead to an increased energy demand.

According to the BP Energy Outlook 2035 the global energy consumption is expected to rise by 37% between 2013 and 2035, where virtually all (96%) of the projected growth is in non-OECD countries [5]. The increasing demand in energy has economic impact, as well. If there is more demand for a product, while supply does not change that much, the product will get more expensive. This basic market mechanism is also true for energy. As an example we show a plot of the annual average price for a barrel of oil in Figure 1.1. We see that prices went up during the oil crisis in the 1970s, when some countries stopped producing and trading oil for a while. The second era of higher oil prices started at the beginning of this millennium. Due to the increasing demand from new growing

1. Energy

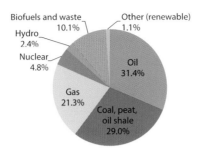

Figure 1.2: The primary energy consumption of the world by source in 2012. The total supply was 155,505 TWh (data from [6]).

economies, the oil prices increased significantly.

A second challenge that we are facing is related to the fact that our energy infrastructure heavily depends on fossil fuels like oil, coal and gas, as shown in Figure 1.2. Fossil fuels are nothing but millions and millions of years of solar energy stored in the form of chemical energy. The problem is that humans deplete these fossil fuels much faster than they are generated through the photosynthetic process in nature. Therefore fossil fuels are not a sustainable energy source. The more fossil fuels we consume, the less easily extractable gas and oil resources will be available. Already now we see that more and more oil and gas is produced with *unconventional* methods, such as extracting oil from tar sands in Alberta, Canada and producing gas with hydraulic fracturing [7], such as in large parts of the United States. These new methods use much more energy to get the fossil fuels out of the ground. Further, offshore drilling is put in regions with ever larger water depths, which leads to new technological risks as we have seen in the Deepwater Horizon oil spill in the Gulf of Mexico in 2010.

A third challenge is that by burning fossil fuels we produce the so-called greenhouse gases such carbon dioxide (CO_2). The additional carbon dioxide created by human activities is stored in our oceans and atmosphere. Figure 1.3 shows the increase in carbon dioxide concentration in the Earth's atmosphere up to 2015. According to the International Panel on Climate Change (IPCC) Fifth Assessment Report (AR5),

> The atmospheric concentrations of carbon dioxide, methane, and nitrous oxide have increased to levels unprecedented in at least the last 800,000 years. Carbon dioxide concentrations have increased by 40% since pre-industrial times, primarily from fossil fuel emissions and secondarily from net land use change emissions. The ocean has absorbed about 30% of the emitted anthropogenic carbon dioxide, causing ocean acidification [11].

Further, in the AR5 it is stated that:

> Human influence on the climate system is clear. This is evident from the increasing greenhouse gas concentrations in the atmosphere, positive radiative forcing, observed warming, and understanding of the climate system [11].

and

> Human influence has been detected in warming of the atmosphere and the ocean, in changes in the global water cycle, in reductions in snow and ice, in global mean

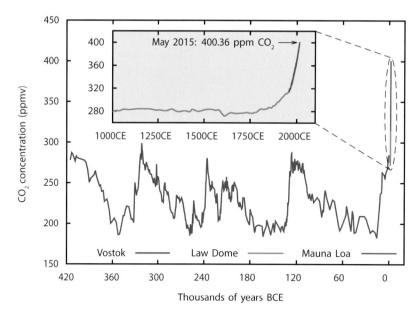

Figure 1.3: Atmospheric CO_2 concentration in the last 420,000 years (detailed view since the year 1000 shown in the inset). The drastic rise of the CO_2 concentration since the onset of the industrial revolution (ca. 1750) is clearly visible. The figure combines data from the Antarctic Vostok [8] and Law Dome [9] ice cores and updated data from the Mauna Loa Observatory in Hawaii [10].

sea level rise, and in changes in some climate extremes. This evidence for human influence has grown since AR4. It is *extremely likely* that human influence has been the dominant cause of the observed warming since the mid-20th century [11].

Hence, it seems very clear that the increase in carbon dioxide is responsible for global warming and climate change, which can have drastic consequences on the habitats of many people.

Since the beginning of the industrial revolution, humankind has been heavily dependent on fossil fuels. Within a few centuries, we are exhausting solar energy that was incident on Earth for hundreds of millions of years, converted into chemical energy by photosynthetic processes and stored in the form of gas, coal and oil.

Before the industrial revolution, the main source of energy was wood and other biomasses, which is a secondary form of solar energy. The energy source was replenished in the same characteristic time as the energy being consumed. In the pre-industrial era, humankind was basically living on a secondary form of solar energy. However, also back then the way we consumed energy was not fully sustainable. For example, deforestation due to increasing population density was already playing a role at the end of the first millennium.

1. Energy

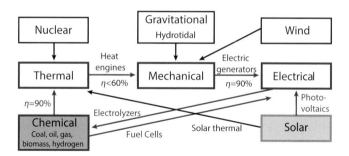

Figure 1.4: The different energy carriers and how we utilise them (adapted from L Freris and D Infield, Renewable Energy in Power Systems (copyright John Wiley & Sons Inc, Chichester, United Kingdom, 2008)) [12].

1.3 Methods of energy conversion

Figure 1.4 shows different energy sources and the ways we utilize them. We see that usually the chemical energy stored in fossil fuels is converted to usable forms of energy via heat by burning, with an efficiency of about 90%. Using heat engines, thermal energy can be converted into mechanical energy. Heat engines have a conversion efficiency of up to 60%. Their efficiency is ultimately limited by the Carnot efficiency limit that we will discuss in Chapter 10. The vast majority of the current cars and trucks works on this principle. Mechanical energy can be converted into electricity using electric generators with an efficiency of 90% or even higher. Most of the world's electricity is generated using *turbogenerators* that are connected to a steam turbine, where coal is the major energy source. This process is explained in more detail in our discussion on solar thermal electric power in Chapter 22. Along all the process steps of making electricity out of fossil fuels, at least 50% of the initial available chemical energy is lost in the various conversion steps.

Chemical energy can be directly converted into electricity using a fuel cell. The most common fuel used in fuel cell technology is hydrogen. Typical conversion efficiencies of fuel cells are 60%. A regenerative fuel cell can operate in both directions and also convert electrical energy into chemical energy. Such an operation is called *electrolysis*; typical conversion efficiencies for hydrogen electrolysis of 50-80% have been reported. We will discuss electrolysis in more detail in Chapter 23.

In *nuclear power plants*, energy is released as heat during *nuclear fission* reactions. The heat generates steam which drives a steam turbine and subsequently an electric generator just as in most fossil fuel power plants.

1.3.1 Renewable energy carriers

All the energy carriers discussed above are either fossil or nuclear fuels. They are not renewable because they are not "refilled" by nature, at least not in a useful amount of time. In contrast, *renewable energy carriers* are energy carriers that are replenished by natural processes at a rate comparable or faster than their rate of consumption by humans. Consequently, hydro, wind and solar energy are renewable energy sources.

Hydroelectricity is an example of an energy conversion technology that is not based on heat generated by fossil or nuclear fuels. The potential energy of rain falling in mountainous areas or elevated plateaus is converted into electrical energy via a *water turbine*. With *tidal pools* the potential energy stored in the tides can also be converted to mechanical energy and subsequently electricity. The kinetic energy of *wind* can be converted into mechanical energy using windmills.

Finally, the energy contained in sunlight, called *solar energy*, can be converted into electricity as well. If this energy is converted into electricity directly using devices based on semiconductor materials, we call it *photovoltaics* (PV). The term *photovoltaic* is derived from the greek word φως (phos), which means light, and volt, which refers to electricity and is a reverence to the Italian physicist Alessandro Volta (1745–1827) who invented the battery. As we will see in this book, typical efficiencies of the most commercial *solar modules* are in the range of 15-20%.

The energy carried with sunlight can also be converted into heat. This application is called *solar thermal energy* and is discussed in detail in Chapter 22. Examples are the heating of water flowing through a black absorber material that is heated in the sunlight. This heat can be used for water heating, heating of buildings or even cooling. If concentrated solar power systems are used, temperatures of several hundreds of degrees are achieved; this is sufficient to generate steam and hence drive a steam turbine and a generator to produce electricity.

Next to generating heat and electricity, solar energy can be converted into chemical energy as well. This is what we refer to as *solar fuels*. For producing solar fuels, photovoltaics and regenerative fuel cells can be combined. In addition, sunlight can also be directly converted into fuels using photoelectrochemical devices. We will discuss solar fuels in Chapter 23.

We just have seen that solar energy can be converted into electricity, heat and chemical energy. The Sun is the energy source for almost all the processes happening on the surface of our planet: wind is a result of temperature difference in the atmosphere induced by solar irradiation; waves are generated by the wind; clouds and rain are initially formed by the evaporation of water due to sunlight. As the Sun is the only real energy source we have, we need to move to an era in which we start to utilize the energy provided by the sun directly for satisfying our energy needs. The aim of this book is to teach the reader how solar energy can be utilized directly.

1.3.2 Electricity

As we see in Figure 1.5 (a), 17% of all the world's final energy is used as electricity, which is a form of energy that can be easily and cheaply transported with relatively small losses through an electric grid. It is important to realize that without electricity modern society as we know it would not be possible. Electricity has been practically used for more than 100 years now. It provides us the energy to cook food, wash, do the laundry, illuminate buildings and streets, and for countless other applications. The access to electricity strongly determines our living standard. Despite this importance of electricity, in 2009 still about 1.3 billion people had no access to electricity.

As we see in Figure 1.5 (b), about 67% of the electricity is generated using fossil fuels, where coal is the dominant contributor. As coal emits about twice as much CO_2 per gen-

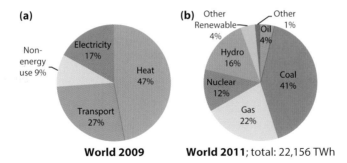

Figure 1.5: (a) The final energy consumption by energy service [13] and (b) the energy carriers used for electricity generation [14] (©OECD/IEA 2012, Insights Series 2012: Policies for renewable heat, IEA Publishing. Licence: www.iea.org/t&c/termsandconditions).

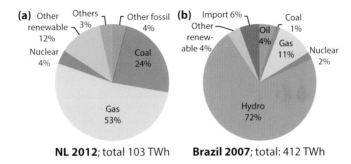

Figure 1.6: The energy mix used for electricity production in (a) the Netherlands [15] and (b) Brazil [16] (Data from Centraal Bureau voor de Statistiek (CBS; Statistics Netherlands) and used according to Creative Commons license: https://creativecommons.org/licenses/by/3.0/nl).

erated kWh as natural gas, coal power plants are a major contributor to global warming. Nuclear is responsible for 12% of the World's electricity generation. With 16%, hydroelectricity is by far the largest contributor among the renewable energy sources.

Of all the generated electricity, about 40% is used for residential purposes and 47% is used by industry. 13% is lost in transmission. In 2007, transport did not play a significant role in the electricity consumption. However, this is expected to change as electric cars become more important.

Figure 1.6 shows which energy carriers are mainly used for electricity generation in the Netherlands and Brazil. We see that in the Netherlands, electricity generation heavily depends on the local gas resources, whereas in Brazil hydroelectricity is the most important resource.

1.4 Exercises

1.1 How many mega joules (MJ) are equivalent to 2.5 kWh?

1.2 To get a feeling of the concepts of power and energy, let us look at the power and energy generated by the human body. In 1994, the Spanish cyclist Miguel Indurain set a world hour record of 53,040 metres in 1 hour. Spanish scientists measured an average power of 509.5 watts produced by him during that hour. What is the energy generated by him in that hour expressed in kWh?

1.3 By using fossil fuels, what is the maximum efficiency that you can achieve from converting chemical energy to electrical energy?

(a) 34%

(b) 49%

(c) 75%

(d) 90%

1.4 Below different energy conversion processes are described.

Person An active person consumes 2,500 kcal of energy in one day.

Lightning A bolt of lightning strikes the ground. It has a voltage of 100 MV and carries a current of 100 kA for 30 μs.

BBQ During a barbeque 1 kg of coal (consisting mostly of carbon) is burned in 1 hour. Assume the following combustion reaction:

$$C + O_2 \rightarrow CO_2 + 4\,eV.$$

Tea To make tea an electric heater is used to boil (i.e. heat from 20 °C to 100 °C) 1 kg of water in three minutes.

House The flat roof of a house (6 x 8 m^2) absorbs sunlight for a year. Assume the house is in the Netherlands where the annual irradiation is about 1,000 kWh/m^2 year.

Battery A 1.5 V battery with a capacity of 2,300 mAh is charged in 3 hours.

Humankind The whole world population (7 billion people) consuming on average 1,500 watts per person for one year.

Solar energy Solar energy reaching planet Earth in one hour. Assume a solar constant of 1,361 W/m^2.

(a) For each of the processes above, answer the following questions.
   i. In what form is the energy before and after the process?
   ii. How much energy (in joules) is converted?
   iii. What is the average conversion power (in watts)?

(b) Order the processes from low to high energy.

(c) Order the processes from low to high power.

Hint: Make simplifying assumptions if needed and use the following background information.
Elementary charge: $q = 1.6022 \times 10^{-19}$ C
Avogadro constant: $N = 6.0022 \times 10^{23}$ mol^{-1} [number of atoms in 12 g of carbon]
Specific heat of water: $c = 4.184$ J/g °C
Radius of Earth: $r = 6,378$ km
1 kcal: the amount of energy it takes to heat 1 kg of water by 1 °C.

2
Status and prospects of PV technology

In this chapter we give a brief overview on the current status of PV technology and discuss its prospects.

Figure 2.1 shows the worldwide cumulative installed PV power, which is exponentially increasing in time. The vertical axis represents the cumulative installed power capacity expressed in GW_p. The letter p denotes *peak power*, which is the maximum power a PV module can deliver if it is illuminated with the standardized AM1.5 solar spectrum that we introduce in Section 5.5. By far the largest share is installed in Europe. It is followed by the Asia Pacific Region, where most of the PV power is installed in Japan. For China we observe a very strong increase in installed PV power since 2010. By the end of 2012 the 100 GW_p threshold was passed for the first time [17]. By the end of 2013, already almost 140 GW_p was installed around the globe [18]. Of all the installed PV power at the end of 2013, almost one third was installed in 2013 alone!

In Figure 2.2 the annually installed capacity of PV modules in recent years is shown. We see that the number of installed PV systems between 2000 and 2011 has grown almost exponentially, with an average growth of 60%. The strongest growth was between 2007 and 2008 with a growth of 143%. In these years, by far the most PV systems were installed in Europe. However, since 2011 the number of installed systems in Europe has been going down rapidly, while it strongly increases in the other regions of the world. While in 2011 74% of all PV systems were installed in Europe, in 2013, this was only 29%. It will be interesting to see how this development continues in the coming years.

In Figure 2.3 the installed PV power in several countries at the end of 2013 is shown. About 26% of the total PV capacity is installed in Germany. This is a result of the German government's progressive feed-in tariff policy that was introduced in 2000 [19].[1] Considering that Germany lies within an area with a relatively low radiation level that is comparable

---
[1] We will discuss the *feed-in tariff* scheme in Chapter 21.

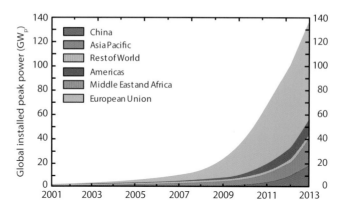

Figure 2.1: The globally installed PV capacity (data from [18] and reproduced with kind permission from the European Photovoltaic Industry Association EPIA).

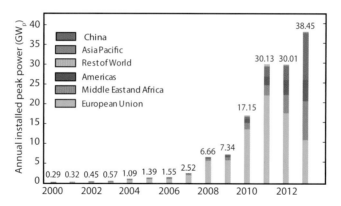

Figure 2.2: The annual installed PV capacity in recent years (data from [18] and reproduced with kind permission from the European Photovoltaic Industry Association EPIA).

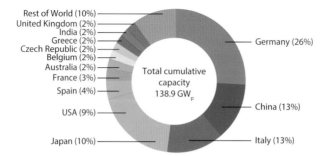

Figure 2.3: Fraction of PV installations for different countries by the end of 2013 (data from [18] and reproduced with kind permission from the European Photovoltaic Industry Association EPIA).

2. Status and prospects of PV technology

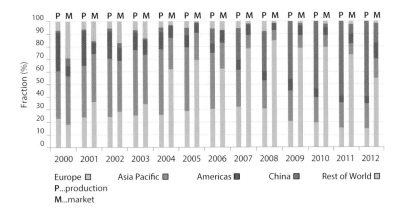

Figure 2.4: Development of the market and production shares of different PV markets since 2000 (data from [17] and reproduced with kind permission from the European Photovoltaic Indusctry Association (EPIA)).

to that of Alaska [20], the large contribution of solar electricity to Germany's electricity production indicates the promising potential of solar energy for the sunnier parts of the world.

A very strong increase is also observed in Italy, which accounts for 13% of the world wide PV capacity. China, with a contribution of 13%, is the fastest growing market at the moment. In 2010, China only contributed 2% to the global PV capacity. Within the top six, we also find the United States, Japan and Spain. Their PV capacity contributes between 4% (Spain) and 9% (USA). Also Japan shows a strong growth in PV installations. After the *Fukushima Daiichi nuclear disaster* on 11 March 2011 the Japanese government introduced some progressive feed-in tariffs to promote and accelerate the introduction of renewable energy conversion technologies.

While PV was mainly a local affair at the beginning of this century, the situation changed strongly around 2009, which is illustrated in Figure 2.4. This figure shows the evolution of the world wide supply and demand of PV modules in the various regions around the world. We see that in 2000 the biggest market was Japan with a total share of 40%. In 2000, Germany introduced the *Erneuerbare Energie Gesetz* (Renewable Energy act) which induced a strong growth of the German and hence the European PV market. By 2008, Europe had a market share of more than 80%. Back then, PV was mainly a European industry. Starting from 2009, the domestic PV markets in China, the Americas (mainly US) and Asia Pacific (mainly Japan) have increased very rapidly and are catching up quickly with Europe.

Figure 2.4 also shows the supply side. Up to 2005 we see that the Asia Pacific and the Europe production shares were slowly increasing, as their growth was faster than that of the other regions. Since then the picture has changed drastically! The Chinese production share has increased very strongly to an amount of about 60% in 2012, which can be explained with huge investments by the Chinese government in order to scale up PV module manufacturing in China.

In 2000, the PV markets were essentially local, meaning the European companies produced for the European market etc. The local demands and supplies in Asia, the Americas and Europe were in balance. More recently, the market has become a global market. As a

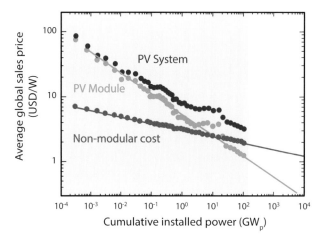

Figure 2.5: The learning curve for PV modules and PV systems (data from [21]).

result, in 2012 no local balance between supply and demand existed anymore. While the majority of the demand is in Europe, the majority of the production is in China.

The demand is also strongly stimulated by the decreasing cost price of PV technology. Figure 2.5 shows the *learning curve* of PV technology. The learning curve shows, in a graphical way, how the cost price develops with increasing experience, where the experience is expressed by the cumulatively installed PV capacity. With more PV produced – and hence also with time – the PV industry gets more experienced. On the one hand, the industry learns to increase the *energy conversion efficiency* without increasing the cost via betterr understanding the production process and hence increasing the *production yield*. On the other hand, industry also learns to produce more efficiently, which means that the manpower required per production unit can be reduced. Also, the materials and energy required for producing the PV modules becomes less and less per production unit. In addition, also upscaling reduces the cost. Learning curves usually show an exponentially decreasing cost price, until the technology or product is fully developed.

In Figure 2.5, the average global sales price of a PV module versus the cumulative installed power up to 20 GW is shown. Note, that the points up to 20 GW (up to 2009) in the grey area are real data points, while the points in the white area are extrapolations of the general trend. It is important to note that the sales prices, except for some fluctuations, follow a largely exponential decay. Currently, the average retail price of PV modules is below 1 US dollar per watt-peak. However, the cost price of a PV system is not only determined by the module. The red dots show the decrease in the cost price of complete PV systems. While in the early days of PV technology, the system price was dominated by the module price, currently, the cost of the *balance of system*, i.e. the *non-modular components* of PV systems, are getting more and more dominant. By non-modular components, we refer to components such as the racking, wiring, inverter, batteries for stand-alone systems, and also the maintenance costs. All these components are discussed in detail in Chapter 19. The difference between the red and green lines corresponds to the non-modular costs, which are dropping more slowly than that of the PV modules.

2. Status and prospects of PV technology

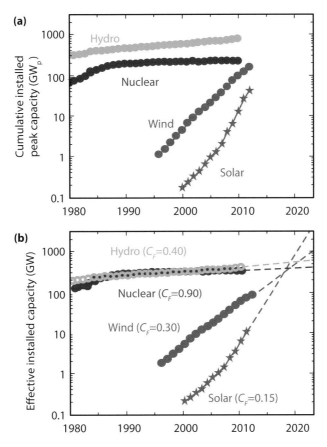

Figure 2.6: (a) Development of the installed capacity (in GW) of several non-fossil electricity generation technologies since 1980. (b) The same graph corrected by the capacity factor C_F and extrapolated until 2020.

As a consequence, PV technologies with higher energy conversion efficiencies have an advantage, because they require less area to deliver the same PV power. As the area is directly linked to the non-modular costs, technologies with higher efficiencies require less modular costs which has a positive effect on the cost price of the complete PV system. Consequently, the c-Si PV technology, with module efficiencies ranging from 14% up to 20% has an advantage with respect to thin-film technologies, that have lower efficiencies.

In Chapter 1 we have seen that hydropower is responsible for 16% of the total worldwide electricity production while 12% of the electricity is generated in nuclear power plants. How do these numbers compare to solar electricity? This question is answered in Figure 2.6 (a), where the installed capacity (in GW) of several electricity generation technologies is shown on a logarithmic scale. The figure only considers electricity generation technologies that are not dependent on fossil fuels. We see that the installed nuclear power capacity is hardly growing any more, while the installed hydropower is still slightly growing with time. Wind is growing at a much faster rate of 20% per year. Solar has by far the largest growth rate with an annual increase of installed capacity exceeding 40% since 2008.

However, it is not fair to compare the installed power between technologies like this, because the numbers shown in the graph represent the maximum (peak) power the different technologies can generate instead of the average power they have delivered in reality. The relationship between the totally installed power and the power generated on average is called the *capacity factor* C_F. Of the technologies shown in Figure 2.6 (a) nuclear has by far the highest capacity factor with $C_F(\text{nuclear}) = 90\%$ followed by hydropower with $C_F(\text{hydro}) = 40\%$. For wind electricity we assume $C_F(\text{wind}) = 30\%$ and for solar electricity $C_F(\text{solar}) = 15\%$. The low capacity factor for PV systems can be explained by the fact that for most geographical locations, almost half of the solar day is devoid of solar radiation at night time.

Figure 2.6 (b) shows the effective installed power corrected with the capacity factors. Currently solar energy generates about an order of magnitude less electricity than wind energy and more than two orders of magnitude less than hydro and nuclear electricity. Seeing the development in recent years, nonetheless we can claim that the trend in the growth of solar energy will continue in the coming years. If we therefore extrapolate the trends of the last decade until 2020 we see that the installed power of solar energy will exceed nuclear, wind and hydropower by then. It is just a matter of time before solar electricity will be the most important electricity generation technology that is not based on the combustion of fossil fuels.

Of course, we have to justify why solar electricity can grow much faster than the other technologies shown in Figure 2.6. First, solar radiation is available everywhere on Earth and it is available in great abundance. The amount of solar energy incident on Earth is about 10,000 times larger than the *total* energy[2] consumption of mankind. As hydroelectricity is powered by water that is evaporated by the Sun and falls on the ground as rain, it is a secondary form of solar energy. Also, wind arises from temperature and pressure differences in the atmosphere and hence is a secondary form of solar energy. As a consequence, solar energy is by far the largest available form of renewable energy.

Secondly, hydro- and nuclear electricity are *centralized* electricity generation technologies. For hydropower plants, big dams are needed. Also nuclear power plants have large power rates of about 1 GW. Building new hydro- and nuclear power plants requires large public or private investments. While solar electricity can be generated in large PV parks or solarthermal power plants (see Chapter 22), it also has a unique advantage: PV systems can be installed decentralized on every roof. Electricity consumers can generate at least a part of their required electricity on their own homes, which makes them partially independent of the electricity market. In addition, the cost price of PV systems has dropped below grid parity in many parts of the world [22]. This means that, averaged during the lifetime of the PV system, PV generated electricity is cheaper than electricity from the grid.

We believe that the installation of decentralized PV systems will be the big force behind the solar revolution in the coming years. It will change the energy landscape much faster than most people think, which is justified in Figure 2.6 (b). As more and more people become aware of these facts, it is more likely that the growth will be further enhanced than be slowed down.

---

[2] We really mean the *total* human energy consumption and not only electricity!

Exercises

For all the exercises below, assume that the worldwide total amount of electricity generated is 20,200 TWh per year.

2.1 In 2010, the worldwide installed hydroelectricity power was 1,010 GW. Assume a capacity factor for hydropower of 40%. What percentage of the total electricity generation worldwide was covered by hydropower in 2010?

2.2 In 2010, the worldwide installed nuclear power was 380 GW. Assume a capacity factor for nuclear power of 90%. What percentage of the total electricity generation worldwide was covered by nuclear power in 2010?

2.3 In 2012, the worldwide installed wind power was 280 GW. Assume a capacity factor for wind power of 30%. What percentage of the total electricity generation worldwide was covered by wind energy in 2012?

2.4 In 2013, the worldwide installed solar power was 140 GW. Assume a capacity factor for solar power of 15%. What percentage of the total electricity generation worldwide was covered by solar energy in 2013?

3
The working principle of a solar cell

In this chapter we present a very simple model of a solar cell. Many notions presented in this chapter will be new but nonetheless the general idea of how a solar cell works should be clear. All the aspects presented in this chapter will be discussed in greater detail in the following chapters.

The working principle of solar cells is based on the *photovoltaic effect*, i.e. the generation of a potential difference at the junction of two different materials in response to electromagnetic radiation. The photovoltaic effect is closely related to the photoelectric effect, where electrons are emitted from a material that has absorbed light with a frequency above a material-dependent threshold frequency. In 1905, Albert Einstein understood that this effect can be explained by assuming that the light consists of well defined energy quanta, called *photons*. The energy of such a photon is given by

$$E = h\nu, \tag{3.1}$$

where h is Planck's constant and ν is the frequency of the light. For his explanation of the photoelectric effect Einstein received the Nobel Prize in Physics in 1921 [23].

The photovoltaic effect can be divided into three basic processes:

1. Generation of charge carriers due to the absorption of photons in the materials that form a junction

Absorption of a photon in a material means that its energy is used to excite an electron from an initial energy level E_i to a higher energy level E_f, as shown in Figure 3.1 (a). Photons can only be absorbed if electron energy levels E_i and E_f are present so that their difference equals the photon energy, $h\nu = E_f - E_i$. In an ideal semiconductor electrons can populate energy levels below the so-called *valence band* edge, E_V, and above the so-called *conduction band* edge, E_C. Between those two bands no allowed energy states exist which could be

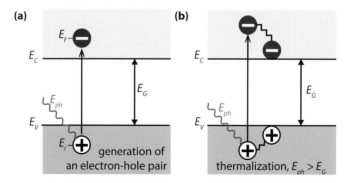

Figure 3.1: (a) Illustrating the absorption of a photon in a semiconductor with bandgap E_G. The photon with energy $E_{\text{ph}} = h\nu$ excites an electron from E_i to E_f. At E_i a hole is created. (b) If $E_{\text{ph}} > E_G$, a part of the energy is thermalized.

populated by electrons. Hence, this energy difference is called the *bandgap*, $E_G = E_C - E_V$. If a photon with an energy smaller than E_G reaches an ideal semiconductor, it will not be absorbed but will traverse the material without interaction.

In a real semiconductor, the valence and conduction bands are not flat, but vary depending on the so-called *k*-vector that describes the momentum of an electron in the semiconductor. This means that the energy of an electron is dependent on its momentum because of the periodic structure of the semiconductor crystal. If the maximum of the valence band and the minimum of the conduction band occur at the same *k*-vector, an electron can be excited from the valence to the conduction band without a change in the momentum. Such a semiconductor is called a *direct bandgap* material. If the electron cannot be excited without changing its momentum, we refer to it as an *indirect bandgap* material. The electron can only change its momentum by momentum exchange with the crystal, i.e. by receiving momentum from or giving momentum to vibrations of the crystal lattice. The absorption coefficient in a direct bandgap material is much higher than in an indirect bandgap material, thus the absorbing semiconductor, often just called the *absorber*, can be much thinner [24].

If an electron is excited from E_i to E_f, a void is created at E_i. This void behaves like a particle with a positive elementary charge and is called a *hole*. The absorption of a photon therefore leads to the creation of an electron-hole pair, as illustrated in Figure 3.2 ❶. The *radiative energy* of the photon is *converted* to the *chemical energy* of the electron-hole pair. The maximal conversion efficiency from radiative energy to chemical energy is limited by thermodynamics. This *thermodynamic limit* lies between 67% for non-concentrated sunlight and 86% for fully concentrated sunlight [25].

The basic physics required for describing semiconductors is presented in Chapter 6.

2. Subsequent separation of the photo-generated charge carriers in the junction

Usually, the electron-hole pair will recombine, i.e. the electron will fall back to the initial energy level E_i, as illustrated in Fig. 3.2 ❷. The energy will then be released either as photon (*radiative recombination*) or transferred to other electrons or holes or lattice vibrations (*non-

3. The working principle of a solar cell

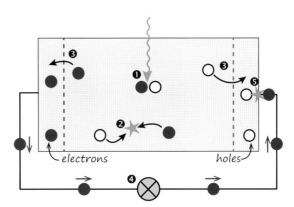

Figure 3.2: A very simple solar cell model. ❶ Absorption of a photon leads to the generation of an electron-hole pair. ❷ Usually, the electrons and holes will recombine. ❸ With semipermeable membranes the electrons and the holes can be separated. ❹ The separated electrons can be used to drive an electric circuit. ❺ After the electrons have passed through the circuit, they will recombine with holes.

radiative recombination). If one wants to use the energy stored in the electron-hole pair for performing work in an external circuit, *semipermeable membranes* must be present on both sides of the absorber, such that electrons can only flow out through one membrane and holes can only flow out through the other membrane [25], as illustrated in Figure 3.2 ❸. In most solar cells, these membranes are formed by n- and p-type materials.

A solar cell has to be designed such that the electrons and holes can reach the membranes before they recombine, i.e. the time it requires the charge carriers to reach the membranes must be shorter than their lifetime. This requirement limits the thickness of the absorber.

We will discuss generation and recombination of electrons and holes in detail in Chapter 7.

3. Collection of the photo-generated charge carriers at the terminals of the junction

Finally, the charge carriers are extracted from the solar cells with electrical contacts so that they can perform work in an external circuit (Fig. 3.2 ❹). The *chemical energy* of the electron-hole pairs is finally converted to *electric energy*. After the electrons have passed through the circuit, they will recombine with holes at a metal-absorber interface, as illustrated in Figure 3.2 ❺.

Loss mechanisms

The two most important *loss mechanisms* in single bandgap solar cells are the inability to convert photons with energies below the bandgap to electricity and thermalization of photon energies exceeding the bandgap, as illustrated in Figure 3.1 (b). These two mechanisms alone amount to the loss of about half the incident solar energy in the conversion process [26]. Thus, the maximal energy conversion efficiency of a single-junction solar cell is considerably below the thermodynamic limit. This *single bandgap limit* was first calculated by Shockley and Queisser in 1961 [27].

A detailed overview of loss mechanisms and the resulting efficiency limits is discussed in Chapter 10.

Part II

PV fundamentals

4
Electrodynamic basics

In this chapter we introduce the basics of electrodynamics that are required for solar cell physics. First, we introduce the electromagnetic wave equations. The existence of these equations explains the existence of electromagnetic waves, such as light. From there we develop the equations describing the interaction of an electromagnetic wave with interfaces between two materials; in this way we naturally derive the basics of optics.

Later in the chapter we introduce the Poisson equation and the continuity equations that are very important for semiconductor physics, which we discuss in Chapter 6.

4.1 The electromagnetic theory

While electricity and magnetism have been known since ancient times, it took until the nineteenth century to realize that these two phenomena are two sides of the same coin, namely *electromagnetism*. We can easily see this by recalling that electric fields are generated by charges while magnetic fields are generated by currents, i.e. . moving charges. Let us now assume that we are within an array of charges. Since charges create an electric field, we will experience such a field. Now we start moving with a constant velocity. This is equivalent to saying that the array of charges moves with respect to us. Since moving charges are a current, we now experience a magnetic field. Thus, when changing from one frame of reference into another that moves with respect to the first one with a constant velocity, electric fields are transformed into magnetic fields and *vice versa*.

Between 1861 and 1862, the Scottish physicist James Clerk Maxwell published works in which he managed to formulate the complete electromagnetic theory by a set of equations, the *Maxwell equations*. A modern formulation of these equations is given in Appendix A.1. The transformation of the electric and magnetic fields between different frames of reference is correctly described by Albert Einstein's theory of special relativity, published in 1905.

One of the most important predictions of the Maxwell equations is the presence of electromagnetic waves. A derivation is given in Appendix A.2. Maxwell soon realized that the speed of these waves is (within experimental accuracy) the same as the speed of light, which was already known. He brilliantly concluded that light is an electromagnetic wave.

In the 1880s the German physicist Heinrich Hertz experimentally confirmed that electromagnetic waves can be generated and have the same speed as light. His work laid the foundation for radio communication that has shaped the modern world.

The electromagnetic theory can perfectly describe how light propagates. However, it fails to explain how light is emitted and absorbed by matter. For this purpose, quantum mechanics is required.

4.2 Electromagnetic waves

As shown in Appendix A.2, electromagnetic waves are described by

$$\left(\frac{\partial^2}{\partial x^2} + \frac{\partial^2}{\partial y^2} + \frac{\partial^2}{\partial z^2}\right)\xi - \frac{n^2}{c_0^2}\left(\frac{\partial^2 \xi}{\partial t^2}\right) = 0 \tag{4.1a}$$

for the electric field $\xi(\mathbf{r}, t)$, where c_0 denotes the speed of light *in vacuo* and n is the refractive index of the material. In a similar manner we can derive the wave equation for the *magnetic field* ζ,

$$\left(\frac{\partial^2}{\partial x^2} + \frac{\partial^2}{\partial y^2} + \frac{\partial^2}{\partial z^2}\right)\zeta - \frac{n^2}{c_0^2}\left(\frac{\partial^2 \zeta}{\partial t^2}\right) = 0. \tag{4.1b}$$

The simplest solution to the wave equations (4.1) is the plane harmonic wave, where light of constant wavelength λ propagates in one direction. Without loss of generality, we assume that the wave travels along the z direction. The electric and magnetic fields in this case are

$$\xi(\mathbf{r}, t) = \xi_0 e^{ik_z z - i\omega t}, \tag{4.2a}$$

$$\zeta(\mathbf{r}, t) = \zeta_0 e^{ik_z z - i\omega t}, \tag{4.2b}$$

where ξ_0 and ζ_0 are constant vectors (the amplitudes), k_z is the *wave number* and ω is the *angular frequency*. By substituting Eq. (4.2a) into Eq. (4.1a) we find that k_z and ω are connected to each other via

$$k_z^2 - \frac{n^2}{c_0^2}\omega^2 = 0. \tag{4.3}$$

Thus,

$$k_z = \frac{n\omega}{c}. \tag{4.4}$$

The angular frequency, measured in *radians per second* is related to the *frequency* of the wave ν, measured in *hertz*, via

$$\omega = 2\pi\nu = \frac{2\pi}{T}, \tag{4.5}$$

4. Electrodynamic basics

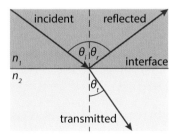

Figure 4.1: Scheme of light, reflected and refracted by a boundary.

where $T = 1/\nu$ is the period, measured in *seconds*. The wave number k_z has the unit of an inverse length. It is related to the *wavelength* λ via

$$\lambda = \frac{2\pi}{k_z} = \frac{2\pi c}{n\omega} = \frac{c}{n\nu}. \tag{4.6}$$

We note that ν and ω are independent of n, while k and λ change when the wave travels to media with different n.

Electromagnetic waves have some extraordinary properties that are derived in detail in Appendix A.3:

- The electric and magnetic field vectors are perpendicular to each other and also perpendicular to the propagation vector,

$$\mathbf{k} \cdot \boldsymbol{\xi}_0 = \mathbf{k} \cdot \boldsymbol{\zeta}_0 = \boldsymbol{\xi}_0 \cdot \boldsymbol{\zeta}_0 = 0. \tag{4.7}$$

- The electric and magnetic fields are perpendicular to the propagation direction, hence electromagnetic waves are *transverse waves*.

- The electric and magnetic vectors have a constant, material-dependent ratio. If the electric field is along the x-direction and the magnetic field is along the y-direction, this ratio is given by

$$\zeta_{y,0} = \frac{n}{c\mu_0}\xi_{x,0} = \frac{n}{Z_0}\xi_{x,0}, \tag{4.8}$$

where

$$Z_0 = c\mu_0 = \sqrt{\frac{\mu_0}{\epsilon_0}} = 376.7\,\Omega \tag{4.9}$$

is the *impedance of free space*.

4.3 Optics of flat interfaces

In this section we repeat the major relations that describe an electromagnetic wave traversing an interface between a medium 1 and medium 2, as illustrated in Figure 4.1. We assume the two media to be non-absorptive. Therefore, only the real parts of the refractive indices are present. We denote them by n_1 and n_2.

A part of the incident light is *reflected*, where the angle of the scattered light θ_r is equal to the incident angle θ_i,

$$\theta_r = \theta_i. \tag{4.10}$$

The other part enters medium 2, where the angle of the *refracted* light θ_t is related to θ_i via *Snell's law*,

$$n_1 \sin \theta_i = n_2 \sin \theta_t. \tag{4.11}$$

Relations between the magnitudes of the incident, reflected and refracted fields are given by *Fresnel equations*. We have to distinguish between *parallel* and *perpendicular polarized light*. Parallel or perpendicular polarized means that the *electric field* is parallel or perpendicular to the plane of incidence, respectively. For *perpendicular* polarized light the Fresnel equations are given by

$$t_s = \left(\frac{\tilde{\xi}_{0t}}{\tilde{\xi}_{0i}}\right)_s = \frac{2 n_1 \cos \theta_i}{n_1 \cos \theta_i + n_2 \cos \theta_t}, \tag{4.12a}$$

$$r_s = \left(\frac{\tilde{\xi}_{0r}}{\tilde{\xi}_{0i}}\right)_s = \frac{n_1 \cos \theta_i - n_2 \cos \theta_t}{n_1 \cos \theta_i + n_2 \cos \theta_t}, \tag{4.12b}$$

where the s stands for *senkrecht*, which is German for perpendicular. The relations for *parallel* polarized light are

$$t_p = \left(\frac{\tilde{\xi}_{0t}}{\tilde{\xi}_{0i}}\right)_p = \frac{2 n_1 \cos \theta_i}{n_1 \cos \theta_t + n_2 \cos \theta_i}, \tag{4.13a}$$

$$r_p = \left(\frac{\tilde{\xi}_{0r}}{\tilde{\xi}_{0i}}\right)_p = \frac{n_1 \cos \theta_t - n_2 \cos \theta_i}{n_1 \cos \theta_t + n_2 \cos \theta_i}. \tag{4.13b}$$

The intensities are proportional to the square of the electric field, $I \propto \xi^2$. For unpolarized light, we have to take the mean values of the two polarizations. For the *reflectivity* R we thus obtain

$$R = \frac{1}{2}\left(r_s^2 + r_p^2\right). \tag{4.14}$$

For normal incidence this leads to

$$R(\theta_i = 0) = \left(\frac{n_1 - n_2}{n_1 + n_2}\right)^2. \tag{4.15}$$

Because of *conservation of energy* the sum of R and the *transmittance* T must be 1,

$$R + T = 1. \tag{4.16}$$

By combining Eq. (4.14) with (4.16) and doing some calculations we find

$$T = 1 - R = \frac{n_2 \cos \theta_t}{n_1 \cos \theta_i} \frac{1}{2}\left(t_p^2 + t_s^2\right), \tag{4.17}$$

which leads to

$$T(\theta_i = 0) = \frac{4 n_1 n_2}{(n_1 + n_2)^2} \tag{4.18}$$

… for normal incidence.

A very important consequence of Snell's law is *total reflection*. If $n_2 > n_1$, there is a *critical angle* at which light can no longer leave the layer with n_2,

$$\sin\theta_{\text{crit}} = \frac{n_1}{n_2}. \tag{4.19}$$

Hence if $\theta_2 \geq \theta_{\text{crit}}$, no light will be transmitted, but everything will be reflected back into the layer. For a silicon–air interface ($n_{\text{Si}} \approx 4.3$), we find $\theta_{\text{crit}} = 13.4°$. For the supporting layers used in solar cells, the critical angle is much larger. For a silicon–glass interface ($n_{\text{glass}} \approx 1.5$), we find $\theta_{\text{crit}} = 20.4°$. And for an interface between silicon and zinc oxide, which is a transparent conducting oxide often used in solar cell technology, ($n_{\text{ZnO}} \approx 2$), the critical angle would be $\theta_{\text{crit}} = 30.3°$.

4.4 Optics in absorptive media

Let us recap what we have seen in this chapter so far: Starting from the Maxwell equations we derived the wave equations and looked at their properties for the special case of plane waves. After that we looked at the behaviour of electromagnetic waves at the interfaces between two media. For the whole discussion so far we implicitly assumed that the media is non-absorbing.

The working principle of solar cells is based on the fact that light is *absorbed* in an absorber material and that the absorbed light is used for exciting charge carriers that can be used to drive an electric circuit. Therefore we will use this section to discuss how absorption of light in a medium can be described mathematically.

In general, the optical properties of an absorbing medium are described by a *complex electric permittivity* $\tilde{\epsilon}$,

$$\tilde{\epsilon} = \epsilon' + i\epsilon''. \tag{4.20}$$

Since the refractive index is given as the square root of $\tilde{\epsilon}$, it is complex too,

$$\tilde{n} = \sqrt{\tilde{\epsilon}} = n + i\kappa. \tag{4.21}$$

Here, κ denotes the imaginary part of the refractive index. From Eq. (4.4) it becomes clear that in our case the wavenumber also becomes complex,

$$\tilde{k}_z = \frac{\tilde{n}\omega}{c} = \frac{n\omega}{c} + i\frac{\kappa\omega}{c} = k'_z + ik''_z. \tag{4.22}$$

Let us now substitute Eq. (4.22) into Eq. (4.2a),

$$\xi_x(z,t) = \xi_{x,0} \cdot e^{i\tilde{k}_z z - i\omega t} = \xi_{x,0} \cdot e^{-k''_z z} e^{ik'_z z - i\omega t}. \tag{4.23}$$

We thus see that the electric field is attenuated exponentially, $\exp(-k''_z z)$ when travelling through the absorbing medium. The intensity of the electromagnetic field is proportional to the square of the electric field,

$$I(\mathbf{r},t) \propto |\boldsymbol{\xi}(\mathbf{r},t)|^2. \tag{4.24}$$

Therefore we find for the attenuation of the intensity of the electromagnetic field

$$I(z) = I_0 \exp(-2k_z'' z) = I_0 \exp(-\alpha z), \tag{4.25}$$

where α is the *absorption coefficient*. It is related to the other properties via

$$\alpha = 2k_z'' = 2\frac{\kappa \omega}{c} = \frac{4\pi \kappa}{\lambda_0}, \tag{4.26}$$

where $\lambda_0 = 2\pi c/\omega$ is the wavelength *in vacuo*.

Equation (4.25) is known as the *Lambert-Beer law*. A magnitude that is often used to judge the absorptivity of a material at a certain wavelength, is the *penetration depth* δ_p,

$$\delta_p = \frac{1}{\alpha}. \tag{4.27}$$

At this depth, the intensity has decayed to a fraction of $1/e$ of the initial value.

In general, the complex refractive index and hence the absorption coefficient are not material constants but vary with the frequency. Especially α may change by several orders of magnitude across the spectrum, making the material very absorptive at one wavelength but almost transparent at other wavelengths. Absorption spectra will be discussed thoroughly later on when looking at various photovoltaic materials in Part III.

4.5 Continuity and Poisson equations

At the end of this chapter we want to mention two equations that are very important for our treatise of semiconductor physics in Chapter 6.

4.5.1 Poisson equation

The first equation is the *Poisson equation* that relates the density of electric charges $\rho(\mathbf{r})$ to the electrical potential $V(\mathbf{r})$. For its derivation, we start with the first Maxwell equation (A.1a). Using Eq. (A.2a) we obtain

$$\frac{\partial \xi_x}{\partial x} + \frac{\partial \xi_y}{\partial y} + \frac{\partial \xi_z}{\partial z} = \frac{\rho}{\epsilon \epsilon_0}, \tag{4.28}$$

where ξ_x, ξ_y and ξ_z are the components of the electric field vector, $\boldsymbol{\xi} = (\xi_x, \xi_y, \xi_z)$. Further, we here assume that we are in an electrostatic situation, i.e. there are no moving charges. Hence, the electric field is *rotation free* as we know from the second Maxwell equation (A.1b). Vector calculus teaches us that the electric field is then connected to the electric potential V via

$$\boldsymbol{\xi} = -\left(\frac{\partial V}{\partial x}, \frac{\partial V}{\partial y}, \frac{\partial V}{\partial z}\right). \tag{4.29}$$

By combining Eqs. (4.28) with (4.29) we find the *Poisson equation*

$$\left(\frac{\partial^2}{\partial x^2} + \frac{\partial^2}{\partial y^2} + \frac{\partial^2}{\partial z^2}\right) V = -\frac{\rho}{\epsilon \epsilon_0}. \tag{4.30}$$

4. Electrodynamic basics

In Chapter 6 we will only use the one-dimensional form given by

$$\frac{d^2 V}{dz^2} = -\frac{\rho}{\epsilon\epsilon_0}. \tag{4.31}$$

4.5.2 Continuity equation

Charge is a conserved quantity. The total amount of charge inside a volume Y can only be changed via charges flowing through the boundary surface A of this volume. This can be expressed mathematically by the equation

$$\frac{dQ_Y}{dt} + \oiint_A \mathbf{J}\, dA = 0, \tag{4.32}$$

where **J** is the current density vector and Q_Y is the total charge contained within the volume Y. It is given by

$$Q_Y = \iiint_Y \rho\, dY. \tag{4.33}$$

Equation (4.32) is the *integral* formulation of the continuity equation. It is equivalent to the *differential* formulation that is given by

$$\frac{\partial \rho}{\partial t} + \left(\frac{\partial J_x}{\partial x} + \frac{\partial J_y}{\partial y} + \frac{\partial J_z}{\partial z}\right) = 0, \tag{4.34}$$

where J_x, J_y and J_z are the components of the current density vector, $\mathbf{J} = (J_x, J_y, J_z)$.

4.6 Exercises

4.1 Solar light is coming from air ($n_1 = 1$) to the upper glass layer of a solar cell ($n_2 = 1.5$).

   (a) The light has an angle of incidence of $\theta_i = 0°$. What is the reflectivity at the air–glass interface? Assume that the solar light is randomly polarized.

   (b) Imagine now an angle of incidence change of $\theta_i = 30°$. Calculate the reflection in this case. Assume again that the solar light is randomly polarized.

4.2 Assume a light beam propagating from a medium with refractive index $n_1 = 1.5$ to a medium with refractive index $n_2 = 1$. Determine the critical angle.

4.3 Let us assume that solar light reaches a silicon solar cell at an angle of incidence of $\theta_i = 0°$. The refractive index of silicon is assumed to be $n_{Si} = 3.5$. The refractive index of air is $n_{air} = 1$. What percentage of light would be lost due to reflection at the air–silicon interface? Assume that the solar light is randomly polarized.

5
Solar radiation

In this chapter we discuss the aspects of solar radiation, which are important for solar energy. After defining the most important radiometric properties in Section 5.2, we discuss blackbody radiation in Section 5.3 and the wave-particle duality in Section 5.4. Equipped with this knowledge, we then investigate the different solar spectra in Section 5.5. However, prior to these discussions we give a short introduction about *the Sun*.

5.1 The Sun

The Sun is the central star of our solar system. It consists mainly of hydrogen and helium. Some basic facts are summarized in Table 5.1 and its structure is sketched in Figure 5.1. The mass of the Sun is so large that it contributes 99.68% of the total mass of the solar system. In the centre of the Sun the pressure–temperature conditions are such that *nuclear*

Table 5.1: Some facts on the Sun.

Mean distance from the Earth	149,600,000 km (the astronomic unit, AU)
Diameter	1,392,000 km (109 × that of the Earth)
Volume	1,300,000 × that of the Earth
Mass	1.993×10^{27} kg (332,000 × that of the Earth)
Density (at its centre)	>10^5 kg m^{-3} (over 100 × that of water)
Pressure (at its centre)	over 1 billion atmospheres
Temperature (at its centre)	about 15,000,000 K
Temperature (at the surface)	6,000 K
Energy radiation	3.8×10^{26} W
The Earth receives	1.7×10^{18} W

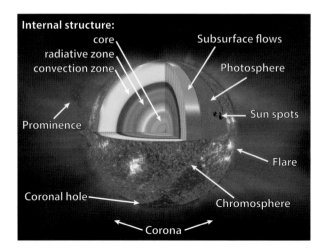

Figure 5.1: The layer structure of the Sun (adapted from a figure obtained from NASA [28]).

fusion can take place. In the major nuclear reaction, the *proton–proton reaction*, via a number of steps, four protons react to produce:

- 1 helium core (two protons and two neutrons);
- 2 positrons (the anti-particles of electrons);
- 2 neutrinos;
- electromagnetic radiation.

The positrons are annihilated with electrons, leading to additional radiation. The mass of the helium core is 0.635% less than that of four protons; the energy difference is converted into energy according to Einstein's equation

$$E = mc_0^2. \tag{5.1}$$

The total power is about 3.8×10^{26} W. Therefore, every second approximately 4 million tons of mass are converted into energy. However, the power density at the centre of the Sun is estimated by theoretical assumptions only to be about 275 W/m^3. As we have seen in Chapter 1, the average power of an adult male is 115.7 W. If we assume his average volume to be 70 L, i.e. 0.07 m^3, the average power density of the human body is 1650 W, hence much higher than the fusion power in the centre of the Sun!

The neutrinos hardly interact with matter and thus can leave the solar core without any hindrance. Every second, about 6.5×10^{10} cm^{-2} pass through the Earth and hence also through our bodies. Neutrinos carry about 2% of the total energy radiated by the Sun.

The remainder of the radiation is released as electromagnetic radiation. The core of the Sun is so dense that radiation cannot travel freely but is continuously absorbed and re-emitted, such that it takes the radiation between 10,000 and 170,000 years to travel to the solar surface. The surface of the Sun is called the photosphere. It has a temperature of about 6,000 K. It behaves very closely to a blackbody (see Section 5.3) and is the source of the solar radiation that hits the Earth. The total irradiance of the solar radiation at the mean

5. Solar radiation

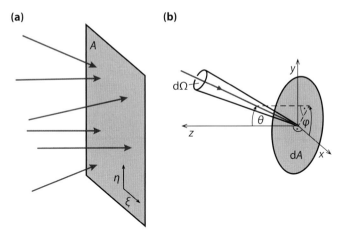

Figure 5.2: (a) Illustrating a surface A irradiated by light from various directions and (b) a surface element dA that receives radiation from a solid angle element $d\Omega$ under an angle θ with respect to the surface normal.

Earth–sun distance on a plane perpendicular to the direction of the Sun, outside the Earth's atmosphere, is referred to as the *solar constant*. Its value is approximately 1,361 W/m².

5.2 Radiometric properties

Radiometry is the branch of optics concerned with the *measurement of light*. Since photovoltaics deals with sunlight that is converted into electricity it is very important to discuss how the "amount of energy" of the light can be expressed physically and mathematically.

In solar science, it is not the total amount of the energy that is important, but the amount of energy per unit time, i.e. the *power P*. For our discussion we assume a surface A that is irradiated by light, as illustrated in Figure 5.2 (a). To obtain the total power that is incident on the surface, we have to integrate over the whole surface. Further we have to take into account that light is incident from all the different directions, which we parameterize with the spherical coordinates (θ, ϕ). The *polar angle* θ is defined with respect to the normal of the surface element dA and ϕ is the *azimuth*, as sketched in Figure 5.2 (b). Thus, we also have to integrate over the hemisphere from which light can be incident on the surface element dA. We therefore obtain

$$P = \int_A \int_{2\pi} L_e \cos\theta \, d\Omega \, dA, \tag{5.2}$$

where

$$d\Omega = \sin\theta \, d\theta \, d\phi \tag{5.3}$$

is a *solid angle* element. The quantity L_e is called the *radiance* and it is one of the most fundamental radiometric properties. The subscript e for L_e and all the other radiometric properties indicates that these are *energetic* properties, i.e. they are related to energies or powers. The physical dimension of L_e is

$$[L_e] = \text{Wm}^{-2}\text{sr}^{-1}.$$

The factor $\cos\theta$ expresses the fact that the surface element dA itself is not the relevant property but the projection of dA to the normal of the direction (θ, ϕ). This is also known as the *Lambert cosine law*.

We can express Eq. (5.2) as integrals of the surface coordinates (ξ, η) and the direction coordinates (θ, ϕ), which reads as

$$P = \int_A \int_{2\pi} L_e(\xi, \eta; \theta, \phi) \cos\theta \sin\theta \, d\theta \, d\phi \, d\xi \, d\eta. \tag{5.4}$$

Since sunlight consists of a spectrum of different frequencies (or wavelengths), it is useful to use *spectral properties*. These are given by

$$P_\nu = \frac{dP}{d\nu}, \qquad\qquad P_\lambda = \frac{dP}{d\lambda}, \tag{5.5}$$

$$L_{e\nu} = \frac{dL_e}{d\nu}, \qquad\qquad L_{e\lambda} = \frac{dL_e}{d\lambda}, \tag{5.6}$$

etc. Their physical dimensions are

$$[P_\nu] = \text{W Hz}^{-1} = \text{Ws}, \qquad\qquad [P_\lambda] = \text{Wm}^{-1},$$

$$[L_{e\nu}] = \text{Wm}^{-2}\text{sr}^{-1}\text{s}, \qquad\qquad [L_{e\lambda}] = \text{Wm}^{-2}\text{sr}^{-1}\text{m}^{-1},$$

Since wavelength and frequency are connected to each other via $\nu\lambda = c$, P_ν and P_λ are related via

$$P_\nu = \frac{dP}{d\nu} = \frac{dP}{d\lambda}\frac{d\lambda}{d\nu} = P_\lambda\left(-\frac{c}{\nu^2}\right), \tag{5.7}$$

and similarly for $L_{e\nu}$ and $L_{e\lambda}$. The $-$ sign is because of the changing direction of integration when switching between ν and λ and usually is omitted.

The spectral power in wavelength thus can be obtained via

$$P_\lambda = \int_A \int_{2\pi} L_{e\lambda} \cos\theta \, d\Omega \, dA, \tag{5.8}$$

and analogously for P_ν. The radiance is given by

$$L_e = \frac{1}{\cos\theta} \frac{\partial^4 P}{\partial A \, \partial \Omega}, \tag{5.9}$$

and similarly for $L_{e\nu}$ and $L_{e\lambda}$.

Another very important radiometric property is the *irradiance* I_e that tells us the power density at a certain point (ξ, η) of the surface. It is often also called the *(spectral) intensity* of the light. It is given as the integral of the radiance over the solid angle,

$$I_e = \int_{2\pi} L_e \cos\theta \, d\Omega = \int_{2\pi} L_e(\xi, \eta; \theta, \phi) \cos\theta \sin\theta \, d\theta \, d\phi. \tag{5.10}$$

The *spectral irradiance* $I_{e\nu}$ or $I_{e\lambda}$ is calculated similarly. The physical dimensions are

$$[I_e] = \text{Wm}^{-2}, \qquad [I_{e\nu}] = \text{Wm}^{-2}\text{s}, \qquad [I_{e\lambda}] = \text{Wm}^{-2}\text{m}^{-1}.$$

The irradiance also is given as
$$I_e = \frac{\partial^2 P}{\partial A}, \tag{5.11}$$
and similarly for $I_{e\nu}$ and $I_{e\lambda}$. Irradiance refers to radiation that is received by the surface. For radiation emitted by the surface, we instead speak of *radiant emittance*, M_e, $M_{e\nu}$, and $M_{e\lambda}$.

As we discussed earlier, the energy of a photon is proportional to its frequency, $E_{ph} = h\nu = hc/\lambda$. Thus, the spectral power P_λ is proportional to the *spectral photon flow* $\Psi_{ph,\lambda}$,
$$P_\lambda = \Psi_{ph,\lambda} \frac{hc}{\lambda}, \tag{5.12}$$
and similarly for P_ν and $\Psi_{ph,\nu}$. The total photon flow Ψ_{ph} is related to the spectral photon flow via
$$\Psi_{ph} = \int_0^\infty \Psi_{ph,\nu}\, d\nu = \int_0^\infty \Psi_{ph,\lambda}\, d\lambda. \tag{5.13}$$
The physical dimensions of the (spectral) photon flow are
$$[\Psi_{ph}] = s^{-1}, \qquad [\Psi_{ph,\nu}] = 1, \qquad [\Psi_{ph,\lambda}] = s^{-1} m^{-1}.$$

The *(spectral) photon flux* Φ_{ph} is defined as the photon flow per area,
$$\Phi_{ph} = \frac{\partial^2 \Psi_{ph}}{\partial A}, \tag{5.14}$$
and similarly for $\Phi_{ph,\nu}$ and $\Phi_{ph,\lambda}$. The physical dimensions are
$$[\Phi_{ph}] = s^{-1} m^{-2}, \qquad [\Phi_{ph,\nu}] = m^{-2}, \qquad [\Phi_{ph,\lambda}] = s^{-1} m^{-2} m^{-1}.$$

By comparing Eqs. (5.11) and (5.14) and looking at Eq. (5.12), we find
$$I_{e\lambda} = \Phi_{ph,\lambda} \frac{hc}{\lambda}, \tag{5.15}$$
and analogously for $I_{e\nu}$ and $\Phi_{ph,\nu}$.

5.3 Blackbody radiation

If we take a piece of metal, for example, and start heating it up, it will start to glow, first in a reddish colour, then getting more and more yellowish as we increase the temperature even further. It thus emits electromagnetic radiation that we call *thermal radiation*. Understanding this phenomenon theoretically and correctly predicting the emitted spectrum was one of the most important topics of physics in the late nineteenth century.

For discussing thermal radiation, the concept of the *blackbody* is very useful. A blackbody, which does not exist in nature, absorbs all the radiation that is incident on it, regardless of wavelength and angle of incidence. Its reflectivity therefore is 0. Of course, since it will also emit light according to its equilibrium temperature, it does not need to appear black to the eye.

Two approximations for the blackbody spectrum were presented around the turn of the century: First, in 1896, Wilhelm Wien empirically derived the following expression for the spectral blackbody radiance:

$$L_{e\lambda}^{W}(\lambda; T) = \frac{C_1}{\lambda^5} \exp\left(-\frac{C_2}{\lambda T}\right), \tag{5.16}$$

where λ and T are the wavelength and the temperature, respectively. While this approximation gives good results for short wavelengths, it fails to predict the emitted spectrum at long wavelengths, thus in the infrared.

Secondly, in 1900 and in a more complete version in 1905, Lord Rayleigh and James Jeans, derived the equation

$$L_{e\lambda}^{RJ}(\lambda; T) = \frac{2ck_B T}{\lambda^4}, \tag{5.17}$$

where $k_B \approx 1.381 \times 10^{-23}$ J/K is the Boltzmann constant. The derivation of this equation was based on electrodynamic arguments. While $L_{e\lambda}^{RJ}$ is in good agreement to measured values at long wavelengths, it diverges to infinity for short wavelength. Further, the radiant emittance, which is obtained via integration over all wavelengths, diverges towards infinity. This so-called *ultraviolet catastrophe* demonstrates that Rayleigh and Jeans did not succeed in developing a model that could adequately describe thermal radiation.

In 1900, Max Planck found an equation that interpolates between the Wien approximation and the Rayleigh–Jeans law,

$$L_{e\lambda}^{BB}(\lambda; T) = \frac{2hc^2}{\lambda^5} \frac{1}{\exp\left(\frac{hc}{\lambda k_B T}\right) - 1}, \tag{5.18a}$$

where $c \approx 2.998 \times 10^8$ m/s is the speed of light *in vacuo* and $h \approx 6.626 \times 10^{-34}$ m²kg/s, which is now known as the *Planck constant*. Via Eq. (5.7) we find the *Planck law* expressed as a function of the frequency ν,

$$L_{e\nu}^{BB}(\nu; T) = \frac{2h\nu^3}{c^2} \frac{1}{\exp\left(\frac{h\nu}{k_B T}\right) - 1}. \tag{5.18b}$$

It is remarkable to see that the Planck law contains three fundamental constants, c, k_B, and h, which are among the most important constants in physics.

Figure 5.3 shows the spectrum of a blackbody with a 6,000 K temperature and the Wien approximation and the Rayleigh–Jeans law. We indeed see that the Wien approximation fits well at short wavelengths, while the Rayleigh–Jeans law matches well at long wavelengths but completely fails at short wavelengths.

Both the Wien approximation (Eq. (5.16)) and the Rayleigh–Jeans law (Eq. (5.17)) can be directly derived from the Planck law:

For short wavelengths,

$$\exp\left(\frac{hc}{\lambda k_B T}\right) \gg 1,$$

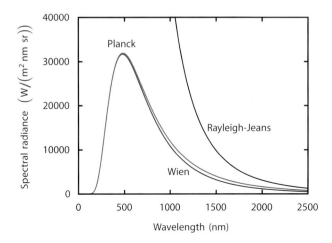

Figure 5.3: The blackbody spectrum at 6,000 K as calculated with the Wien approximation, the Rayleigh–Jeans law and the Planck law.

such that the -1 can be ignored and we arrive at the Wien approximation with $C_1 = 2hc^2$ and $C_2 = hc/k_B$.

For long wavelengths we can use the approximation

$$\exp\left(\frac{hc}{\lambda k_B T}\right) - 1 \approx \frac{hc}{\lambda k_B T},$$

which directly results in the Rayleigh–Jeans law.

The total radiant emittance of a blackbody is given by

$$M_e^{BB}(T) = \int_0^\infty \int_0^{2\pi} \int_0^{\pi/2} L_{e\lambda}^{BB}(\lambda; T) \cos\theta \sin\theta \, d\theta \, d\phi \, d\lambda = \sigma T^4, \tag{5.19}$$

where

$$\sigma = \frac{2\pi^5 k_B^4}{15 c^2 h^3} \approx 5.670 \times 10^{-8} \, \frac{\text{W}}{\text{m}^2 \text{K}^4} \tag{5.20}$$

is the *Stefan–Boltzmann* constant. Equation (5.19) is known as the *Stefan-Boltzmann law*. As a matter of fact, it had already been discovered in 1879 and 1884 by Jožef Stefan and Ludwig Boltzmann, respectively, i.e. about twenty years prior to the derivation of Planck's law. This law is very important because it tells us that if the temperature of a body (in K) is doubled, it emits 16 times as much power. Small temperature variations thus have a large influence on the total emitted power.

Another important property of blackbody radiation is Wien's displacement law, which states that the wavelength of maximal radiance is inversely proportional to the temperature,

$$\lambda_{\max} T = b \approx 2.898 \times 10^{-3} \, \text{mK}. \tag{5.21}$$

Figure 5.4 shows the spectra for three different temperatures. Note the strong increase in radiance with temperature and also the shift of the maximum to shorter wavelengths.

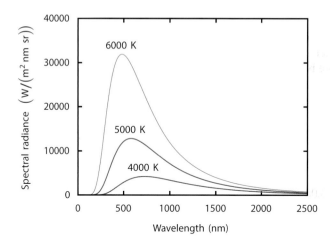

Figure 5.4: The blackbody spectrum at three different temperatures.

5.4 Wave-particle duality

In Planck's law, as stated in Eqs. (5.18), the constant h appeared for the first time. Its product with the frequency, $h\nu = hc/\lambda$ has the unit of energy. Planck himself did not see the implications of h. It was Einstein, who understood in 1905 that Planck's law actually has to be interpreted such that light comes in quanta of energy with the size

$$E_{\mathrm{ph}} = h\nu. \tag{5.22}$$

Nowadays, these quanta are called *photons*. In terms of classical mechanics we could say that *light shows the behaviour of particles*.

On the other hand, we have seen in Chapter 4 that light also shows *wave character* which becomes obvious when looking at the propagation of light through space or at reflection and refraction at a flat interface. It also was discovered that other particles, such as electrons, show wave-like properties.

This behaviour is called *wave-particle* duality and is a very intriguing property of *quantum mechanics* that was discovered and developed in the first quarter of the twentieth century. Many discussion were held on how this duality was to be interpreted – but this is out of the focus of this book. So we will just accept that depending on the situation, light might behave as a wave or as a particle.

5.5 Solar spectra

As we already mentioned in Chapter 3, only photons of appropriate energy can be absorbed and hence generate electron-hole pairs in a semiconductor material. Therefore, it is important to know the spectral distribution of the solar radiation, i.e. the number of photons of a particular energy as a function of the wavelength λ. Two quantities are used to describe the solar radiation spectrum, namely the *spectral irradiance* $I_{e\lambda}$ and the *spectral photon flux* $\Phi_{\mathrm{ph}}(\lambda)$. We defined these quantities in Section 5.2.

5. Solar radiation

The surface temperature of the Sun is about 6,000 K. If it was a perfect blackbody, it would emit a spectrum as described by Eqs. (5.18), which give the spectral radiance. To calculate the spectral *irradiance* a blackbody with the size and position of the Sun would have on Earth, we have to multiply the spectral radiance with the solid angle of the Sun as seen from Earth,

$$I_{e\lambda}^{BB}(T; \lambda) = L_{e\lambda}^{BB}(T; \lambda)\Omega_{\text{Sun}}. \quad (5.23)$$

We can calculate Ω_{Sun} with

$$\Omega_{\text{Sun}} = \pi \left(\frac{R_{\text{Sun}}}{\text{AU} - R_{\text{Earth}}}\right)^2. \quad (5.24)$$

Using R_{Sun} = 696,000 km, an astronomical unit AU = 149,600,000 km, and R_{Earth} = 6,370 km, we find

$$\Omega_{\text{Sun}} \approx 68.5 \, \mu\text{sr}. \quad (5.25)$$

The blackbody spectrum is illustrated in Figure 5.5. The spectrum outside the atmosphere of Earth is already very different. It is called the AM0 spectrum, because no (or "zero") atmosphere is traversed. AM0 also is shown in Figure 5.5. The irradiance at AM0 is $I_e(\text{AM0}) = 1361 \, \text{Wm}^{-2}$.

When solar radiation passes through the atmosphere of Earth, it is attenuated. The most important parameter that determines the solar irradiance under clear sky conditions is the distance that the sunlight has to travel through the atmosphere. This distance is the shortest when the Sun is at the zenith, i.e. directly overhead. The ratio of an actual path length of the sunlight to this minimal distance is known as the *optical air mass*. When the Sun is at its zenith the optical air mass is unity and the spectrum is called the air mass 1 (AM1) spectrum. When the Sun is at an angle θ with the zenith, the air mass is given by

$$\text{AM} := \frac{1}{\cos\theta}. \quad (5.26)$$

For example, when the Sun is 60° from the zenith, i.e. 30° above the horizon, we receive an AM2 spectrum. Depending on the position on the Earth and the position of the Sun in the sky, terrestrial solar radiation varies both in intensity and spectral distribution. The attenuation of solar radiation is due to scattering and absorption by air molecules, dust particles and/or aerosols in the atmosphere. Especially, water vapour (H_2O), oxygen (O_2) and carbon dioxide (CO_2) cause absorption. Since this absorption is wavelength-selective, it results in gaps in the spectral distribution of solar radiation as apparent in Figure 5.5. Ozone absorbs radiation with wavelengths below 300 nm. Depletion of ozone from the atmosphere allows more ultraviolet radiation to reach the Earth, with consequent harmful effects upon biological systems. CO_2 molecules contribute to the absorption of solar radiation at wavelengths above 1 µm. By changing the CO_2 content in the atmosphere the absorption in the infrared is enhanced, which has consequences for our climate.

Solar cells and photovoltaic modules are produced by many different companies and laboratories. Further, many different solar cell technologies are investigated and sold. It is therefore of utmost importance to define conditions that allow a comparison of all different solar cells and PV modules. These conditions are the *standard test conditions* (STC), characterized by an irradiance of 1,000 Wm^{-2}, an AM1.5 spectrum and a cell temperature

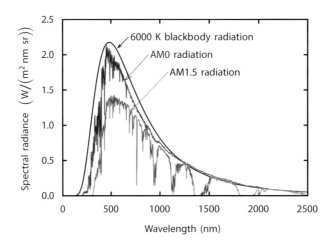

Figure 5.5: Different solar spectra: the blackbody spectrum of a blackbody at 6,000 K, the extraterrestrial AM0 spectrum and the AM1.5 spectrum.

of 25° C. The AM1.5 spectrum is a reference solar spectral distribution, being defined in the International Standard IEC 60904-3 [29]. This spectrum is based on the solar irradiance received on a Sun-facing plane surface tilted at 37° to the horizontal. Both direct sunlight, diffuse sunlight and the wavelength-dependent albedo of light bare soil are being taken into account. Albedo is the part of the solar radiation that is reflected by the Earth's surface and depends on the reflectivity of the environment. The total irradiance of the AM1.5 spectrum is 1,000 Wm^{-2} and is close to the maximum received at the surface of the Earth on a cloudless day. Both STC and the AM1.5 spectrum are used all over the world in both industry and (test) laboratories. The power generated by a PV module at STC is thus expressed in the unit watt peak, Wp.

The actual amount of solar radiation that reaches a particular place on the Earth is extremely variable. In addition to the regular daily and annual variation due to the apparent motion of the Sun, irregular variations have to be taken into account that are caused by local atmospheric conditions, such as clouds. These conditions particularly influence the direct and diffuse components of solar radiation. The direct component of solar radiation is that part of the sunlight that directly reaches the surface. Scattering of the sunlight in the atmosphere generates the diffuse component. Albedo may also be present in the total solar radiation. We use the term global radiation to refer to the total solar radiation, which is made up of these three components.

The design of an optimal photovoltaic system for a particular location depends on the availability of the solar insolation data at the location. Solar irradiance integrated over a period of time is called solar irradiation. For example, the average annual solar irradiation in the Netherlands is 1,000 kWh/m^2, while in the Sahara the average value is 2,200 kWh/m^2, thus more than twice as high. We will discuss these issues in more detail in Chapters 17 and 18.

5.6 Exercises

5.1 The radius of the Sun is 6.96×10^8 m and the distance between the Sun and the Earth is roughly 1.50×10^{11} m. You may assume that the Sun is a perfect sphere and that the irradiance arriving on the Earth is the value for AM0, 1,350 W/m². Calculate the temperature at the surface of the Sun.

5.2 The power output of a solar module under STC is 250 W_p. The module has an area of 2 m², while the solar spectrum provides 1,000 W/m² under STC.

  (a) How much radiative power does the module receive in watts?

  (b) What is the overall efficiency of the module?

5.3 4×10^{18} photons from a monochromatic light source with a wavelength of 500 nm are incident in 1 second on a surface with area of 20 cm². Calculate the photon flux (in $s^{-1}m^{-2}$).

5.4 What is the ratio between the path length of sunlight through the atmosphere if the Sun is directly overhead ($\theta = 0°$) and when the Sun is $\theta = 30°$ overhead?

5.5 What is the ratio between the path length at an air mass 1.5 and the path length at an air mass 2?

5.6 Figure 5.6 shows the AM1.5 solar spectrum illustrated by the yellow region. A rough approximation of the AM1.5 solar spectrum is represented by the blue region. The spectral irradiance of this region is divided into two spectral ranges,

$$I_{e\lambda} = 1.00 \times 10^9 \text{ Wm}^{-2}\text{m}^{-1} \text{ for } 250 \text{ nm} < \lambda < 1,000 \text{ nm},$$
$$I_{e\lambda} = 0.25 \times 10^9 \text{ Wm}^{-2}\text{m}^{-1} \text{ for } 1,000 \text{ nm} < \lambda < 2,000 \text{ nm}.$$

  (a) Calculate the irradiance I_e.

  (b) Calculate the photon flux.

5.7 Solar simulators are used to study the performance of solar cells and modules. A solar simulator is a lamp which has to simulate the solar spectrum under standard test conditions (STC). Figure 5.7 shows the spectral irradiance of a solar simulator. The spectral irradiance density is divided into two spectral ranges,

$$I_{e\lambda} = 4.0 \times 10^{15} \text{ Wm}^{-2}\text{m}^{-2} \times \lambda - 1.2 \times 10^9 \text{ Wm}^{-2}\text{m}^{-1}$$
$$\text{for } 300 \text{ nm} < \lambda < 800 \text{ nm},$$
$$I_{e\lambda} = -4.0 \times 10^{15} \text{ Wm}^{-2}\text{m}^{-2} \times \lambda + 5.2 \times 10^9 \text{ Wm}^{-2}\text{m}^{-1}$$
$$\text{for } 800 \text{ nm} < \lambda < 1300 \text{ nm},$$

where the wavelength λ is expressed in metres.

  (a) Calculate the total irradiance of the solar simulator.

  (b) What is the photon flux of the solar simulator?

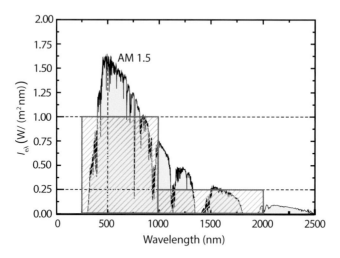

Figure 5.6: Simplified AM1.5 spectrum.

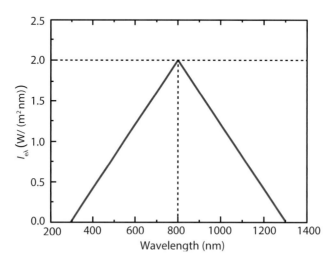

Figure 5.7: Spectral power density of a solar simulator.

6
Basic semiconductor physics

6.1 Introduction

In this chapter we start with the discussion of some important concepts from *semiconductor physics*, which are required to understand the operation of solar cells. After giving a brief introduction into semiconductor physics in this chapter, we will discuss the most important *generation and recombination* mechanisms in Chapter 7. Finally, we will focus on the physics of *semiconductor junctions* in Chapter 8.

The first successful solar cell was made from crystalline silicon (c-Si), which still is by far the most widely used PV material. Therefore we shall use c-Si as an example to explain the concepts of semiconductor physics that are relevant to solar cell operation. This discussion will give us a basic understanding of how solar cells based on other semiconductor materials work.

The *central semiconductor parameters* that determine the design and performance of a solar cell are:

1. Concentrations of doping atoms, which can be of two different types: *donor atoms*, which donate free electrons or *acceptor atoms*, which accept electrons. The concentrations of donor and acceptor atoms are denoted by N_D and N_A, respectively, and determine the width of the space-charge region of a junction, as we see in Chapter 8.

2. The mobility μ and the diffusion coefficient D of charge carriers is used to characterize the transport of carriers due to drift and diffusion, respectively, which we will discuss in Section 6.5.

3. The lifetime τ and the diffusion length L of the *excess carriers* characterize the recombination-generation processes, discussed in Chapter 7.

4. The band gap energy E_g, and the complex refractive index $n - ik$, where k is linked

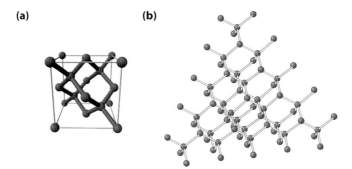

Figure 6.1: (a) A diamond lattice unit cell representing a unit cell of single crystal Si [30], (b) the atomic structure of a part of single crystal Si.

to the absorption coefficient α, characterize the ability of a semiconductor to absorb electromagnetic radiation.

6.2 Atomic structure

The *atomic number* of silicon is 14, which means that 14 electrons are orbiting the nucleus. In ground state configuration, two electrons are in the first shell, both in the 1s orbital. Further, eight electrons are in the second shell, two in the 2s and six in the 2p orbitals. Hence, four electrons are in the third shell, which is the outermost shell for a Si atom. Only these four electrons interact with other atoms, for example via forming chemical bonds. They are called the *valence electrons*.

Two Si atoms are bonded together when they share each other's valence electron. This is the so-called *covalent bond* that is formed by two electrons. Since Si atoms have four valence electrons, they can be covalently bonded to four other Si atoms. In the crystalline form each Si atom is covalently bonded to four neighbouring Si atoms, as illustrated in Figure 6.1.

In the ground state, two valence electrons live in the 3s orbital and the other two are present in the three 3p orbitals (p_x, p_y and p_z). In this state only the two electrons in the 3p orbitals can form bonds as the 3s orbital is full. In a silicon crystal, where every atom is *symmetrically* connected to four others, the Si atoms are present as so-called sp_3 *hybrids*. The 3p and 3s orbitals are mixed forming 4 sp_3 orbitals. Each of these four orbitals is occupied by one electron that can form a covalent bond with a valence electron from a neighbouring atom.

All bonds have the same length and the angles between the bonds are equal to 109.5°. The number of bonds that an atom has with its immediate neighbours in the atomic structure is called the *coordination number* or *coordination*. Thus, in single crystal silicon, the coordination number for all Si atoms is four, we can also say that Si atoms are fourfold coordinated. A *unit cell* can be defined, from which the crystal lattice can be reproduced by duplicating the unit cell and stacking the duplicates next to each other. Such a regular atomic arrangement is described as a structure with *long range order*.

6. Basic semiconductor physics

A diamond lattice unit cell represents the real lattice structure of monocrystalline silicon. Figures 6.1 (a) and (b) show the arrangement of the unit cell and the atomic structure of single crystal silicon, respectively. One can determine from Figure 6.1 (a) that there are eight Si atoms in the volume of the unit cell. Since the lattice constant of c-Si is 543.07 pm, one can easily calculate that the density of atoms is approximately 5×10^{22} cm^{-3}. Figure 6.1 (a) shows the crystalline Si atomic structure with no foreign atoms. In practice, a semiconductor sample always contains some impurity atoms. When the concentration of impurity atoms in a semiconductor is insignificant we refer to such a semiconductor as an *intrinsic semiconductor*.

At practical operational conditions, e.g. at room temperature,[1] there are always some of the covalent bonds broken. The breaking of the bonds results in liberating the valence electrons from the bonds and making them mobile through the crystal lattice. We refer to these electrons as free electrons (henceforth simply referred to as electrons). The position of a missing electron in a bond, which can be regarded as positively charged, is referred to as a hole. This situation can be easily visualized by using the *bonding model* illustrated in Figure 6.2.

In the bonding model the atomic cores (atoms without valence electrons) are represented by circles and the valence or bonding electrons are represented by lines interconnecting the circles. In case of c-Si, one Si atom has four valence electrons and four nearest neighbours. Each of the valence electrons is equally shared with the nearest neighbour. There are therefore eight lines terminating on each circle. In an ideal Si crystal at 0 K all valence electrons take part in forming covalent bonds between Si atoms and therefore no free electrons are present in the lattice. This situation is schematically shown in Figure 6.2 (a).

At temperatures higher than 0 K the bonds start to break due to the absorption of thermal energy. This process results in the creation of mobile electrons and holes. Figure 6.2 (b) shows a situation where a covalent bond is broken and one electron departs from the bond leaving a so-called *hole* behind. A single line between the atoms in Figure 6.2 (b) represents the remaining electron of the broken bond. When a bond is broken and a hole created, a valence electron from a neighbouring bond can "jump" into this empty position and restore the bond. The consequence of this transfer is that at the same time the jumping electron creates an empty position in its original bond. The subsequent "jumps" of a valence electron can be viewed as a motion of the hole – a positive charge representing the empty position – in the opposite direction to the motion of the valence electron through the bonds.

Since the breaking of a covalent bond leads to the formation of an electron-hole pair, in intrinsic semiconductors the concentration of electrons is equal to the concentration of holes. In intrinsic silicon at 300 K approximately 1.5×10^{10} cm^{-3} broken bonds are present. This number then also gives the concentration of holes, p, and electrons, n. Hence, at 300 K, $n = p = 1.5 \times 10^{10}$ cm^{-3}. This concentration is called the *intrinsic carrier concentration* and is denoted as n_i.

---

[1] In semiconductor physics most of the time a temperature of 300 K is assumed.

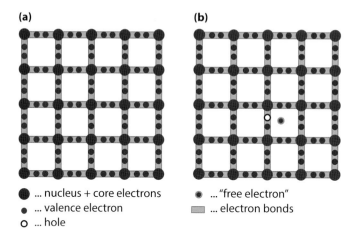

Figure 6.2: The bonding model for c-Si. (a) No bonds are broken. (b) A bond between two Si atoms is broken resulting in a free electron and hole.

6.3 Doping

The concentrations of electrons and holes in c-Si can be manipulated by *doping*. Doping of silicon means that atoms of other elements replace Si atoms in the crystal lattice. The substitution has to be carried out by atoms with three or five valence electrons. The most used elements to dope c-Si are boron (B) and phosphorus (P), with atomic numbers of 5 and 15, respectively.

The process of doping action can best be understood with the aid of the bonding model and is illustrated in Figure 6.3. When introducing a phosphorus atom into the c-Si lattice, four of the five phosphorus atom valence electrons will readily form bonds with the four neighbouring Si atoms. The fifth valence electron cannot take part in forming a bond and becomes rather weakly bound to the phosphorus atom. It is easily liberated from the phosphorus atom by absorbing the thermal energy which is available in the c-Si lattice at room temperature. Once free, the electron can move throughout the lattice. In this way the phosphorus atom that replaces a Si atom in the lattice *"donates"* a free (mobile) electron into the c-Si lattice. The impurity atoms that enhance the concentration of electrons are called *donors*. We denote the concentration of donors by N_D.

An atom with three valence electrons such as boron cannot form bonds with all four neighbouring Si atoms when it replaces a Si atom in the lattice. However, it can readily *"accept"* an electron from a nearby Si-Si bond. The thermal energy of the c-Si lattice at room temperature is sufficient to enable an electron from a nearby Si-Si bond to attach itself to the boron atom and complete the bonding to the four Si neighbours. In this process a hole is created that can move around the lattice. The impurity atoms that enhance the concentration of holes are called *acceptors*. We denote the concentration of acceptors by N_A.

Note that by substituting Si atoms with only one type of impurity atoms the concentration of only one type of mobile charge carrier is increased. Charge neutrality of the material is nevertheless maintained because the sites of the bonded and thus fixed impur-

6. Basic semiconductor physics

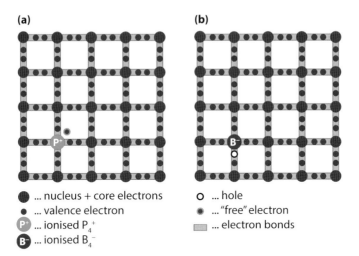

Figure 6.3: The doping process illustrated using the bonding model. (a) A phosphorus (P) atom replaces a Si atom in the lattice resulting in the positively-ionized P atom and a free electron. (b) A boron (B) atom replaces a Si atom resulting in the negatively ionized B atom and a hole.

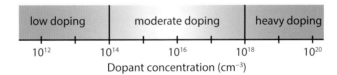

Figure 6.4: The range of doping levels used in c-Si.

ity atoms become charged. The donor atoms become positively ionized and the acceptor atoms become negatively ionized.

The possibility to control the electrical conductivity of a semiconductor by doping is one of the most important semiconductor features. The electrical conductivity in semiconductors depends on the concentration of electrons and holes as well as their mobility. The concentration of electrons and holes is influenced by the amount of the doping atoms that are introduced into the atomic structure of the semiconductor. Figure 6.4 shows the range of doping that is used in case of c-Si. We denote a semiconductor as *p*-type or *n*-type when holes or electrons, respectively, dominate its electrical conductivity. When one type of charge carriers has a higher concentration than the other type, these carriers are called majority carriers (holes in the *p*-type and electrons in the *n*-type), while the other type with lower concentration are then called minority carriers (electrons in the *p*-type and holes in the *n*-type).

6.4 Carrier concentrations

6.4.1 Intrinsic semiconductors

Any operation of a semiconductor device depends on the concentration of carriers that transport charge inside the semiconductor and hence cause electrical currents. In order to determine and to understand device operation it is important to know the precise concentration of these charge carriers. In this section the concentrations of charge carriers inside a semiconductor are derived assuming the semiconductor is under *thermal equilibrium*. The term equilibrium is used to describe the unperturbed state of a system, to which no external voltage, magnetic field, illumination, mechanical stress, or other perturbing forces are applied. In the equilibrium state, the observable parameters of a semiconductor do not change with time.

In order to determine the carrier concentration one has to know the function of density of allowed energy states of electrons and the occupation function of the allowed energy states. The density of energy states function, $g(E)$, describes the number of allowed states per unit volume and energy. Usually it is abbreviated with *density of states* function (DoS). The occupation function is the Fermi–Dirac distribution function, $f(E)$, which describes the ratio of states filled with an electron to total allowed states at given energy E. In an isolated Si atom, electrons are allowed to have only discrete energy values. The periodic atomic structure of single crystal silicon results in the ranges of allowed energy states for electrons that are called *energy bands,* and the excluded energy ranges, forbidden gaps or *band gaps*. Electrons that are liberated from the bonds determine the charge transport in a semiconductor. Therefore, we further discuss only those bands of energy levels, which concern the valence electrons. Valence electrons, which are involved in the covalent bonds, have their allowed energies in the *valence band* (VB) and the allowed energies of electrons liberated from the covalent bonds form the *conduction band* (CB). The valence band is separated from the conduction band by a band of forbidden energy levels. The maximum attainable valence-band energy is denoted E_V, and the minimum attainable conduction-band energy is denoted E_C. The energy difference between the edges of these two bands is called the band gap energy or band gap, E_G, and it is an important material parameter:

$$E_G = E_C - E_V. \tag{6.1}$$

At room temperature (300 K), the band gap of crystalline silicon is 1.12 eV. A plot of the allowed electron energy states as a function of position is called the energy band diagram; an example is shown in Figure 6.5 (a).

The density of energy states at an energy E in the conduction band close to E_C and in the valence band close to E_V are given by

$$g_C(E) = 4\pi \left(\frac{2m_n^*}{h^2}\right)^{\frac{3}{2}} \sqrt{E - E_C}, \tag{6.2a}$$

$$g_V(E) = 4\pi \left(\frac{2m_p^*}{h^2}\right)^{\frac{3}{2}} \sqrt{E - E_V}, \tag{6.2b}$$

where m_n^* and m_p^* are the *effective masses* of electrons and holes, respectively. As the electrons and holes move in the periodic potential of the c-Si crystal, the mass has to be replaced by the effective mass, which takes the effect of a periodic force into account. The

6. Basic semiconductor physics

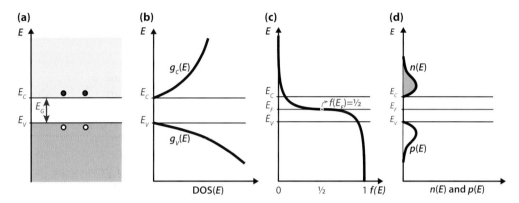

Figure 6.5: (a) The basic energy band diagram with electrons and holes indicated in the conduction and valence bands, respectively. (b) The density of states (DOS) functions g_C in the conduction band and g_V in the valence band. (c) The Fermi–Dirac distribution. (d) The electron and hole densities in the conduction and valence bands, respectively, obtained by combining (b) and (c).

effective mass is also averaged over different directions to take anisotropy into account. Both g_C and g_V have a parabolic shape, which is also illustrated in Figure 6.5 (b).

The Fermi–Dirac distribution function is given by

$$f(E) = \frac{1}{1 + \exp\left(\frac{E-E_F}{k_B T}\right)}, \tag{6.3}$$

where k_B is Boltzmann's constant ($k_B = 1.38 \times 10^{-23}$ J/K) and E_F is the so-called *Fermi energy*. $k_B T$ is the *thermal energy*, at 300 K it is 0.0258 eV. The Fermi energy – also called *Fermi level* – is the electrochemical potential of the electrons in a material and in this way it represents the averaged energy of electrons in the material. The Fermi–Dirac distribution function is illustrated in Figure 6.5 (c). Figure 6.6 illustrates the Fermi–Dirac distribution at different temperatures.

The carriers that contribute to charge transport are electrons in the conduction band and holes in the valence band. The concentration of electrons in the conduction band and the total concentration of holes in the valence band is obtained by multiplying the density of states function with the distribution function and integrating across the whole energy band, as illustrated in Figure 6.5 (d):

$$n(E) = g_C(E) f(E), \tag{6.4a}$$
$$p(E) = g_V(E) \left[1 - f(E)\right]. \tag{6.4b}$$

The total concentration of electrons and holes in the conduction band and valence band, respectively, is then obtained via integration,

$$n = \int_{E_C}^{E_{\text{top}}} n(E) dE, \tag{6.5a}$$

$$p = \int_{E_{\text{bottom}}}^{E_V} p(E) dE. \tag{6.5b}$$

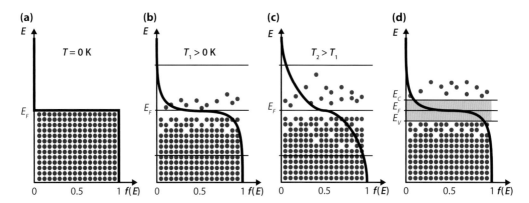

Figure 6.6: The Fermi–Dirac distribution function. (a) For $T = 0$ K, all allowed states below the Fermi level are occupied by two electrons. (b, c) At $T > 0$ K not all states below the Fermi level are occupied and there are some states above the Fermi level that are occupied. (d) In an energy gap between bands no electrons are present.

Substituting the density of states and the Fermi–Dirac distribution function into Eq. (6.5) the resulting expressions for n and p are obtained after solving the equations. The full derivation can be found for example in [24]:

$$n = N_C \exp\left(\frac{E_F - E_C}{k_B T}\right) \quad \text{for} \quad E_C - E_F \geq 3 k_B T, \tag{6.6a}$$

$$p = N_V \exp\left(\frac{E_V - E_F}{k_B T}\right) \quad \text{for} \quad E_F - E_V \geq 3 k_B T, \tag{6.6b}$$

where N_C and N_V are the *effective* densities of the conduction band states and the valence band states, respectively. They are defined as

$$N_C = 2 \left(\frac{2\pi m_n^* k_B T}{h^2}\right)^{\frac{3}{2}} \quad \text{and} \quad N_V = 2 \left(\frac{2\pi m_p^* k_B T}{h^2}\right)^{\frac{3}{2}} \tag{6.7}$$

For crystalline silicon, we have at 300 K

$$N_C = 3.22 \times 10^{19} \text{ cm}^{-3}, \tag{6.8a}$$

$$N_V = 1.83 \times 10^{19} \text{ cm}^{-3}. \tag{6.8b}$$

When the requirement that the Fermi level lies in the band gap more than $3 k_B T$ from either band edge is satisfied, the semiconductor is referred to as a *nondegenerate* semiconductor.

If an intrinsic semiconductor is in equilibrium, we have $n = p = n_i$. By multiplying the corresponding sides of Eqs. (6.6) we obtain

$$np = n_i^2 = N_C N_V \exp\left(\frac{E_V - E_C}{k_B T}\right) = N_C N_V \exp\left(-\frac{E_g}{k_B T}\right), \tag{6.9}$$

which is independent of the position of the Fermi level and thus valid for doped semiconductors as well. When we denote the position of the Fermi level in the intrinsic material

6. Basic semiconductor physics

E_{Fi} we may write

$$n_i = N_C \exp\left(\frac{E_{Fi} - E_C}{k_B T}\right) = N_V \exp\left(\frac{E_V - E_{Fi}}{k_B T}\right). \tag{6.10}$$

From Eq. (6.10) we can easily find the position of E_{Fi} to be

$$E_{Fi} = \frac{E_C + E_V}{2} + \frac{k_B T}{2} \ln\left(\frac{N_V}{N_C}\right) = E_C - \frac{E_g}{2} + \frac{k_B T}{2} \ln\left(\frac{N_V}{N_C}\right). \tag{6.11}$$

The Fermi level E_{Fi} lies close to the midgap $[(E_C + E_V)/2]$; a slight shift is caused by the difference in the densities of the valence and conduction band.

6.4.2 Doped semiconductors

It has already been mentioned in Section 6.3 that the concentrations of electrons and holes in c-Si can be manipulated by doping. The concentration of electrons and holes is influenced by the amount of impurity atoms that substitute silicon atoms in the lattice. Under the assumption that the semiconductor is uniformly doped and in equilibrium, a simple relationship between the carrier and dopant concentrations can be established. We assume that at room temperature the dopant atoms are ionized. Inside a semiconductor the local charge density is given by

$$\rho = q\left(p + N_D^+ - n - N_A^-\right), \tag{6.12}$$

where q is the elementary charge ($q \approx 1.602 \times 10^{-19}$ C). N_D^+ and N_A^- denote the density of the *ionized donor and acceptor* atoms, respectively. As every ionized atom corresponds to a free electron (hole), N_D^+ and N_A^- tell us the concentration of electrons and holes due to doping, respectively.

Under equilibrium conditions, the local charge of the uniformly doped semiconductor is zero, which means that the semiconductor is charge-neutral everywhere. We thus can write:

$$p + N_D^+ - n - N_A^- = 0. \tag{6.13}$$

As previously discussed, the thermal energy available at room temperature is sufficient to ionize almost all the dopant atoms. We therefore may assume

$$N_D^+ \approx N_D \quad \text{and} \quad N_A^- \approx N_A, \tag{6.14}$$

and hence

$$p + N_D - n - N_A = 0, \tag{6.15}$$

which is the common form of the *charge neutrality equation*.

Let us now consider an n-type material. At room temperature almost all donor atoms N_D are ionized and donate an electron into the conduction band. Under the assumption that $N_A = 0$, Eq. (6.15) becomes

$$p + N_D - n = 0. \tag{6.16}$$

Further, assuming that

$$N_D \approx N_D^+ \approx n, \tag{6.17}$$

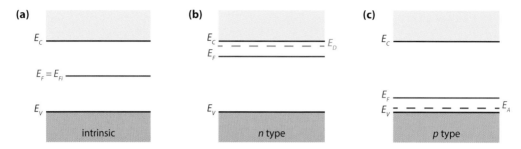

Figure 6.7: A shift of the position of the Fermi energy in the band diagram and the introduction of the allowed energy level into the bandgap due to the doping.

we can expect that the concentration of holes is lower than that of electrons, and becomes very low when N_D becomes very large. From Eq. (6.9), we can calculate the concentration of holes in the n-type material more accurately,

$$p = \frac{n_i^2}{n} \approx \frac{n_i^2}{N_D} \ll n. \tag{6.18}$$

In the case of a p-type material almost all acceptor atoms N_A are ionized at room temperature. Therefore, they accept an electron and leave a hole in the valence band. Under the assumption that $N_D = 0$, Eq. (6.15) becomes

$$p - n - N_A = 0. \tag{6.19}$$

Further, when assuming that

$$N_A \approx N_A^- \approx p, \tag{6.20}$$

we can expect that the concentration of electrons is lower than that of holes. From Eq. (6.9), we can calculate the concentration of electrons in the p-type material more accurately,

$$n = \frac{n_i^2}{p} \approx \frac{n_i^2}{N_A} \ll p. \tag{6.21}$$

Inserting donor and acceptor atoms into the lattice of crystalline silicon introduces allowed energy levels into the forbidden bandgap, as illustrated in Fig. 6.7. For example, the fifth valence electron of the P atom does not take part in forming a bond, is rather weakly bound to the atom and is easily liberated from the P atom. The energy of the liberated electron lies in the CB. The energy levels, which we denote E_D, of the weakly-bound valence electrons of the donor atoms have to be positioned close to the CB, as shown in Fig. 6.7 (b). This means that an electron, which occupies the E_D level, is localized in the vicinity of the donor atom. Similarly, the acceptor atoms introduce allowed energy levels E_A close to the VB, as in Fig. 6.7 (c).

Doping also influences the position of the Fermi energy. When we increase the electron concentration by increasing the donor concentration the Fermi energy will increase, which is represented by bringing the Fermi energy closer to the CB in the band diagram. In the p-type material the Fermi energy is moved closer to the VB. These changes in the Fermi energy positions are illustrated in Figure 6.7.

6. Basic semiconductor physics

The position of the Fermi level in an n-type semiconductor can be calculated with Eqs. (6.6a); in a p-type semiconductor Eqs. (6.6b) and (6.20) can be used:

$$E_C - E_F = k_B T \ln\left(\frac{N_C}{N_D}\right) \qquad \text{for } n\text{-type,} \qquad (6.22a)$$

$$E_F - E_V = k_B T \ln\left(\frac{N_V}{N_A}\right) \qquad \text{for } p\text{-type.} \qquad (6.22b)$$

Example

This example demonstrates how much the concentration of electrons and holes can be manipulated by doping. A c-Si wafer is uniformly doped with 1×10^{17} cm^{-3} P atoms. P atoms act as donors and therefore at room temperature the concentration of electrons is almost equal to the concentration of donor atoms:

$$n = N_D^+ \approx N_D = 10^{17} \text{ cm}^{-3}.$$

The concentration of holes in the n-type material is calculated from Eq. (6.17),

$$p = \frac{n_i^2}{n} = \frac{(1.5 \times 10^{10})^2}{10^{17}} = 2.25 \times 10^3 \text{ cm}^{-3}.$$

We notice that there is a difference of 14 orders of magnitude between n (10^{17} cm^{-3}) and p (2.25×10^3 cm^{-3}). It is now obvious why electrons in n-type materials are called the majority carriers and holes the minority carriers. We can calculate the change in the Fermi energy due to the doping. Let us assume that the reference energy level is the bottom of the conduction band, $E_C = 0$ eV. Using Eq. (6.11) we calculate the Fermi energy in the intrinsic c-Si:

$$E_{Fi} = E_C - \frac{E_g}{2} + \frac{k_B T}{2} \ln\left(\frac{N_V}{N_C}\right) = -\frac{1.12}{2} + \frac{0.0258}{2} \ln\left(\frac{1.83 \times 10^{19}}{3.22 \times 10^{19}}\right) = -0.57 \text{ eV}.$$

The Fermi energy in the n-type doped c-Si wafer is calculated from Eq. (6.6a):

$$E_F = E_C + k_B T \ln\left(\frac{n}{N_C}\right) = 0.0258 \times \ln\left(\frac{10^{17}}{3.22 \times 10^{19}}\right) = -0.15 \text{ eV}.$$

We notice that the doping with P atoms has resulted in the shift of the Fermi energy towards the CB. Note that when $n > N_C$, $E_F > E_C$ and the Fermi energy lies in the CB.

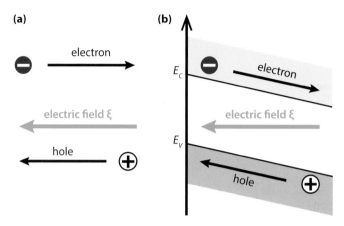

Figure 6.8: Visualization of (a) the direction of carrier fluxes due to an electric field; and (b) the corresponding band diagram.

6.5 Transport properties

In contrast to the equilibrium conditions, under operational conditions a net electrical current flows through a semiconductor device. The electrical currents are generated in a semiconductor due to the transport of charge by electrons and holes. The two basic transport mechanisms in a semiconductor are *drift* and *diffusion*.

6.5.1 Drift

Drift is charged particle motion in response to an electric field. In an electric field the force acts on the charged particles in a semiconductor, which accelerates the positively charged holes in the direction of the electric field and the negatively charged electrons in the opposite direction. Because of collisions with the thermally vibrating lattice atoms and ionized impurity atoms, the carrier acceleration is frequently disturbed. The resulting motion of electrons and holes can be described by average drift velocities $\mathbf{v}_{dn}$ and $\mathbf{v}_{dp}$ for electrons and holes, respectively. In the case of low electric fields, the average drift velocities are directly proportional to the electric field ξ as expressed by

$$\mathbf{v}_{dn} = -\mu_n \xi, \tag{6.23a}$$
$$\mathbf{v}_{dp} = \mu_p \xi. \tag{6.23b}$$

The proportionality factor is called mobility μ. It is a central parameter that characterizes electron and hole transport due to drift. Although the electrons move in the opposite direction to the electric field, because the charge of an electron is negative the resulting electron drift current is in the same direction as the electric field. This is illustrated in Figure 6.8.

The electron and hole drift current densities are then given as

$$\mathbf{J}_{n,\text{drift}} = -qn\mathbf{v}_{dn} = qn\mu_n \xi, \tag{6.24a}$$
$$\mathbf{J}_{p,\text{drift}} = qp\mathbf{v}_{dp} = qp\mu_p \xi. \tag{6.24b}$$

6. Basic semiconductor physics

Combining Eqs. (6.24a) and (6.24b) leads to the total drift current,

$$\mathbf{J}_{\text{drift}} = q(p\mu_p + n\mu_n)\xi. \tag{6.25}$$

Mobility is a measure of how easily the charge particles can move through a semiconductor material. For example, for c-Si with a doping concentration N_D or N_A at 300 K, the mobilities are

$$\mu_n \approx 1360 \text{ cm}^2\text{V}^{-1}\text{s}^{-1},$$
$$\mu_p \approx 450 \text{ cm}^2\text{V}^{-1}\text{s}^{-1},$$

respectively. As mentioned earlier, the motion of charged carriers is frequently disturbed by collisions. When the number of collisions increases, the mobility decreases. Increasing the temperature increases the collision rate of charged carriers with the vibrating lattice atoms, which results in a lower mobility. Increasing the doping concentration of donors or acceptors leads to more frequent collisions with the ionized dopant atoms, which also results in a lower mobility. The dependence of mobility on doping and temperature is discussed in more detail in standard textbooks for [24, 31].

6.5.2 Diffusion

Diffusion is a process whereby particles tend to spread out from regions of high particle concentration into regions of low particle concentration as a result of random thermal motion. The driving force of diffusion is a *gradient* in the particle concentration. In contrast to the drift transport mechanism, the particles need not be charged to be involved in the diffusion process. Currents resulting from diffusion are proportional to the gradient in particle concentration. For electrons and holes, they are given by

$$\mathbf{J}_{n,\text{ diff}} = qD_n \nabla n, \tag{6.26a}$$
$$\mathbf{J}_{p,\text{ diff}} = -qD_p \nabla p. \tag{6.26b}$$

Combining Eqs. (6.26a) and (6.26b) leads to the total diffusion current,

$$\mathbf{J}_{\text{diff}} = q(D_n \nabla n - D_p \nabla p). \tag{6.27}$$

The proportionality constants, D_n and D_p are called the electron and hole *diffusion coefficients*, respectively. The diffusion coefficients of electrons and holes are linked with the mobilities of the corresponding charge carriers by the *Einstein relationship* that is given by

$$\frac{D_n}{\mu_n} = \frac{D_p}{\mu_p} = \frac{k_B T}{q}. \tag{6.28}$$

Figure 6.9 visualizes the diffusion process as well as the resulting directions of particle fluxes and current.

Combining Eqs. (6.25) and (6.27) leads to the total current,

$$\begin{aligned}\mathbf{J} &= \mathbf{J}_{\text{drift}} + \mathbf{J}_{\text{diff}} \\ &= q(p\mu_p + n\mu_n)\xi + q(D_n \nabla n - D_p \nabla p).\end{aligned} \tag{6.29}$$

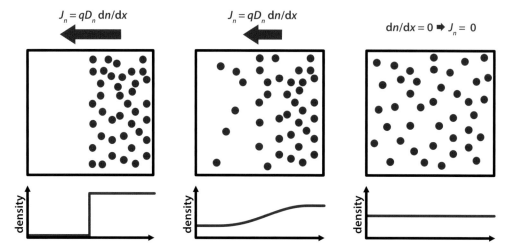

Figure 6.9: Visualization of electron diffusion.

Example

To obtain some idea about values of diffusion coefficients, let us assume a c-Si wafer at room temperature, doped with donors, $N_D = 10^{14}\,\text{cm}^{-3}$. According to Eq. (6.28),

$$D_N = \frac{k_B T}{q}\mu_n = 0.0258\,\text{V} \times 1360\,\text{cm}^2\text{V}^{-1}\text{s}^{-1} = 35\,\text{cm}^2\text{s}^{-1}.$$

6.5.3 Continuity equations

We have seen in Section 4.5 that charge is a conserved quantity, i.e. it fulfils the continuity equation (Eq. (4.34)). Now we will formulate the continuity equation in terms of the electron and hole concentrations. Therefore we have to take drift, diffusion, and recombination as well as generation processes into account; the latter are discussed in Chapter 7.

For electrons, the rate at which the concentration changes is given by

$$\frac{\partial n}{\partial t} = \left.\frac{\partial n}{\partial t}\right|_{\text{drift}} + \left.\frac{\partial n}{\partial t}\right|_{\text{diff}} + \left.\frac{\partial n}{\partial t}\right|_{R} + \left.\frac{\partial n}{\partial t}\right|_{G}, \tag{6.30a}$$

where R and G denote recombination and generation, respectively. For holes, we obtain

$$\frac{\partial p}{\partial t} = \left.\frac{\partial p}{\partial t}\right|_{\text{drift}} + \left.\frac{\partial p}{\partial t}\right|_{\text{diff}} + \left.\frac{\partial p}{\partial t}\right|_{R} + \left.\frac{\partial p}{\partial t}\right|_{G}. \tag{6.30b}$$

In these equations $\partial n/\partial t$ ($\partial p/\partial t$) is the time rate change in the electron (hole) concentration.

Without any generation or recombination, the number of electrons and holes is conserved, i.e. Eq. 4.34 can be applied. We thus obtain

$$\left.\frac{\partial n}{\partial t}\right|_{\text{drift}} + \left.\frac{\partial n}{\partial t}\right|_{\text{diff}} = \frac{1}{q}\nabla \times \mathbf{J}_n, \tag{6.31a}$$

$$\left.\frac{\partial p}{\partial t}\right|_{\text{drift}} + \left.\frac{\partial p}{\partial t}\right|_{\text{diff}} = -\frac{1}{q}\nabla \times \mathbf{J}_p, \tag{6.31b}$$

where $\mathbf{J}_n$ ($\mathbf{J}_p$) is the electron (hole) current density.

The equations can be written in a more compact form when introducing substitutions,

$$\left.\frac{\partial n}{\partial t}\right|_{R} = -R_n, \tag{6.32a}$$

$$\left.\frac{\partial p}{\partial t}\right|_{R} = -R_p, \tag{6.32b}$$

$$\left.\frac{\partial n}{\partial t}\right|_{G} = G_n, \tag{6.33a}$$

$$\left.\frac{\partial p}{\partial t}\right|_{G} = G_p. \tag{6.33b}$$

R_n (R_p) denotes the net thermal recombination-generation rate of electrons (holes), G_n (G_p) is the generation rate of electrons (holes) due to other processes, such as photogeneration. We will discuss these processes in detail in Chapter 7. Substituting Eqs. (6.31), (6.32) and (6.33) into Eq. (6.30) finally leads to

$$\frac{\partial n}{\partial t} = \frac{1}{q}\nabla \times \mathbf{J}_n - R_n + G_n, \tag{6.34a}$$

$$\frac{\partial p}{\partial t} = -\frac{1}{q}\nabla \times \mathbf{J}_p - R_p + G_p. \tag{6.34b}$$

6.6 Exercises

6.1 Which statement is *not* true about a semiconductor at room temperature and in the dark?

(a) The band gap of the semiconductor is within a range of 0.5 eV to 3 eV.

(b) In a semiconductor only a few electrons fill the conduction band.

(c) In a semiconductor electrons almost fully fill the valence band.

(d) In a semiconductor the valence band is fully filled with electrons.

6.2 What are the 'free' carriers in the different electronic bands of a semiconductor?

(a) Electrons in conduction band and holes in valence band.

(b) Holes in conduction band and electrons in valence band.

(c) Both electrons and holes in conduction band when an electric field is applied.

(d) Both electrons and holes in valence band when an electric field is applied.

6.3 What is the maximum number of electrons that can occupy the 3p energy state in an atom?

(a) 2 electrons.

(b) 6 electrons.

(c) Only 1 electron can occupy this state according to the Pauli exclusion principle.

(d) 8 electrons.

6.4 According to the molecular description of the band gap, which of the following statements are true?

(a) The anti-bonding level is a lower energy state than the bonding level.

(b) The anti-bonding level is a higher energy state than the bonding level.

(c) The closer the two neighbouring atoms making a molecular orbital are together, the smaller the energy splitting between the bonding and anti-bonding levels.

(d) The bonding state of a molecular orbital represents the conduction band.

6.5 In case of *p*-doping of Si, the energy of the acceptor state...

(a) ...is located in the Si bandgap, relatively close to the conduction band.

(b) ...is located out of the Si bandgap, relatively close to the conduction band edge.

(c) ...is located in the Si bandgap, relatively close to the valence band.

(d) ...is located out of the Si bandgap, relatively close to the valence band edge.

6.6 Considering Si as the bulk material, which of the dopant materials below can be used in order to achieve *n* doping?

(a) Ge.

(b) In.

(c) Ga.

(d) As.

6.7 A photon with energy $E_{ph} = 1.35$ eV is absorbed in a semiconductor creating one electron-hole pair. At the same time, the energy lost due to thermal relaxation of the electron and hole is 0.27 eV. What is the bandgap of the semiconductor?

6. Basic semiconductor physics

6.8 Silicon is doped with 10^{16} arsenic atoms per cm^3. Assume intrinsic carrier concentration equal to 1.5×10^{10} cm^{-3} at room temperature. What is the minority carrier concentration at room temperature ($T = 300$ K)?

6.9 Which of the following statements is false regarding *diffusion* of charge carriers in semiconductors?

(a) Diffusion occurs only in the presence of an electric field.

(b) During diffusion, net flow of carriers takes place from high concentration to low concentration regions.

(c) Over time, carriers will diffuse randomly throughout the cell, until concentrations of different regions are uniform.

(d) Diffusion occurs faster at higher temperatures.

6.10 Which of the following statements is false regarding the *drift* of charge carriers?

(a) It is the dominant carrier transport mechanism when an electric field is applied in the semiconductor.

(b) Holes move in the direction opposite to that of the applied field.

(c) During drift, the carrier transport is characterized by their respective electron/hole mobilities.

(d) During drift, electrons and holes move in opposite directions.

6.11 An isolated piece of a *p*-type ...

(a) ...is positively charged, due to a hole excess.

(b) ...maintains charge neutrality.

(c) ...is negatively charged, due to an electron excess.

6.12 What is the electron configuration of phosphorus (atomic number 15) in its ground state?

(a) $1s^2 2s^2 2p^6 3s^0 3p^5$.

(b) $1s^2 2s^2 2p^6 3s^1 3p^4$.

(c) $1s^2 2s^2 2p^6 3s^2 3p^3$.

(d) $1s^2 2s^2 2p^6 3s^3 3p^2$.

6.13 If we assume there is no light absorption, then the conductivity of an intrinsic semiconductor...

(a) ...decreases when temperature increases.

(b) ...is zero at $T = 0$ K.

(c) ...is generally lower than the conductivity of an insulator.

(d) ...is not affected in case of doping.

6.14 The diffusion coefficient of electrons in silicon is $D_n = 36$ cm^{2}s^{-1}. In a silicon layer, the electron density drops linearly from $n = 2.7 \times 10^{16}$ cm^{-3} down to $n = 10^{15}$cm^{-3} over a distance of 2 µm. What is the electron diffusion current density $J_{n,\,\text{diff}}$ induced by such a density gradient?

6.15 From the previous data, what electric field over the gradient zone of 2 µm would be required to compensate the electron diffusion flux with an electron drift flux? The electron mobility in silicon is $\mu_n = 1,350$ cm^{2}V^{-1}s^{-1} and the direction of drift to compensate diffusion flux is directed from lower density to higher density.

7
Generation and recombination of electron-hole pairs

7.1 Introduction

Assume that a piece of a semiconductor is illuminated by a light pulse, which leads to excitations of electrons from the valence band to the conduction band and consequently to the creation of holes in the valence band. This illumination will disturb the semiconductor from the state of thermal equilibrium. In the valence band an excess concentration of holes $p > p_0$ is present, where p_0 denotes the equilibrium concentration. Similarly, in the conduction band the electron concentration is larger than the equilibrium concentration, $n > n_0$, which means that *excess carriers are present*. It is clear that in this *non-equilibrium* state Eq. (6.9) is no longer fulfilled. Instead, now an inequality is valid,

$$np > n_i^2. \qquad (7.1)$$

After the pulse stops, the excess electrons will *recombine* with holes until the equilibrium state is reached again. Depending on the properties of the semiconductor, different types of recombination can and will occur. In this chapter we discuss the most important of these mechanisms.

The recombination rate strongly determines the performance of the solar cells. On the one hand, it will reduce the current that can be collected and hence utilized from the solar cell. However, the photogeneration rate is often several magnitudes higher than the recombination rate, such that the effect of recombination on the solar cell current is negligible. On the other hand, the recombination rate strongly determines the saturation current density; the more recombination, the higher the saturation current density. As we will see in Chapter 8, a high saturation current density has a detrimental effect on the solar cell

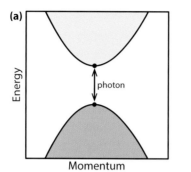 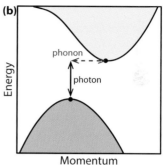

Figure 7.1: Illustrating the dispersion diagram of (a) an direct bandgap semiconductor and (b) an indirect bandgap semiconductor.

voltage, and hence on the energy conversion efficiency.

Before we can actually start with the treatment on the different generation and recombination mechanisms, we have to distinguish between *direct* and *indirect* semiconductors. Figure 7.1 shows the energy–momentum space of the electrons, which also is called the *electronic dispersion diagram*. On the vertical axis the energy state in the electronic bands is plotted. On the horizontal axis the momentum of the charge carrier is shown. This momentum is also called the *crystal momentum*, and is related to the *wave vector* **k** of the electron. It is important to realize that the position of the valence and conduction band may differ in different directions of the lattice coordination. We can understand this by realizing that the crystal can look very different if we look at it from different directions. Hence, the energy levels in which electrons and holes can propagate across the crystal depend on the direction. The dispersion diagram of silicon is discussed in more detail in Chapter 12.

For a direct band gap material the highest point of the valence band is vertically aligned with the lowest point of the conduction band, as shown in Figure 7.1 (a). This means that exciting an electron from the valence to the conduction band requires only the energy provided by a photon without any additional momentum transfer. In contrast, for an indirect band gap the highest point of the valence band is not aligned with the lowest point of the conduction band, as shown in Figure 7.1 (b). Therefore, exciting an electron from the valence to the conduction band requires energy provided by a photon and momentum provided from vibrations of the crystal lattice. Just as light can be described as wave and as particle, the lattice vibrations can also be described as waves (the vibrations) and as particles, which we call *phonons*. A phonon therefore is a quantized mode of lattice vibrations. Transferring momentum from the lattice to the electron can be described as an electron that absorbs a phonon and hence changes its momentum.

It is clear that the excitation of an electron induced by photon absorption is more likely to happen for direct band gap materials than for indirect band gap materials and hence the absorption coefficient for direct band gap materials is significantly higher than for indirect band gap materials. The same principle makes the reverse process of radiative recombination more likely to happen in a direct band gap material. In an indirect band gap material additional momentum is required to make the electron and hole recombine.

7. Generation and recombination of electron-hole pairs

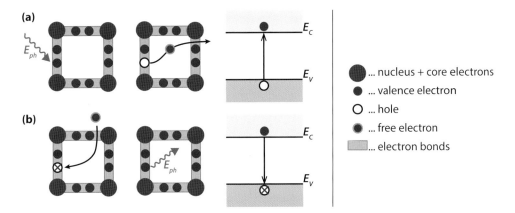

Figure 7.2: Visualization of bandgap-to-bandgap (a) generation; and (b) recombination processes using the bonding model and the energy band diagram.

Crystalline silicon is an indirect band gap material. In such a material, the radiative recombination is inefficient and recombination will be dominated by the Auger recombination mechanism discussed in Section 7.4. For direct band gap materials such as gallium arsenide under moderate illumination conditions, radiative recombination will be the dominant loss mechanism of charge carriers. For very high illumination conditions, Auger recombination starts to play a role as well. Gallium arsenide based solar cells are discussed in Chapter 13.

7.2 Bandgap-to-bandgap processes

Generation and recombination processes that happen from bandgap to bandgap are also called *direct* generation and recombination. They are much more likely to happen in direct bandgap materials, as no change in momentum is required for an electron that is excited into the conduction band. These processes are most usually *radiative*, which means that a photon is absorbed when an electron-hole pair is created, and a photon is emitted if electron-hole pairs recombine directly. In this section, we will also introduce important concepts for generation and recombination, such as the *minority carrier lifetime*.

7.2.1 Radiative generation

When light penetrates into a material it will be (partially) absorbed as it propagates through the material. If the photon energy is higher than the bandgap energy of the semiconductor, it is sufficient to break bonds and to excite a valence electron into the conduction band, leaving a hole behind in the valance band; hence electron-hole pairs are created. This process is called *photogeneration* and illustrated in Figure 7.2 (a).

The absorption profile in the material depends on the absorption coefficient of the material, which is wavelength dependent. The most frequent approach to calculate the absorption profile of photons in semiconductor devices is by using *Lambert-Beer's law* that we introduced in Eq. (4.25), where it was formulated for the decay of intensity. Here we formulate it with the *photon flux* $\Phi_{\text{ph},\lambda}(x)$, which decreases exponentially with the distance x travelled through the absorber,

$$\Phi_{\text{ph},\lambda}(x) = \Phi^0_{\text{ph},\lambda} \exp\left[-\alpha(\lambda)x\right], \tag{7.2}$$

where $\Phi^0_{\text{ph},\lambda}$ is the incident photon flux and $\alpha(\lambda)$ is the absorption coefficient. The photon flux is defined as the number of photons per unit area, unit time and unit wavelength. As we have seen in Chapter 5, it is related to the spectral irradiance $I_{e\lambda}$ of the solar radiation via

$$\Phi^0_{\text{ph},\lambda} = I_{e\lambda} \frac{\lambda}{hc}. \tag{7.3}$$

The spectral generation rate $G_{L,\lambda}(x)$, which is the number of electron-hole pairs generated at a depth x in the film per second unit volume and unit wavelength, by photons of wavelength λ, is calculated according to

$$G_{L,\lambda}(x) = \eta_g \Phi^0_{\text{ph},\lambda} \alpha(\lambda) \exp[-\alpha(\lambda)x], \tag{7.4}$$

where we assume zero reflection. η_g is the generation quantum efficiency, usually assumed equal to unity. This assumption means that every photon generates one and only one electron-hole pair. The optical generation rate $G_L(x)$ is calculated from the spectral generation rate by integrating over the desired wavelength range,

$$G_L(x) = \int_{\lambda_1}^{\lambda_2} G_{L,\lambda}(x)\, d\lambda. \tag{7.5}$$

It has the unit $[G_L] = \text{cm}^{-3}\text{s}^{-1}$. The optical generation rate is related to the absorption profile $A(x)$ in the film via

$$G_L(x) = \eta_g A(x). \tag{7.6}$$

Hence,

$$A(x) = \int_{\lambda_1}^{\lambda_2} \Phi^0_{\text{ph},\lambda} \alpha(\lambda) \exp\left[-\alpha(\lambda)x\right] d\lambda. \tag{7.7}$$

Because of the photogeneration, excess electrons and holes will be created. The rate of the generated excess electrons' concentration is equal to the rate of the generated excess hole concentration per second; therefore we can write

$$\left.\frac{\partial n}{\partial t}\right|_{\text{light}} = \left.\frac{\partial p}{\partial t}\right|_{\text{light}} = G_L. \tag{7.8}$$

Example

Let us calculate the total absorption in a $d = 300$ μm thick c-Si wafer for light with an incident irradiance of $1,000\, \text{Wm}^{-2}$. For simplicity, we assume that all the light has a wavelength of 500 nm. The optical constants of c-Si at this wavelength are: refractive index is $n = 4.293$, extinction coefficient $k = 0.045$ and absorption coefficient $\alpha = 1.11 \times 10^4\, \text{cm}^{-1}$.

First we calculate the photon flux at 500 nm corresponding to the irradiance of $1000\, \text{Wm}^{-2}$.

Using Eq. (7.3) we obtain

$$\Phi^0_{ph,\lambda} = I_{e\lambda}\frac{\lambda}{hc} = \frac{1000\,\text{Wm}^{-2} \times 500 \times 10^{-9}\,\text{m}}{6.625 \times 10^{-34}\,\text{Js} \times 2.998 \times 10^{8}\,\text{ms}^{-1}} = 2.5 \times 10^{21}\,\text{m}^{-2}\text{s}^{-1}.$$

Using Eq. (4.15) we calculate how many incident photons are reflected from the surface,

$$R = \left|\frac{\tilde{n}_0 - \tilde{n}_1}{\tilde{n}_0 + \tilde{n}_1}\right|^2 = \left|\frac{1.0 - 0.0i - (4.293 - 0.045i)}{1.0 - 0.0i + 4.293 - 0.045i}\right|^2 = 0.38.$$

Using Lambert–Beer's law (Eq. (7.2)) we calculate the photon flux at the backside of the wafer, i.e. at 300 µm distance from the surface. We take the reflected light into account by adapting Eq. (7.2),

$$\Phi_{ph,\lambda}(d) = \Phi^0_{ph,\lambda}(1-R)\exp[-\alpha(\lambda)d]$$
$$= 2.5 \times 10^{21}\,\text{m}^{-2}\text{s}^{-1} \times (1 - 0.38)\exp\left(-1.11 \times 10^6\,\text{m}^{-1} \times 300 \times 10^{-6}\,\text{m}\right) \approx 0.$$

The total absorption in the wafer is the difference between the photon flux at the surface after reflection and the photon flux at the back of the wafer,

$$A = \Delta\Phi_{ph,\lambda} = \Phi^0_{ph,\lambda}(1-R) - \Phi_{ph,\lambda}(d)$$
$$= 2.5 \times 10^{21}\,\text{m}^{-2}\text{s}^{-1} \times (1 - 0.38) - 0 = 1.55 \times 10^{21}\,\text{m}^{-2}\text{s}^{-1}.$$

When we assume that all the absorbed photons generate one electron-hole pair ($\eta_g = 1$), we can calculate the photocurrent density corresponding to the absorbed photon flux,

$$J_{ph} = qA = 1.602 \times 10^{-19}\,\text{C} \times 1.55 \times 10^{21}\,\text{m}^{-2}\text{s}^{-1} = 248.31\,\text{Cm}^{-2}\text{s}^{-1} = 248.31\,\text{Am}^{-2}.$$

7.2.2 Direct recombination

We will now discuss direct recombination which mainly occurs in direct bandgap semiconductors, such as gallium arsenide. It is illustrated in Figure 7.2 (b). In this section we roughly follow the derivation by Sze [31].

Let us first look at the situation at *thermal equilibrium*. If the temperature is higher than 0 K, the crystal lattice is vibrating. This vibrational energy will be sufficient to break bonds from time to time, which leads to the generation of electron-hole pairs at a generation rate G_{th}, where the *th* stands for *thermal*. As we are in thermal equilibrium, the expression

$$np = n_i^2 \tag{7.9}$$

must be valid. Hence, recombination takes place at the same rate as generation,

$$R_{th} = G_{th}. \tag{7.10}$$

We may assume that the direct recombination rate is proportional to the concentration of electrons in the conduction band and to the concentration of the available holes in the valence band,

$$R^* = \beta np, \tag{7.11}$$

where β is a proportionality factor. For the thermal recombination we have

$$R_{\text{th}} = \beta n_0 p_0. \tag{7.12}$$

We now look at a situation where the semiconductor is illuminated such that a constant generation rate G_L is present throughout the volume of the semiconductor. In this situation excess electrons and holes are created. As the electron and hole concentrations increase, the recombination rate will also increase according to Eq. (7.11). At some point, the generation and recombination rates will be the same, such that n and p do not change any more. This situation is called the *steady state* situation. The total recombination and generation rates are given by

$$R^* = \beta n p = \beta (n_0 + \Delta n)(p_0 + \Delta p), \tag{7.13}$$
$$G = G_{\text{th}} + G_L, \tag{7.14}$$

where n_0 and p_0 are the equilibrium concentrations. Δn and Δp are the excess carrier concentrations that are given by

$$\Delta n = n - n_0, \tag{7.15a}$$
$$\Delta p = p - p_0. \tag{7.15b}$$

In steady state R^* and G are equal, hence

$$G_L = R^* - G_{\text{th}} = R_d, \tag{7.16}$$

where R_d denotes the *net* radiative recombination rate. By substituting Eqs. (7.11) and (7.12) into Eq. (7.16), we obtain

$$G_L = R_d = \beta(np - n_0 p_0). \tag{7.17}$$

We now assume the semiconductor to be *n*-type and under *low-level injection*, which means that $\Delta n \ll n$ and $p \ll n$. Under these assumptions the recombination rate becomes

$$R_d \approx \beta n_0 (p - p_0) = \frac{p - p_0}{\tau_{pd}}, \tag{7.18}$$

where

$$\tau_{pd} = \frac{1}{\beta n_0} \tag{7.19}$$

is the *lifetime of the minority holes* in the *n*-type semiconductor. Clearly, if no excess carriers are present, $R_d = 0$. The excess carrier concentration is given as the product of the generation rate and the lifetime,

$$p - p_0 = G_L \tau_{pd}. \tag{7.20}$$

To understand the meaning of the *lifetime*, we consider a situation where the light and hence generation at the rate G_L is suddenly shut off. Without loss of generality we may assume that the light is shut off at the instant $t = 0$. As there is no longer any generation, the excess carrier concentration will change according to the differential equation

$$\frac{dp}{dt} = -\frac{p(t) - p_0}{\tau_{pd}}. \tag{7.21}$$

If we solve this equation with the boundary condition $p(t=0) = p_0 + G_L \tau_{pd}$, we find

$$p(t) = p_0 + G_L \tau_{pd} \exp\left(-\frac{t}{\tau_{pd}}\right). \tag{7.22}$$

We therefore see that the minority carrier lifetime is the time constant at which an excess carrier concentration decays exponentially, if external generation is no longer taking place.

For a p-type semiconductor at low-level injection ($\Delta p \ll p$ and $n \ll p$) we find similar expressions,

$$R_d \approx \beta p_0 (n - n_0) = \frac{n - n_0}{\tau_{nd}}, \tag{7.23}$$

where the lifetime of the electrons is given by

$$\tau_{nd} = \frac{1}{\beta p_0}. \tag{7.24}$$

Let us assume that in a semiconductor several recombination mechanisms are present, with recombination rates $R_1, R_2, \ldots$. The total recombination rate then is given by

$$R_{\text{tot}} = R_1 + R_2 + \cdots \tag{7.25}$$

If we have an n-type semiconductor under low-level injection we may assume

$$R_{\text{tot}} = \frac{p - p_0}{\tau_{p1}} + \frac{p - p_0}{\tau_{p2}} + \cdots = \frac{p - p_0}{\tau_{p,\text{tot}}}. \tag{7.26}$$

Hence, the overall (total) lifetime is related to the lifetimes of the different processes via

$$\frac{1}{\tau_{p,\text{tot}}} = \frac{1}{\tau_{p1}} + \frac{1}{\tau_{p2}} + \cdots \tag{7.27}$$

The more recombination mechanisms are present, the shorter the overall lifetime of the excess minority carriers.

The last aspect that we want to discuss in this section is a situation where the excess carrier generation is not uniform throughout the semiconductor. In this case diffusion of excess carriers takes place in the semiconductor until they recombine with majority carriers. The distance over which the minority carriers diffuse is defined as:

$$L_n = \sqrt{D_n \tau_n} \quad \text{for electrons in a } p\text{-type material,} \tag{7.28a}$$

$$L_p = \sqrt{D_p \tau_p} \quad \text{for holes in an } n\text{-type material,} \tag{7.28b}$$

where D_n and D_p are the diffusion coefficients as introduced in Section 6.5. L_n and L_p are called the *minority carrier diffusion lengths*. They follow from the solution of the steady-state continuity equations (6.34) and indicate the distance over which the minority carrier densities drop by a factor of $1/e$ (where e is the base of the natural logarithm); this is demonstrated for a p-n homojunction in Appendix B. Often these diffusion lengths are interpreted as the typical distance minority carriers diffuse before being annihilated.

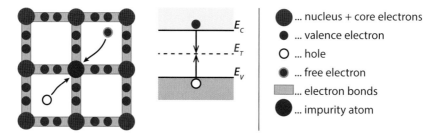

Figure 7.3: Visualization of Shockley–Read–Hall recombination using the bonding model and the energy band diagram.

Example

To get an idea about the diffusion lengths, let us assume the mobility of electrons in a p-type c-Si wafer to be $\mu_n \approx 1{,}250$ cm^{2}V^{-1}s^{-1}, which corresponds to doping of $N_A = 10^{14}$ cm^{-3}, and $\tau_n = 10^{-6}$ s. Further, assume room temperature (300 K). For the given conditions, the electron diffusion length in the p-type c-Si can be calculated from Eq. (7.28a):

$$L_n = \sqrt{D_N \tau_n} = \sqrt{\frac{k_B T}{q} \mu_n \tau_n} = \sqrt{0.0258\,\text{V} \times 1250\,\text{cm}^2\text{V}^{-1}\text{s}^{-1} \times 10^{-6}\text{s}} = 57\,\mu\text{m}.$$

7.3 Shockley-Read-Hall recombination

In the *Shockley–Read–Hall* (SRH) recombination process, which is illustrated in Figure 7.3, the recombination of electrons and holes does not occur directly from bandgap to bandgap. It is facilitated by an *impurity atom* or *lattice defects*. Their concentration is usually small compared to the acceptor or donor concentrations. These recombination centres introduce allowed energy levels (E_T) within the forbidden gap, so-called *trap states*. An electron can be *trapped* at such a defect and consequently recombines with a hole that is attracted by the trapped electron. Though this process seems to be less likely than the direct thermal recombination, it is the dominant recombination-generation process in semiconductors at most operational conditions. The process is typically non-radiative and the excess energy is dissipated into the lattice in the form of heat. The name is a reverence to William B. Shockley, William T. Read and Robert N. Hall, who published the theory of this recombination mechanism in 1952 [32, 33].

We distinguish between two kinds of traps: first, *donor-type* traps that are neutral when they contain an electron and positively charged when they do not. Secondly, *acceptor-type* traps that are negatively charged when they contain an electron and neutral when they do not.

The SRH statistics are based on four processes that are involved in recombination in a single-electron trap:

- r_1: capture of an electron from the conduction band;

7. Generation and recombination of electron-hole pairs

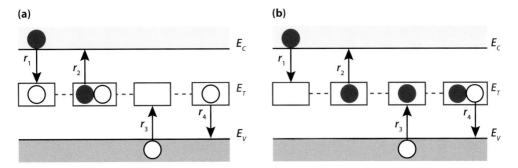

Figure 7.4: Schematic illustration of the processes involved with SRH recombination in a single-electron trap state for (a) a donor-type; and (b) an acceptor-type trap.

- r_2: emission of an electron to the conduction band;
- r_3: capture of a hole from the valence band; and
- r_4: emission of a hole to the valence band.

These processes are illustrated in Figure 7.4 for both donor- and acceptor-type traps. The electron and hole capture rates are proportional to the free carrier concentration, n or p, respectively, the thermal velocity v_{th}, the trap density N_T, the trap occupancy by electron, f, or holes, $1 - f$, and the electron and hole capture cross-section of the traps, σ_n and σ_p. The emission rates are proportional to the trap density and the electron or hole occupancy of the traps, as well as the emission coefficient for electrons or holes, e_n or e_p, respectively. All processes and their rates are listed in Table 7.1.

The *thermal velocity* is the average velocity of the electrons and holes due to thermal movement. It can be obtained by setting the thermal and the kinetic energy equal. Since electrons and holes have three degrees of freedom, we obtain

$$\tfrac{1}{2}m_n^* v_{\text{th},n}^2 = \tfrac{3}{2}k_B T \qquad \tfrac{1}{2}m_p^* v_{\text{th},p}^2 = \tfrac{3}{2}k_B T, \qquad (7.29)$$

where m_n^* and m_p^* are the effective masses of the electrons and holes, respectively. For electrons in silicon and gallium arsenide, the thermal velocity is about 10^7 cm/s. For the following derivation we assume that v_{th} is the same for electrons and holes.

The electron capture cross–section σ_n describes the effectiveness of the trap state to capture an electron. It is a measure of how close an electron has to come to the trap to be captured. It has the unit of area, cm^2. Similarly, σ_p describes the effectiveness of a trap state to capture a hole.

The derivation of the recombination efficacy, below, is valid for both donor- and acceptor-like traps. Therefore, the capture cross-section and emission coefficients in Table 7.1 are generalized by omitting their charge state. Depending on the type of trap considered, the appropriate cross-sections and emission coefficients need to be substituted.

According to the Fermi–Dirac statistics, the carrier distribution in a semiconductor in thermal equilibrium depends on the chemical potential of the carriers, which is referred to as the Fermi level E_F. When the device is illuminated or a bias voltage is applied, the carriers on either side of the band gap are no longer in equilibrium. Yet, they do relax

Table 7.1: Processes associated with single-electron trapping and their rates.

Donor-like traps		Acceptor-like traps	
Process	Rate	Process	Rate
r_1 electron capture	$nv_{th}\sigma_n^+ N_T(1-f)$	r_1 electron capture	$nv_{th}\sigma_n^0 N_T(1-f)$
r_2 electron emission	$e_n^0 N_T f$	r_2 electron emission	$e_n^- N_T f$
r_3 hole capture	$pv_{th}\sigma_p^0 N_T f$	r_3 hole capture	$pv_{th}\sigma_p^- N_T f$
r_4 hole emission	$e_p^+ N_T(1-f)$	r_4 hole emission	$e_p^0 N_T(1-f)$

to a state of quasi-equilibrium with their respective bands. This leads to the definition of the quasi-Fermi levels for electrons and holes, E_{Fn} and E_{Fp}, which determine the carrier concentrations under non-equilibrium conditions. Note that in thermal equilibrium $E_{Fn} = E_{Fp} = E_F$. General expressions for the free electron and hole concentrations n and p, respectively, both under equilibrium and non-equilibrium conditions, read

$$n = N_C \exp\left(\frac{E_{Fn} - E_C}{k_B T}\right), \tag{7.30a}$$

$$p = N_V \exp\left(\frac{E_V - E_{Fp}}{k_B T}\right), \tag{7.30b}$$

where E_C (E_V) is the conduction (valence) band edge and N_C (N_V) the effective density of states in the conduction (valence) band, respectively. According to the Fermi–Dirac statistics the occupation function in thermal equilibrium is given by

$$f(E_T) = \frac{1}{1 + \exp\left(\frac{E_T - E_F}{k_B T}\right)}, \tag{7.31}$$

where E_T is the trap energy.

In thermal equilibrium no net recombination occurs, such that $r_1 = r_2$ and $r_3 = r_4$. Substituting the rate equations from Table 7.1 and Eqs. (7.30–7.31) yields the following expressions for the emission coefficients:

$$e_n = v_{th}\sigma_n N_C \exp\left(\frac{E_T - E_C}{k_B T}\right), \tag{7.32a}$$

$$e_p = v_{th}\sigma_p N_V \exp\left(\frac{E_V - E_T}{k_B T}\right). \tag{7.32b}$$

By substituting N_C and N_V by the intrinsic carrier concentration n_i times an exponential according to Eq. (6.10), we obtain

$$e_n = v_{th}\sigma_n n_i \exp\left(\frac{E_T - E_{Fi}}{k_B T}\right), \tag{7.33a}$$

$$e_p = v_{th}\sigma_p n_i \exp\left(\frac{E_{Fi} - E_T}{k_B T}\right). \tag{7.33b}$$

Consider now a non-equilibrium steady-state situation. It is assumed that the emission coefficients are approximately equal to the emission coefficients under equilibrium. As recombination involves exactly one electron and one hole, at steady state the rate at which the electrons leave the conduction band equals the rate at which the holes leave the valence band. The recombination rate is therefore equal to

$$R_{\text{SRH}} = \frac{dn}{dt} = \frac{dp}{dt} = r_1 - r_2 = r_3 - r_4. \tag{7.34}$$

By substituting the rates from Table 7.1, the expression for the steady-state occupation function can be determined to be

$$f(E_T) = \frac{v_{\text{th}}\sigma_n n + e_p}{V_{\text{th}}\sigma_n n + v_{\text{th}}\sigma_p p + e_n + e_p}. \tag{7.35}$$

Finally, the recombination rate is obtained by substituting Eq. (7.35) in the rate equations as in Eq. (7.34), yielding

$$R_{\text{SRH}} = v_{\text{th}}^2 \sigma_p \sigma_n N_T \frac{np - n_i^2}{v_{\text{th}}\sigma_n n + v_{\text{th}}\sigma_p p + e_n + e_p}, \tag{7.36}$$

where n_i is the intrinsic carrier concentration as in Eq. (6.9).

We can simplify the general expression of Eq. (7.36) when we assume the same capture cross–sections for electrons and holes, $\sigma_n = \sigma_p \equiv \sigma_0$, which yields

$$e_n + e_p = 2v_{\text{th}}\sigma_0 n_i \cosh\left(\frac{E_T - E_{Fi}}{k_B T}\right), \tag{7.37}$$

and hence

$$R_{\text{SRH}} = v_{\text{th}}\sigma N_T \frac{np - n_i^2}{n + p + 2n_i \cosh\left(\frac{E_T - E_{Fi}}{k_B T}\right)}. \tag{7.38}$$

We now look at an *n*-type semiconductor at low injection rate, i.e. the concentration of excess electrons is small compared to the total electron concentration, $n \approx n_0$, where n_0 is the electron concentration under thermal equilibrium. Further, we may assume $n \gg p$. By applying these assumptions to Eq. (7.38) we obtain

$$R_{\text{SRH}} = v_{\text{th}}\sigma N_T \frac{p - p_0}{1 + 2\frac{n_i}{n_0}\cosh\left(\frac{E_T - E_{Fi}}{k_B T}\right)} = c_p N_T (p - p_0) = \frac{p - p_0}{\tau_{p,\text{SRH}}}, \tag{7.39}$$

where c_p is called the *hole capture coefficient*. $\tau_{p,\text{SRH}}$ is the lifetime of holes in an *n*-type semiconductor.

In a similar manner, we can derive for a *p*-type semiconductor at a low injection rate

$$R_{\text{SRH}} = v_{\text{th}}\sigma N_T \frac{n - n_0}{1 + 2\frac{n_i}{p_0}\cosh\left(\frac{E_T - E_{Fi}}{k_B T}\right)} = c_n N_T (n - n_0) = \frac{n - n_0}{\tau_{n,\text{SRH}}}, \tag{7.40}$$

with the electron capture coefficient c_n and the electron lifetime $\tau_{n,\text{SRH}}$.

We see that the lifetime is related to the capture coefficients via

$$\tau_{p,\text{SRH}} = \frac{1}{c_p N_T} \quad \text{and} \quad \tau_{n,\text{SRH}} = \frac{1}{c_n N_T}. \tag{7.41}$$

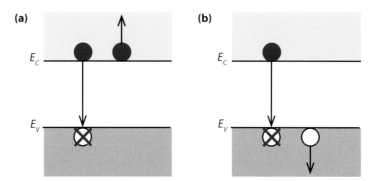

Figure 7.5: Schematic illustration of Auger recombination with (a) two electrons; and (b) two holes involved.

The lifetime of the minority carriers due to Shockley–Read–Hall recombination therefore is indirectly proportional to the trap density N_T. Hence, for a good semiconductor device it is crucial to keep N_T low.

The values of the minority-carrier lifetimes can vary a lot. When the trap concentration in c-Si is very low, τ_n (τ_p) can achieve values around 1 ms. On the other hand, the intentional introduction of gold atoms into Si, which introduce efficient traps into Si, can decrease τ_n (τ_p) to values around 1 ns. Typical minority-carrier lifetimes in most c-Si devices are usually around 1 μs. For an efficient collection of photo–generated carriers in c-Si solar cells the minority–carrier lifetimes should be in the range of tens of milliseconds.

7.4 Auger recombination

We already mentioned that direct recombination is not possible or at least very limited for indirect semiconductors, because both transfer in energy *and* momentum must occur for an electron in the conduction band to recombine with a hole in the valence band. In indirect semiconductors, *Auger recombination* becomes important. In comparison to direct and SRH recombination, which involve two particles, i.e. an electron and a hole, Auger recombination is a *three particle process*, as illustrated in Figure 7.5.

In Auger recombination, momentum and energy of the recombining hole and electron is conserved by transferring energy and momentum an another electron (or hole). If the third particle is an electron, it is excited into higher levels in the electronic band. This excited electron relaxes again, transferring its energy to vibrational energy of the lattice, or *phonon modes*, and finally heat. Similarly, if the third particle is a hole, it is excited into deeper levels of the valence band, from where it rises back to the valence band edge by transferring its energy to phonon modes.

As Auger recombination is a three particle process, the Auger recombination rate R_{Aug} strongly depends on the charge carrier densities for the electrons n and holes p. The recombination rates for electron-electron-hole (eeh) and electron-hole-hole (ehh) processes

7. Generation and recombination of electron-hole pairs

are given by

$$R_{\text{eeh}} = C_n n^2 p, \quad (7.42a)$$

$$R_{\text{ehh}} = C_p n p^2, \quad (7.42b)$$

respectively, where C_n and C_p are the proportionality constants that are strongly dependent on the temperature [31]. R_{eeh} is dominant when the electrons are the majority charge carriers, while R_{ehh} is dominant when the holes are the majority charge carries. Adding them leads to the total Auger recombination rate,

$$R_{\text{Aug}} = R_{\text{eeh}} + R_{\text{ehh}} = C_n n^2 p + C_p n p^2. \quad (7.43)$$

In strongly doped n-type silicon with a donor concentration N_D under low-level injection we can assume that $n \approx N_D$ and hence that the eeh process is dominant. We then can write

$$R_{\text{eeh}} = C_n N_D^2 p. \quad (7.44)$$

Hence, the lifetime can be approximated with

$$\tau_{\text{eeh}} = \frac{1}{C_n N_D^2}. \quad (7.45)$$

Similarly, for strongly doped p-type silicon with acceptor concentration N_A we may assume $p \approx N_A$ and hence the ehh process being dominant,

$$R_{\text{ehh}} = C_p N_A^2 n. \quad (7.46)$$

The lifetime then is

$$\tau_{\text{ehh}} = \frac{1}{C_p N_A^2}. \quad (7.47)$$

As the Auger recombination under these conditions is proportional to the square of the doping levels, the more important it becomes, the higher the doping.

For an n-type semiconductor under high-injection conditions, as it occurs for example for concentrator photovoltaics (CPV) with very high irradiance values, the lifetime can be approximated with

$$\tau_{\text{Auger, hi}} = \frac{1}{(C_n + C_p)\Delta n^2} = \frac{1}{C_a \Delta n^2}, \quad (7.48)$$

where $\Delta n = n - n_0 = p - p_0$ is the excess carrier density [34]. Under these circumstances Auger recombination might also become important for direct bandgap materials such as gallium arsenide.

7.5 Surface recombination

All the recombination mechanisms that we discussed so far are *bulk recombination* mechanisms, which can happen inside the bulk of a semiconductor. For example, impurities can cause trap states within the semiconductor bandgap leading to Shockley–Read–Hall recombination. However, in semiconductor devices, not only is bulk recombination important, but also surface recombination. As we see in Figure 7.6 (a), at a silicon surface

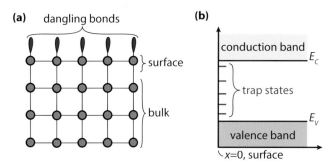

Figure 7.6: (a) Illustrating dangling bonds (surface defects) on a semiconductor surface. (b) The trap states within the bandgap created by the surface defects.

many valence electrons on the surface cannot find a partner to create a covalent bond with. The result is a so-called *dangling bond*, which is a defect. Due to these defects many surface trap states are created within the band gap, as illustrated in Figure 7.6 (b). These defects will induce SRH recombination. In very pure semiconductors, recombination might be dominated by surface recombination. The *surface recombination rate* R_s for an *n*-type semiconductor can be approximated with [31]

$$R_s \approx v_{\text{th}} \sigma_p N_{sT}(p_s - p_0), \tag{7.49}$$

where v_{th} is the thermal velocity in cm/s [see Eq. (7.29)], N_{sT} is the surface trap density in cm^{-2}, and σ_p is the capture cross–section for holes in cm^2. p_s is the hole concentration at the surface and p_0 is the equilibrium hole concentration in the *n*-type semiconductor. For a *p*-type semiconductor, we have to replace σ_p by σ_n, p_s by n_s, and p_0 by n_0.

Note that the product $v_{\text{th}} \sigma N_{sT}$ has the unit of a velocity; it is called the *surface recombination velocity*

$$S_r := v_{\text{th}} \sigma N_{sT}, \tag{7.50}$$

with σ_p or σ_n for an *n*- or *p*-type semiconductor, respectively. A low surface recombination velocity means that little recombination takes place while a (theoretical) value of $S_r = \infty$ would mean that every minority carrier coming to the proximity of the surface recombines.

For high quality solar cells it is crucial to have a low surface recombination velocity S_r, which can be achieved in two different ways: first, S_r can be made low by reducing the trap density N_{sT}. In semiconductor technology, N_{sT} can be reduced with so-called *passivation*. This means that the defect density is reduced by depositing a thin layer of a suitable material onto the semiconductor surface. Because of this layer, the valence electrons on the surface can form covalent bonds, such that N_{sT} is reduced.

Secondly, the excess minority carrier concentration at the surface (p_s or n_s) can be reduced, for example by high doping of the region just underneath the surface in order to create a barrier. Because of this barrier, the minority carrier concentration is reduced and hence the recombination rate R_s. We discuss both ways in more detail during our discussion on crystalline silicon solar cells in Section 12.4.

More detailed discussions for the extreme cases $S_R \rightarrow \infty$ and $S_r = 0$ are found in Appendix C.

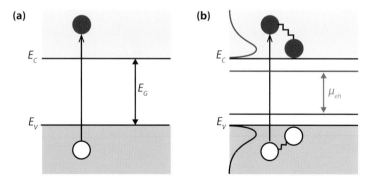

Figure 7.7: Thermalization of photogenerated electron-hole pairs resulting in non-equilibrium charge-carrier concentrations described by the quasi-Fermi levels.

7.6 Carrier concentration in non-equilibrium

When a semiconductor is illuminated additional electrons and holes are generated in the material by the absorption of photons. The *photogenerated carriers* interact with the semiconductor lattice. The extra energy that the electron-hole pairs receive from the photons with energies larger than the band gap of the semiconductor is released into the lattice in the form of heat. After this so–called *thermalization* process, which is very fast and takes approximately 10^{-12} s, the carrier concentrations achieve a steady state. In this non-equilibrium state the electron and hole concentrations differ from those in the equilibrium state. In non-equilibrium states two Fermi distributions are used to describe the electron and hole concentrations. One Fermi distribution with the *quasi-Fermi energy for electrons*, E_{Fn}, describes the occupation of states in the conduction band with electrons. Another Fermi distribution with the *quasi-Fermi energy for holes*, E_{Fp}, describes the occupation of states in the valence band with electrons, and therefore determines also the concentration of holes. Using the band diagram with the quasi-Fermi levels the process of creation of electron-hole pairs and their subsequent thermalization that describe the carrier concentration under illumination is illustrated in Figure 7.7. The difference between the quasi-Fermi levels is the electrochemical energy, μ_{eh}, of the generated electron-hole pairs which represents the measure for the conversion efficiency of solar radiation.

The density of electrons and holes under non-equilibrium conditions is described by

$$n = N_C \exp\left(\frac{E_{Fn} - E_C}{k_B T}\right), \tag{7.51a}$$

$$p = N_V \exp\left(\frac{E_V - E_{Fp}}{k_B T}\right). \tag{7.51b}$$

It then follows that under non-equilibrium conditions

$$np = N_C N_V \exp\left(\frac{E_V - E_C}{k_B T}\right) \exp\left(\frac{E_{Fn} - E_{Fp}}{k_B T}\right) = n_i^2 \exp\left(\frac{E_{Fn} - E_{Fp}}{k_B T}\right). \tag{7.52}$$

By using the quasi-Fermi level formalism for describing the concentration of charge carriers in non-equilibrium conditions, the electron and hole current densities inside a semi-

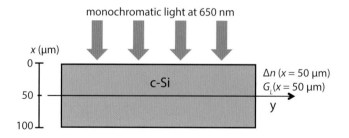

Figure 7.8: c-Si wafer illuminated with a monochromatic light.

conductor, J_N and J_P, can be expressed with

$$J_N = n\mu_n \nabla E_{Fn}, \qquad (7.53a)$$
$$J_P = p\mu_p \nabla E_{Fp}. \qquad (7.53b)$$

One can notice from Eqs. (7.53) that when a quasi-Fermi level varies with position,

$$\frac{dE_{Fn}}{dx} \neq 0 \qquad \text{or} \qquad \frac{dE_{Fp}}{dx} \neq 0,$$

the current is flowing inside the semiconductor. By checking the position dependence of the quasi-Fermi levels in an energy band diagram, one can easily determine whether current flows inside the semiconductor.

7.7 Exercises

7.1 Which of the charge carrier recombination mechanisms below, occurs due to the electron-hole recombination via a defect state in the bandgap?

(a) Radiative recombination.

(b) Auger recombination.

(c) Shockley–Read–Hall (SRH) recombination.

(d) All of the above.

7.2 The *diffusion length* is the average length that a carrier moves between generation and recombination. Calculate the minority diffusion length of a minority carrier having a lifetime of $\tau = 10$ μs and minority carrier diffusivity of $D = 25.6$ cm^2/s.

7.3 The minority carrier lifetime of a material is the average time which a carrier can spend in an excited state after electron-hole generation before it recombines. Calculate the minority carrier lifetime for a single crystalline solar cell having diffusion length of $L = 200$ μm and minority carrier diffusivity of $D = 27$ cm^2/s.

7.4 We have discussed that indirect band gap materials have a lower absorption coefficient than direct band gap materials, due to the fact that the charge carriers need a change in energy *and* momentum in order to be excited. If both Si and Ge are indirect band gap materials, why does Ge have a much higher absorption coefficient than Si in the visible wavelength range?

(a) Because Ge has more valence electrons than Si.

(b) Because Ge has a higher band gap than Si.

(c) Because Si has direct transitions in this part of the spectrum.

(d) Because Ge has direct transitions in this part of the spectrum.

7.5 Consider a 100 µm thick p-doped c-Si wafer illuminated with a monochromatic light at a wavelength of 650 nm as illustrated in Figure 7.8. The optical complex refractive index ($\tilde{n} = n - ik$) of the c-Si at 650 nm is $\tilde{n} = 3.84 - 0.015i$. The incident irradiance is 1,000 W/m². The absorption coefficient α is given by $\alpha = 4\pi k/\lambda$. Calculate:

(a) The absorption coefficient at 650 nm.

(b) The reflectance at interface air/Si (assume $\tilde{n}_{air} = 1$).

(c) The photon flux after reflection at $x = 0$ and $x = 50$ µm.

(d) The generation rate G_L at $x = 50$ µm.

(e) The excess of minority carriers, Δn at $x = 50$ µm in the p-doped wafer.

Assume the following steady-state conditions: sample is uniformly illuminated along y direction as shown in the figure; dominant thermal recombination and generation process and condition of low injection level and finally there is no current flowing through the wafer, which means:

$$\left.\frac{\partial n}{\partial t}\right|_{diff} = \left.\frac{\partial p}{\partial t}\right|_{drift} = 0.$$

7.6 Consider a slab of silicon crystal 10 cm by 10 cm by 10 cm at room temperature, in the dark. Exactly 1.5×10^{19} phosphorus atoms were added to the crystal when it was still molten, during its growth. The effective conduction band density of states of c-Si can be described as: $N_C \approx 6.2 \times 10^{15} \times T^{3/2}$ cm^{-3}. The effective valance band density of states of c-Si can be described as: $N_V \approx 3.5 \times 10^{15} \times T^{3/2}$ cm^{-3}. Assume room temperature conditions and that the density of atoms in c-Si is approximately 5×10^{22} cm^{-3}.

(a) Calculate the density of phosphorus atoms in the c-Si slab (cm^{-3}). Calculate the number of phosphorus atoms per million silicon atoms, i.e. ppm.

(b) Calculate the concentration of electrons (n), holes (p) and the intrinsic concentration of charge carriers (n_i). Assume the bandgap of c-Si is 1.1 eV and that all of the P atoms are ionized.

(c) Calculate the position of the Fermi level (E_F) in respect to the conduction band edge (E_C).

(d) What 'type' is our c-Si and what are the majority carriers?

(e) Answer questions (b) to (d) for the silicon slab after it has been heated to a high temperature (727 °C).

(f) Why is silicon a different 'type' at room temperature compared to the higher temperature? Where do the extra holes/electrons, for example, come from at the high temperature compared to the room temperature?

(g) Draw an energy band diagram and include the position of the Fermi level as a dashed line for both cases. Comment on the position of the Fermi level at room temperature compared to at the high temperature.

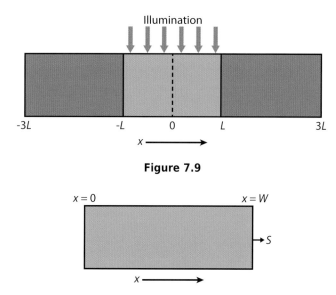

Figure 7.9

Figure 7.10

(h) A drift current density of $J_{drift} = 110 \text{A/cm}^2$ is required in p-type c-Si (hole mobility $\mu_p \approx 470 \text{cm}^2\text{V}^{-1}\text{s}^{-1}$) with an applied electric field of $E = 25 \text{V/cm}$. What doping concentration is required to achieve this current?

7.7 An *n*-type semiconductor is shown in Figure 7.9. Illumination produces a constant excess carrier generation rate, G_0, in the region $-L < x < +L$. Assume that the minority–carrier lifetime is infinite and assume that the excess–minority carrier hole concentration is zero at $x = -3L$ and at $x = 3L$. Find the steady-state excess minority–carrier concentration versus x, for the case of low injection and for zero applied electric field.

7.8 A bar of *p*-type crystalline silicon with width W, is shown in Figure 7.10. The minority carrier diffusion constant is D_n and thermal equilibrium concentration is n_0. In this bar, carriers are uniformly generated at a rate G_L. Assume a steady-state situation while it is also given that no external electric field is present and that for $x = 0$ the minority carrier concentration is zero. At $x = W$ the surface recombination velocity is S.

(a) Find an expression for the excess minority–carrier concentration for the case of infinite minority–carrier lifetime in the silicon and $S = 0$.

(b) Find an expression for the excess minority–carrier concentration for the case of infinite minority carrier lifetime in the silicon and $S > 0$.

(c) Sketch in one plot the excess minority–carrier concentration as a function of position for questions (a) and (b).

8
Semiconductor junctions

Almost all solar cells contain junctions between (different) materials of different doping. Since these junctions are crucial to the operation of the solar cell, we will discuss their physics in this chapter.

A *p-n* junction fabricated in the same semiconductor material, such as c-Si, is an example of a *p-n homojunction*. There are also other types of junctions: A *p-n* junction that is formed by two chemically different semiconductors is called a *p-n heterojunction*. In a *p-i-n* junction the region of the internal electric field is extended by inserting an intrinsic, *i*, layer between the *p*-type and the *n*-type layers. The *i*-layer behaves like a capacitor: it stretches the electric field formed by the *p-n* junction across itself. Another type of the junction is between a *metal* and a *semiconductor*; this is called a MS junction. The Schottky barrier formed at the metal-semiconductor interface is a typical example of the MS junction.

8.1 *p-n* homojunctions

8.1.1 Formation of a space-charge region in the *p-n* junction

Figure 8.1 shows schematically isolated pieces of a *p*-type and an *n*-type semiconductor and their corresponding band diagrams. In both isolated pieces the charge neutrality is maintained. In the *n*-type semiconductor the large concentration of negatively-charged free electrons is compensated by positively-charged ionized donor atoms. In the *p*-type semiconductor holes are the majority carriers and the positive charge of holes is compensated by negatively-charged ionized acceptor atoms. For the isolated *n*-type semicon-

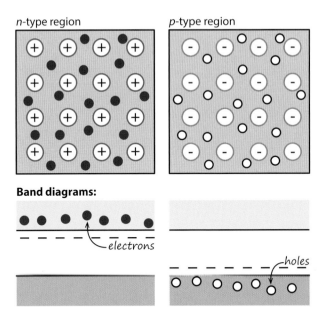

Figure 8.1: Schematic representation of an isolated *n*-type and *p*-type semiconductor and corresponding band diagrams.

ductor we can write

$$n = n_{n0} \approx N_D, \quad (8.1a)$$

$$p = p_{n0} \approx n_i^2 / N_D. \quad (8.1b)$$

For the isolated *p*-type semiconductor we have

$$p = p_{p0} \approx N_A, \quad (8.2a)$$

$$n = n_{p0} \approx n_i^2 / N_A. \quad (8.2b)$$

When a *p*-type and an *n*-type semiconductor are brought together, a very large difference in electron concentration between *n*- and *p*-type regions causes a diffusion current of electrons from the *n*-type material across the *metallurgical junction* into the *p*-type material. The term "metallurgical junction" denotes the interface between the *n*- and *p*-type regions. Similarly, the difference in hole concentration causes a diffusion current of holes from the *p*- to the *n*-type material. Due to this diffusion process the region close to the metallurgical junction becomes almost completely depleted of mobile charge carriers. The gradual depletion of the charge carriers gives rise to a space charge created by the charge of the ionized donor and acceptor atoms that is not compensated by the mobile charges any more. This region of the space charge is called the *space-charge* region or *depleted region* and is schematically illustrated in Fig. 8.2. Regions outside the depletion region, in which the charge neutrality is conserved, are denoted as the quasi-neutral regions.

8. Semiconductor junctions

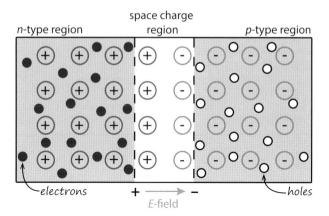

Figure 8.2: Formation of a space-charge region, when n-type and p-type semiconductors are brought together to form a junction. The coloured part represents the space-charge region.

The space charge around the metallurgical junction results in the formation of an internal electric field which forces the charge carriers to move in the opposite direction than the concentration gradient. The diffusion currents continue to flow until the forces acting on the charge carriers, namely the concentration gradient and the internal electrical field, compensate each other. The driving force for the charge transport does not exist any more and no net current flows through the p-n junction.

8.1.2 The *p-n* junction under equilibrium

The p-n junction represents a system of charged particles in diffusive equilibrium in which the electrochemical potential is constant and independent of position. The electro-chemical potential describes an average energy of electrons and is represented by the Fermi energy. Figure 8.3 (a) shows the band diagrams of isolated n- and p-type semiconductors. The band diagrams are drawn such that the *vacuum energy level* E_{vac} is aligned. This energy level represents the energy just outside the atom, if an electron is elevated to E_{vac} it leaves the sphere of influence of the atom. Also the electron affinity χ_e is shown, which is defined as the potential that an electron present in the conduction band requires to be elevated to an energy level just outside the atom, i.e. E_{vac}.

The band diagram of the p-n junction in equilibrium is shown in Figure 8.3 (b). Note, that in the band diagram the Fermi energy is constant across the junction, not the vacuum energy. As the Fermi energy denotes the "filling level" of electrons, this is the level that is constant throughout the junction. To visualize this, we take a look at Fig. 8.4 (a), which shows two tubes of different lengths that are partially filled with water. The filling level is equivalent to the Fermi energy in a solid state material. The vacuum energy would be the upper boundary of the tube; if the water was elevated above this level, it could leave the tube. If the two tubes are connected as illustrated in Fig. 8.4 (b), the water level in both tubes will be the same. However, the length of the tubes might be different; for leaving the first tube, a different energy can be required than for leaving the second tube.

In addition to the Fermi energy being constant across the junction, the band-edge en-

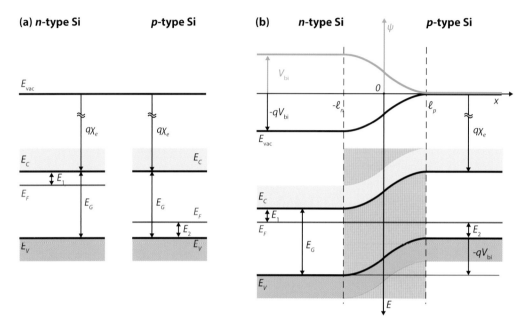

Figure 8.3: (a) The energy band diagrams of an n- and a p-type material that are separated from each other. (b) The energy-band diagram of the p-n junction under equilibrium. The electrostatic potential profile (green curve) is also presented in the figure.

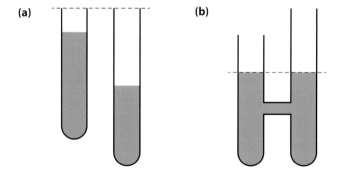

Figure 8.4: (a) Two tubes of different length (upper boundary represents vacuum level) and filling level (representing Fermi energy). (b) If the tubes are connected, the filling level will equalize but the heights of the tube boundaries can be different.

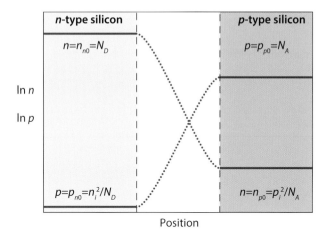

Figure 8.5: Concentrations profile of mobile charge carriers in a *p-n* junction under equilibrium.

ergies E_C and E_V as well as the vacuum energy E_{vac} must be *continuous*. Hence, the bands get *bended*, which indicates the presence of an electric field in this region. Due to the electric field a difference in the electrostatic potential is created between the boundaries of the space-charge region. Across the depletion region the changes in the carrier concentration are compensated by changes in the electrostatic potential. The electrostatic-potential profile ψ is also drawn in Fig. 8.3 (b).

The concentration profile of charge carriers in a *p-n* junction is schematically presented in Fig. 8.5. In the quasi-neutral regions the concentration of electrons and holes is the same as in the isolated doped semiconductors. In the space-charge region the concentrations of majority charge carriers decrease very rapidly. This fact allows us to use the assumption that the space-charge region is depleted of mobile charge carriers. This assumption means that the charge of the mobile carriers represents a negligible contribution to the total space charge in the depletion region. The space charge in this region is fully determined by the ionized dopant atoms fixed in the lattice.

The presence of the internal electric field inside the *p-n* junction means that there is an electrostatic potential difference, V_{bi}, across the space-charge region. We shall determine a profile of the internal electric field and electrostatic potential in the *p-n* junction. First we introduce an approximation, which simplifies the calculation of the electric field and electrostatic-potential. This approximation (*the depletion approximation*) assumes that the space-charge density, ρ, is zero in the quasi-neutral regions and it is fully determined by the concentration of ionized dopants in the depletion region. In the depletion region of the *n*-type semiconductor it is the concentration of positively charged donor atoms, N_D, which determines the space charge in this region. In the *p*-type semiconductor, the concentration of negatively charged acceptor atoms, N_A, determines the space charge in the depletion region. This is illustrated in Fig. 8.6. Further, we assume that the *p-n* junction is a step junction; it means that there is an abrupt change in doping at the metallurgical junction and the doping concentration is uniform both in the *p*-type and the *n*-type semiconductors.

In Fig. 8.6, the position of the metallurgical junction is placed at zero, the width of

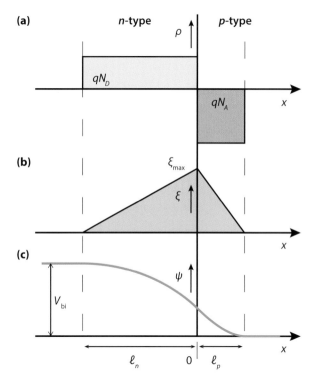

Figure 8.6: (a) The space-charge density $\rho(x)$, (b) the electric field $\xi(x)$, and (c) the electrostatic potential $\psi(x)$ across the depletion region of a n-p junction under equilibrium.

8. Semiconductor junctions

the space-charge region in the *n*-type material is denoted as ℓ_n and the width of the space-charge region in the *p*-type material is denoted as ℓ_p. The space-charge density is described by

$$\rho(x) = qN_D \quad \text{for} \quad -\ell_n \leq x \leq 0, \tag{8.3a}$$
$$\rho(x) = -qN_A \quad \text{for} \quad 0 \leq x \leq \ell_p, \tag{8.3b}$$

where N_D and N_A is the concentration of donor and acceptor atoms, respectively. Outside the space-charge region the space-charge density is zero. The electric field which is calculated from the Poisson's equation, in one dimension can be written as

$$\frac{d^2\psi}{dx^2} = -\frac{d\xi}{dx} = -\frac{\rho}{\epsilon_r\epsilon_0}, \tag{8.4}$$

where ψ is the electrostatic potential, ξ is the electric field, ρ is the space-charge density, ϵ_r is the semiconductor dielectric constant and ϵ_0 is the vacuum permittivity. The vacuum permittivity is $\epsilon_0 = 8.854 \times 10^{-14}$ F/cm and for crystalline silicon $\epsilon_r = 11.7$. The electric field profile can be found by integrating the space-charge density across the space-charge region,

$$\xi = \frac{1}{\epsilon_r\epsilon_0}\int \rho\, dx. \tag{8.5}$$

Substituting the space-charge density with Eqs. (8.3) and using the boundary conditions

$$\xi(-\ell_n) = \xi(\ell_p) = 0, \tag{8.6}$$

we obtain as solution for the electric field

$$\xi(x) = \frac{q}{\epsilon_r\epsilon_0}N_D(\ell_n + x) \quad \text{for} \quad -\ell_n \leq x \leq 0, \tag{8.7a}$$
$$\xi(x) = \frac{q}{\epsilon_r\epsilon_0}N_A(\ell_p - x) \quad \text{for} \quad 0 \leq x \leq \ell_p. \tag{8.7b}$$

At the metallurgical junction, $x = 0$, the electric field is continuous, which requires that the following condition has to be fulfilled

$$N_A\ell_p = N_D\ell_n. \tag{8.8}$$

Outside the space-charge region the material is electrically neutral and therefore the electric field is zero there.

The profile of the electrostatic potential is calculated by integrating the electric field throughout the space-charge region and applying the boundary conditions,

$$\psi = -\int \xi\, dx. \tag{8.9}$$

We define the zero electrostatic potential level at the outside edge of the *p*-type semiconductor. Since we assume no potential drop across the quasi-neutral region the electrostatic potential at the boundary of the space-charge region in the *p*-type material is also zero,

$$\psi(\ell_p) = 0. \tag{8.10}$$

Using Eqs. (8.7) for describing the electric field in the *n*- and *p*-doped regions of the space-charge region, and taking into account that at the metallurgical junction the electrostatic potential is continuous, we can write the solution for the electrostatic potential as

$$\psi(x) = -\frac{q}{2\epsilon_r \epsilon_0} N_D (x+\ell_n)^2 + \frac{q}{2\epsilon_r \epsilon_0}\left(N_D \ell_n^2 + N_A \ell_p^2\right) \quad \text{for} \quad -\ell_n \leq x \leq 0, \quad (8.11a)$$

$$\psi(x) = \frac{q}{2\epsilon_r \epsilon_0} N_A (x-\ell_p)^2 \quad \text{for} \quad 0 \leq x \leq \ell_p. \quad (8.11b)$$

Under equilibrium a difference in electrostatic potential, V_{bi}, develops across the space-charge region. The electrostatic potential difference across the *p-n* junction is an important characteristic of the junction and is denoted as the *built-in voltage* or diffusion potential of the *p-n* junction. We can calculate V_{bi} as the difference between the electrostatic potential at the edges of the space-charge region,

$$V_{bi} = \psi(-\ell_n) - \psi(\ell_p) = \psi(-\ell_n). \quad (8.12)$$

Using Eq. (8.11a) we obtain for the built-in voltage

$$V_{bi} = \frac{q}{2\epsilon_r \epsilon_0}\left(N_D \ell_n^2 + N_A \ell_p^2\right) \quad (8.13)$$

The built-in potential V_{bi} can also be determined with using the energy-band diagram presented in Fig. 8.3 (b).

$$qV_{bi} = E_G - E_1 - E_2. \quad (8.14)$$

Using Eqs. (6.1) and (6.22), which determine the band gap, and the positions of the Fermi energy in the *n*- and *p*-type semiconductor, respectively,

$$E_G = E_C - E_V,$$
$$E_1 = E_C - E_F = k_B T \ln(N_C/N_D),$$
$$E_2 = E_F - E_V = k_B T \ln(N_V/N_A).$$

We can write

$$qV_{bi} = E_G - k_B T \ln\left(\frac{N_V}{N_A}\right) - k_B T \ln\left(\frac{N_C}{N_D}\right)$$
$$= E_G - k_B T \ln\left(\frac{N_V N_C}{N_A N_D}\right). \quad (8.15)$$

Using the relationship between the intrinsic concentration, n_i and the band gap, E_G [see Eq. (6.9)],

$$n_i^2 = N_C N_V \exp\left[-\frac{E_G}{k_B T}\right],$$

we can rewrite Eq. (8.15) and obtain

$$V_{bi} = \frac{k_B T}{q} \ln\left(\frac{N_A N_D}{n_i^2}\right) \quad (8.16)$$

8. Semiconductor junctions

This equation allows us to determine the built-in potential of a *p-n* junction from the standard semiconductor parameters, such as doping concentrations and the intrinsic carrier concentration.

We can calculate the width of the space charge region of the *p-n* junction in the thermal equilibrium starting from Eq. (8.13). Using Eq. (8.8), either ℓ_n or ℓ_p can be eliminated from Eq. (8.13), resulting in expressions for ℓ_n and ℓ_p respectively,

$$\ell_n = \sqrt{\frac{2\epsilon_r \epsilon_0 V_{bi}}{q} \frac{N_A}{N_D} \left(\frac{1}{N_A + N_D} \right)}, \qquad (8.17a)$$

$$\ell_p = \sqrt{\frac{2\epsilon_r \epsilon_0 V_{bi}}{q} \frac{N_D}{N_A} \left(\frac{1}{N_A + N_D} \right)}. \qquad (8.17b)$$

The total space-charge width, W, is the sum of the partial space-charge widths in the *n*- and *p*-type semiconductors. Using Eq. (8.17) we find

$$W = \ell_n + \ell_p = \sqrt{\frac{2\epsilon_r \epsilon_0}{q} V_{bi} \left(\frac{1}{N_A} + \frac{1}{N_D} \right)}. \qquad (8.18)$$

The space-charge region is not uniformly distributed in the *n*- and *p*- regions. The widths of the space-charge region in the *n*- and *p*-type semiconductor are determined by the doping concentrations as illustrated by Eqs. (8.17). Knowing the expressions for ℓ_n and ℓ_p we can determine the maximum value of the internal electric field, which is at the metallurgical junction. By substituting ℓ_p from Eq. (8.17b) into Eq. (8.7b) we obtain the expression for the maximum value of the internal electric field,

$$\xi_{max} = \sqrt{\frac{2q}{\epsilon_r \epsilon_0} V_{bi} \left(\frac{N_A N_D}{N_A + N_D} \right)}. \qquad (8.19)$$

Example

A crystalline silicon wafer is doped with 10^{16} acceptor atoms per cubic centimetre. A 1 micrometer thick emitter layer is formed at the surface of the wafer with a uniform concentration of 10^{18} donors per cubic centimetre. Assume a step p-n junction and that all doping atoms are ionized. The intrinsic carrier concentration in silicon at 300 K is $1.5 \cdot 10^{10}$ cm^{-3}.

Let us calculate the electron and hole concentrations in the p- and n-type quasi-neutral regions at thermal equilibrium. We shall use Eqs. (8.1) and (8.2) to calculate the charge carrier concentrations.

P-type region: $\qquad p = p_{p0} \approx N_A = 10^{16}$ cm^{-3}.

$\qquad n = n_{p0} = n_i^2 / p_{p0} = \left(1.5 \cdot 10^{10}\right)^2 / 10^{16} = 2.25 \cdot 10^4$ cm^{-3}

N-type region: $\qquad n = n_{n0} \approx N_A = 10^{18}$ cm^{-3}.

$\qquad p = p_{n0} = n_i^2 / n_{n0} = \left(1.5 \cdot 10^{10}\right)^2 / 10^{18} = 2.25 \cdot 10^2$ cm^{-3}

We can calculate the position of the Fermi energy in the quasi-neutral n-type and p-type regions, respectively, using Eq. (6.22a). We assume that the reference energy level is the bottom of the conduction band, $E_C = 0$ eV.

N-type region: $E_F - E_C = -k_B T \ln(N_C/n) = -0.0258 \ln\left(3.32 \cdot 10^{19}/10^{18}\right) = -0.09$ eV.

P-type region: $E_F - E_C = -k_B T \ln(N_C/n) = -0.0258 \ln\left(3.32 \cdot 10^{19}/2.24 \cdot 10^{4}\right) = -0.90$ eV.

The minus sign tells us that the Fermi energy is positioned below the conduction band.
The built-in voltage across the p-n junction is calculated using Eq. (8.16),

$$V_{bi} = \frac{k_B T}{q} \ln\left(\frac{N_A N_D}{n_i^2}\right) = 0.0258 \text{ V} \left[\frac{10^{16} \, 10^{18}}{(1.5 \cdot 10^{10})^2}\right] = 0.81 \text{ V}.$$

The width of the depletion region is calculated from Eq. (8.18),

$$W = \sqrt{\frac{2\epsilon_r \epsilon_0}{q} V_{bi} \left(\frac{1}{N_A} + \frac{1}{N_D}\right)} = \sqrt{\frac{2 \cdot 11.7 \cdot 8.854 \cdot 10^{-14}}{1.602 \cdot 10^{-19}} \cdot 0.81 \left(\frac{1}{10^{16}} + \frac{1}{10^{18}}\right)}$$
$$= 3.25 \cdot 10^{-5} \text{ cm} = 0.325 \, \mu m.$$

A typical thickness of c-Si wafers is 300 µm. The depletion region is 0.3 µm which represents 0.1% of the wafer thickness. It is important to realize that almost the whole bulk of the wafer is a quasi-neutral region without an internal electrical field.
The maximum electric field is at the metallurgical junction and is calculated from Eq. (8.19).

$$\xi_{max} = \sqrt{\frac{2q}{\epsilon_r \epsilon_0} V_{bi} \left(\frac{N_A N_D}{N_A + N_D}\right)} = \sqrt{\frac{2 \times 1.602 \cdot 10^{-19}}{11.7 \times 8.854 \times 10^{-14}} \times 0.81 \left(\frac{10^{16} \, 10^{18}}{10^{16} + 10^{18}}\right)} = 5 \times 10^4 \text{ V cm}^{-1}.$$

8.1.3 The *p-n* junction under applied voltage

When an external voltage, V_a, is applied to a *p-n* junction the potential difference between the *n*- and *p*-type regions will change and the electrostatic potential across the space-charge region will become $(V_{bi} - V_a)$. Remember that under equilibrium the built-in potential is negative in the *p*-type region with respect to the *n*-type region. When the applied external voltage is negative with respect to the potential of the *p*-type region, the applied voltage will increase the potential difference across the *p-n* junction. We refer to this situation as *p-n* junction under *reverse-bias voltage*. The potential barrier across the junction is increased under reverse-bias voltage, which results in a wider space-charge region.

Figure 8.7 (a) shows the band diagram of the *p-n* junction under reverse-biased voltage. Under external voltage the *p-n* junction is no longer under equilibrium any more and the concentrations of electrons and holes are described by the quasi-Fermi energy for electrons, E_{Fn}, and the quasi-Fermi energy for holes, E_{Fp}, respectively. When the applied external voltage is positive with respect to the potential of the *p*-type region, the applied voltage will decrease the potential difference across the *p-n* junction. We refer to this situation as *p-n* junction under *forward-bias voltage*. The band diagram of the *p-n* junction under forward-biased voltage is presented in Figure 8.7 (b). The potential barrier across the junction is decreased under forward-bias voltage and the space–charge region becomes narrower.

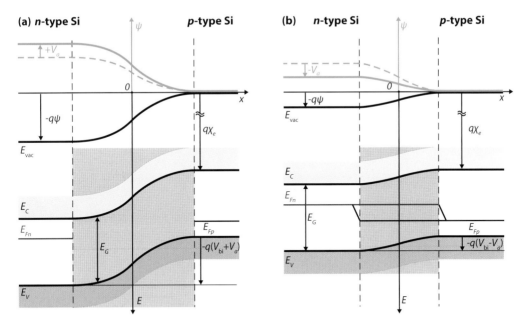

Figure 8.7: Energy band diagram and electrostatic-potential (in green) of a *p-n* junction under (a) reverse bias and (b) forward bias conditions.

The balance between the forces responsible for diffusion (concentration gradient) and drift (electric field) is disturbed. Lowering the electrostatic–potential barrier leads to a higher concentration of minority carriers at the edges of the space-charge region compared to the situation in equilibrium. This process is referred to as minority-carrier *injection*. This gradient in concentration causes diffusion of the minority carriers from the edge into the bulk of the quasi-neutral region.

The diffusion of minority carriers into the quasi-neutral region causes a so-called recombination current density, J_{rec}, since the diffusing minority carriers recombine with the majority carriers in the bulk. The recombination current is compensated by the so-called thermal generation current, J_{gen}, which is caused by the drift of minority carriers. These are present in the corresponding doped regions (electrons in the *p*-type region and holes in the *n*-type region), across the junction. Both the recombination and generation current densities have contributions from electrons and holes. When no voltage is applied to the *p-n* junction, the situation inside the junction can be viewed as the balance between the recombination and generation current densities,

$$J = J_{\text{rec}} - J_{\text{gen}} = 0 \qquad \text{for } V_a = 0 \text{ V.} \qquad (8.20)$$

It is assumed that when a moderate forward-bias voltage is applied to the junction the recombination current density increases with the Boltzmann factor $\exp(eV_a/k_BT)$,

$$J_{\text{rec}}(V_a) = J_{\text{rec}}(V_a = 0) \, \exp\left(\frac{qV_a}{k_BT}\right). \qquad (8.21)$$

This assumption is called the *Boltzmann approximation*.

The generation current density, on the other hand, is almost independent of the potential barrier across the junction and is determined by the availability of the thermally-generated minority carriers in the doped regions,

$$J_{\text{gen}}(V_a) \approx J_{\text{gen}}(V_a = 0). \tag{8.22}$$

The external net current density can be expressed as

$$J(V_a) = J_{\text{rec}}(V_a) - J_{\text{gen}}(V_a) = J_0 \left[\exp\left(\frac{qV_a}{k_B T} \right) - 1 \right], \tag{8.23}$$

where J_0 is the saturation current density of the p-n junction, given by

$$J_0 = J_{\text{gen}}(V_a = 0). \tag{8.24}$$

Equation (8.23) is known as the *Shockley equation* that describes the current–voltage behaviour of an ideal p-n diode. It is a fundamental equation for microelectronics device physics. The saturation current density is also known as *dark current density*; its detailed derivation for the p-n junction is carried out in Appendix B.1. The saturation-current density is given by

$$J_0 = q\, n_i^2 \left(\frac{D_N}{L_N N_A} + \frac{D_P}{L_P N_D} \right). \tag{8.25}$$

The saturation current density depends in a complex way on the fundamental semiconductor parameters. Ideally the saturation current density should be as low as possible and this requires an optimal and balanced design of the p-type and n-type semiconductor properties. For example, an increase in the doping concentration decreases the diffusion length of the minority carriers, which means that the optimal product of these two quantities requires a delicate balance between these two properties.

The recombination of the majority carriers due to the diffusion of the injected minority carriers into the bulk of the quasi-neutral regions results in a lowering of the concentration of the majority carriers compared to the one under equilibrium. The drop in the concentration of the majority carriers is balanced by the flow of the majority carriers from the electrodes into the bulk. In this way the net current flows through the p-n junction under forward-bias voltage. For high reverse-bias voltage, the Boltzmann factor in Eq. (8.23) becomes very small and can be neglected. The net current density is given by

$$J(V_a) = -J_0, \tag{8.26}$$

and represents the flux of thermally generated minority carriers across the junction. The current density-voltage (J-V) characteristic of an ideal p-n junction is shown schematically in Figure 8.8.

8.1.4 The p-n junction under illumination

When a p-n junction is illuminated, additional electron-hole pairs are generated in the semiconductor. The concentration of minority carriers (electrons in the p-type region and holes in the n-type region) strongly increases, leading to flow of the minority carriers across the depletion region into the quasi-neutral regions. Electrons flow from the p-type into the

8. Semiconductor junctions

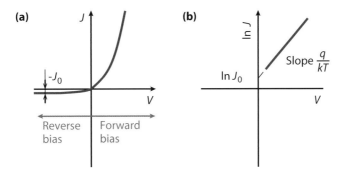

Figure 8.8: *J-V* characteristic of a *p-n* junction; (a) linear plot; and (b) semi-logarithmic plot.

n-type region and holes from the *n*-type into the *p*-type region. The flow of the photogenerated carriers causes the so-called *photogeneration current density*, J_{ph}, which adds to the thermal-generation current, J_{gen}. When no external electrical contact between the *n*-type and the *p*-type regions is established, the junction is in *open circuit condition*. Hence, the current resulting from the flux of photogenerated and thermally-generated carriers has to be balanced by the opposite recombination current. The recombination current will increase through lowering of the electrostatic potential barrier across the depletion region.

The band diagram of the illuminated *p-n* junction under open circuit condition is presented in Figure 8.9 (a). As we have seen in Section 7.6, under non-equilibrium the Fermi level is replaced by quasi-Fermi levels that are different for electrons and holes and denote their electrochemical potential. In the open circuit condition, the quasi-Fermi level of electrons, denoted by E_{Fn}, is higher than the quasi-Fermi level of holes (denoted by E_{Fp}) by an amount of qV_{oc}. This means that a voltmeter will measure a voltage difference of V_{oc} between the contacts of the *p-n* junction. We refer to V_{oc} as the *open circuit voltage*.

The bands in the quasi-neutral regions (i.e. outside the depletion region) are flat. Under the assumption that the charge density in each of the regions is homogeneous, the bands follow a parabolic shape in the depletion region. In this case there is generation anywhere in the device. In the open circuit condition, the external current density is zero. So we have:

$$J_n = J_{n,\text{drift}} + J_{n,\text{diff}} = q\mu_n n E + qD_n \frac{dn}{dx} = q\mu_n n \left(\frac{1}{q}\frac{dE_C}{dx}\right) + k_B T \mu_n \frac{dn}{dx}, \quad (8.27a)$$

$$J_p = J_{p,\text{drift}} + J_{p,\text{diff}} = q\mu_p p E - qD_p \frac{dp}{dx} = q\mu_p p \left(\frac{1}{q}\frac{dE_V}{dx}\right) - k_B T \mu_p \frac{dp}{dx}. \quad (8.27b)$$

The carrier densities are given by:

$$n = N_C \exp\left(-\frac{E_C - E_{Fn}}{k_B T}\right) \quad \text{and} \quad p = N_V \exp\left(-\frac{E_{Fp} - E_V}{k_B T}\right). \quad (8.28)$$

So we find for the derivatives

$$\frac{dn}{dx} = -\frac{n}{k_B T}\left(\frac{dE_C}{dx} - \frac{dE_{Fn}}{dx}\right) \quad \text{and} \quad \frac{dp}{dx} = \frac{p}{k_B T}\left(\frac{dE_V}{dx} - \frac{dE_{Fp}}{dx}\right). \quad (8.29)$$

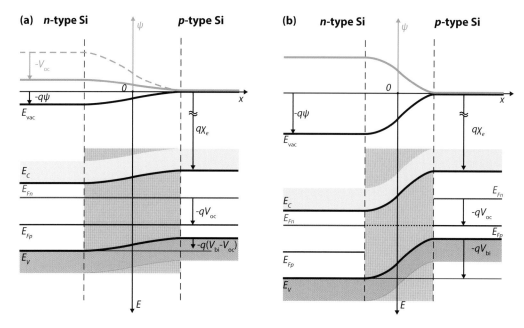

Figure 8.9: Energy band diagram and electrostatic-potential (in green) of an illuminated *p-n* junction under the (a) open-circuit and (b) short-circuit conditions.

We then find for each current component

$$J_n = q\mu_n n \left(\frac{1}{q}\frac{dE_C}{dx}\right) + k_B T \mu_n \frac{dn}{dx} = \mu_n n \frac{dE_{Fn}}{dx}, \quad (8.30a)$$

$$J_p = q\mu_p p \left(\frac{1}{q}\frac{dE_V}{dx}\right) - k_B T \mu_p \frac{dp}{dx} = \mu_p p \frac{dE_{Fp}}{dx}. \quad (8.30b)$$

Hence, the total current is given by

$$J = J_n + J_p = \mu_n n \frac{dE_{Fn}}{dx} + \mu_p p \frac{dE_{Fp}}{dx} = 0. \quad (8.31)$$

Note that the last step implies that the current density at $V = V_{oc}$ is zero. The current density can only be zero if

$$\frac{dE_{Fn}}{dx} \equiv \frac{dE_{Fp}}{dx} \equiv 0, \quad (8.32)$$

which implies that the quasi-Fermi levels are horizontal in the entire band diagram of the solar cell.

Figure 8.9 (b) shows the band diagram of the *short circuited p-n* junction. In this situation, the photogenerated current will also flow through the external circuit. In the short circuit condition the electrostatic-potential barrier is not changed, but from a strong variation of the quasi-Fermi levels inside the depletion region one can determine that the current is flowing inside the semiconductor.

When a load is connected between the electrodes of the illuminated *p-n* junction, only a fraction of the photogenerated current will flow through the external circuit. The electrochemical potential difference between the *n*-type and *p*-type regions will be lowered by a voltage drop over the load. This in turn lowers the electrostatic potential over the depletion region which results in an increase of the recombination current. In the *superposition approximation*, the net current flowing through the load is determined as the sum of the photo- and thermal- generation currents and the recombination current. The voltage drop at the load can be simulated by applying a forward-bias voltage to the junction. Therefore Eq. (8.23), which describes the behaviour of the junction under applied voltage, is included to describe the net current of the illuminated *p-n* junction,

$$J(V_a) = J_{\text{rec}}(V_a) - J_{\text{gen}}(V_a) - J_{\text{ph}} = J_0 \left[\exp\left(\frac{qV_a}{k_B T}\right) - 1 \right] - J_{\text{ph}}. \quad (8.33)$$

Both the dark and illuminated *J-V* characteristics of the *p-n* junction are represented in Figure 8.10. Note that in the figure the superposition principle is reflected. The illuminated *J-V* characteristic of the *p-n* junction is the same as the dark *J-V* characteristic, but it is shifted down by the photogenerated current density J_{ph}. The detailed derivation of the photogenerated current density of the *p-n* junction is carried out in Appendix B.2. Under a uniform generation rate, G, its value is

$$J_{\text{ph}} = qG(L_N + W + L_P), \quad (8.34)$$

where L_N and L_P are the minority-carriers diffusion length for electrons and holes, respectively, and W is the width of the depletion region. It means that only carriers generated in the depletion region and in the regions up to the minority-carrier diffusion length from the depletion region can contribute to the photogenerated current. When designing the thickness of a solar cell, Eq. (8.34) must be considered. The thickness of the absorber should not be greater than the region from which the carriers contribute to the photogenerated current.

8.2 Heterojunctions

In the previous section we discussed the physics of junctions between an *n*-doped and a *p*-doped semiconductor of the same material. In these junctions, that are called *homojunctions*, the bandgap and the electron affinity are the same at both sides of the junction. Of course, junctions between different materials can also be made. These junctions are called *heterojunctions*. Heterojunctions are very important for solar cells; in fact, as of 2014, the best solar cells based on crystalline silicon have heterojunctions of crystalline and amorphous silicon, as we will see in Chapter 12. In this section we will look at the most important features of heterojunctions.

We distinguish between four types of heterojunctions: *n-P*, *p-N*, *n-N*, and *p-P*, where the lower case letter denotes the material with the lower bandgap and the upper-case letter denotes the material with the larger bandgap. In this section, we only consider *n-P* heterojunctions. Band diagrams of the other cases are for example shown in [24].

Figure 8.11 (a) shows the band diagrams of an *n*-type semiconductor and of a *P*-type semiconductor; the latter has a larger bandgap. As a reference level in this graph the vacuum level is used. We see that not only are the bandgaps E_{Gn} and E_{GP} different, but also

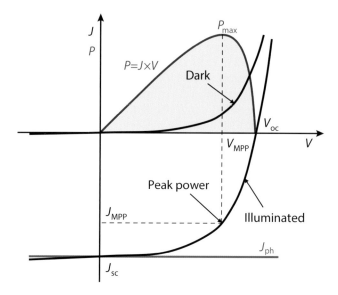

Figure 8.10: *J-V* characteristics of a *p-n* junction in the dark and under illumination.

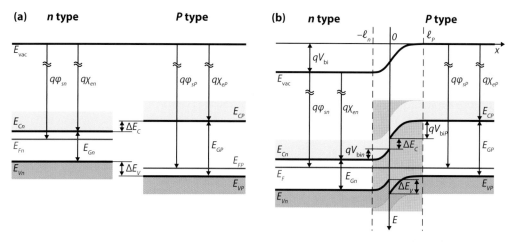

Figure 8.11: Energy band diagrams of (a) an *n*-type and a *P*-type semiconductor with a larger bandgap than the *n*-type material; and (b) an *n-P* heterojunction.

8. Semiconductor junctions

the electron affinities χ_{en} and χ_{eP}, which denote the potential difference between the conduction band edges and the vacuum level. In addition the work functions ϕ_{sn} and ϕ_{sP} are indicated, which are defined as the potential differences between the vacuum level and the Fermi energies. As we can see in this figure, offsets between the edges of the conduction band and the valence band exist; we denote them with ΔE_C and ΔE_V, respectively. These offsets are related to the other relevant parameters via

$$\Delta E_C = q(\chi_n - \chi_P), \text{ and} \tag{8.35}$$
$$\Delta E_C + \Delta E_V = E_{GP} - E_{Gn} = \Delta E_G. \tag{8.36}$$

In the ideal situation, discontinuities with the same ΔE_C and ΔE_V will exist if the two semiconductors form an interface. This behaviour is known as the *electron affinity rule*.

The band diagram of the n-P junction is shown in Figure 8.11 (b). For drawing the band diagram, we use two constraints. First, in equilibrium the Fermi energy is constant across the junction, as this was the case for homojunctions. Secondly, the vacuum energy should be continuous across the junction. Because of the band offsets of the two materials, the conduction and valence bands are not continuous but have discontinuities according to the electron affinity rule.

The built-in voltage of the heterojunction is given by the difference of the work functions,

$$V_{bi} = \phi_{sP} - \phi_{sn}. \tag{8.37}$$

It also can be expressed as [24]

$$qV_{bi} = -\Delta E_C + \Delta E_g + k_B T \ln\left(\frac{N_{Vn}}{p_{n0}}\right) - k_B T \ln\left(\frac{N_{VP}}{p_{P0}}\right), \tag{8.38}$$

where p_{n0} and p_{P0} are the hole concentrations in the n and P materials, respectively. N_{Vn} and N_{VP} are the effective density of state functions in the valence bands of the n and P materials, respectively.

Just as for the homojunction, the electric field is maximal at the junction and decays linearly throughout the depletion region. For the n- and P-type depletion regions it is given as

$$\xi_n(x) = \frac{qN_{dn}}{\epsilon_0 \epsilon_n}(\ell_n + x) \qquad (-\ell_n \leq x < 0), \tag{8.39a}$$
$$\xi_P(x) = \frac{qN_{aP}}{\epsilon_0 \epsilon_P}(\ell_P - x) \qquad (0 < x < \ell \leq \ell_P), \tag{8.39b}$$

with the doping concentrations N_{dn} and N_{aP}, the dielectric constants ϵ_n and ϵ_P, and the depletion widths ℓ_n and ℓ_P for the n- and P-type materials, respectively. The net charge in the n-type region is equivalent to the net charge in the P-type region,

$$N_{dn}\ell_n = N_{aP}\ell_P. \tag{8.40}$$

The widths of the depletion regions in the n- and P-type zones are given by [24]

$$\ell_n = \sqrt{\frac{2\epsilon_0\epsilon_n\epsilon_P N_{aP} V_{bi}}{qN_{dn}(\epsilon_n N_{dn} + \epsilon_P N_{aP})}}, \tag{8.41a}$$

$$\ell_P = \sqrt{\frac{2\epsilon_0\epsilon_n\epsilon_P N_{dn} V_{bi}}{qN_{aP}(\epsilon_n N_{dn} + \epsilon_P N_{aP})}}. \tag{8.41b}$$

For the total depletion width we find

$$W = \ell_n + \ell_P = \sqrt{\frac{2\epsilon_0\epsilon_n\epsilon_P (N_{dn} + N_{aP})^2 V_{bi}}{qN_{dn}N_{aP}(\epsilon_n N_{dn} + \epsilon_P N_{aP})}}. \tag{8.42}$$

As we have seen in Figure 8.11, the band diagrams of heterojunctions are much more complex than that of homojunctions because of the different bandgaps and electron affinities. In particular, the discontinuities at the edges of the valence and conduction bands can lead to the formation of barriers for the electrons, the holes, or both. Depending on the barrier heights and the applied voltage, different transport mechanisms might be dominant: these are diffusion, quantum-mechanic tunnelling, and thermionic emission, which is discussed in more detail in Section 8.3.

One possible issue of heterojunctions is *lattice mismatch*. The two semiconductor materials that are forming the junction might have different lattice constants, which can lead to the creation of dislocations and hence interface defects, which are detrimental to the performance of the junction. Lattice mismatch is discussed in more detail in Section 13.2.

8.3 Metal-semiconductor junctions

Almost all solar cells have metal-semiconductor junctions at the back and in most cases also at the front. These junctions are very important for connecting the solar cells to the external electric circuit, which consists of cables that – of course – also use metals as conducting materials. In this section we will briefly summarize the physics of metal-semiconductor junctions. We roughly follow Sze [31] and will discuss the main concepts for n-type semiconductors. From these results, the equations that are valid for p-type semiconductors can easily be derived.

Depending on the material properties of the metal and the semiconductor, we distinguish between two types of metal-semiconductor junctions: the *rectifying* type and the *nonrectifying* type, which is also called the *ohmic* type. Figure 8.12 (a) shows the separated band diagrams of a metal and an n-type semiconductor next to each other. A metal is characterized by the fact that a band is partially filled with electrons, i.e. the Fermi energy E_F is in the middle of this band. An important parameter for the metal is the work function ϕ_m (given in volts), which is defined as the energy that is required to remove an electron from the Fermi level to a position just outside the material, i.e. the *vacuum level*. The semiconductor is characterized by two parameters: the *electron affinity* χ, defined as the distance between the vacuum level and the lower edge of the conduction band, and the semiconductor work function ϕ_s.

8. Semiconductor junctions

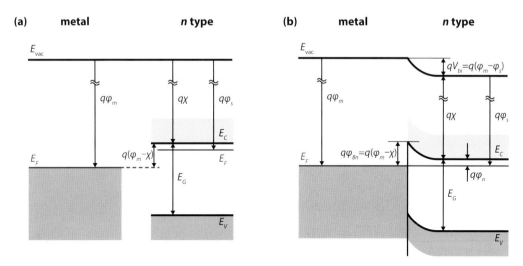

Figure 8.12: (a) The band diagrams of a metal and an n-type semiconductor that are separated from each other. (b) The band diagram of a junction between a metal and an n-type semiconductor.

If the metal and the semiconductor are brought together and form an ideal contact, two requirements must be fulfilled: at thermal equilibrium the Fermi energy must be constant throughout the junction and the vacuum level must be continuous. As a consequence, a barrier forms between the metal and the semiconductor, as depicted in Figure 8.12 (b). For an n-type semiconductor, the height of this barrier is given by

$$q\phi_{Bn} = q\phi_m - q\chi. \tag{8.43a}$$

In the case of a p-type semiconductor we find

$$q\phi_{Bp} = E_G - q(\phi_m - \chi). \tag{8.43b}$$

As a result of Eqs. (8.43), the sum of the barrier heights of n- and p-type substrates of a semiconductor is equal to its bandgap,

$$q(\phi_{Bn} + \phi_{Bp}) = E_G. \tag{8.44}$$

Note that this result is independent of the metal. An electron that wants to travel from the conduction band of the semiconductor to the metal experiences a built-in voltage,

$$V_{\text{bi}} = \phi_{Bn} - \phi_n, \tag{8.45}$$

where ϕ_n is the distance between the lower edge of the conduction band and the Fermi level of the n-type semiconductor.

Let us now take a look at how an external applied voltage affects the band diagrams of metal-semiconductor junctions, which is depicted in Figure 8.13. If no voltage is applied, as shown in Figure 8.13 (a), we get the same situation as that already depicted in Figure 8.12. If we apply a positive voltage to the metal with respect to the n-type semiconductor, the barrier height for an electron that wants to travel from the semiconductor to the

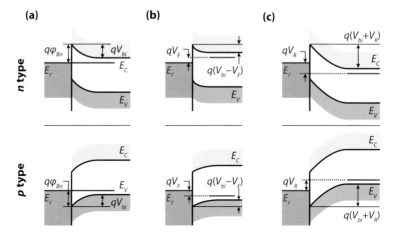

Figure 8.13: The band diagrams of metal-semiconductor junctions with n-type and p-type substrates under (a) no external bias; (b) forward bias; and (c) reverse bias.

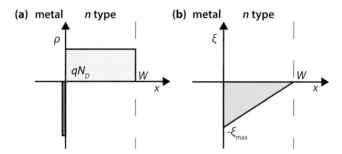

Figure 8.14: (a) The charge distribution and (b) the electric field band in a metal-semiconductor junction.

metal decreases to $q(V_{bi} - V_F)$, as depicted in Figure 8.13 (b). Hence it becomes easier for electrons to overcome the barrier. On the other hand, if a negative voltage is applied, the barrier becomes larger with $q(V_{bi} + V_R)$, as shown in Fig. 8.13 (c). In this case, transport of electrons from the semiconductor to the metal becomes more difficult. For a junction with a p-type substrate the situation is similar, only the polarization of the voltage is the other way round.

Figures 8.14 (a) and (b) show the charge distribution and the field across the junction, respectively. As the metal is assumed to be a perfect conductor and only in a very narrow region close to the metal-semiconductor interface, charge transferred from the semiconductor is present. In the semiconductor, a space-charge region exists that extends from the interface into the semiconductor for a width W. This means that inside this space charge region the charge density is $\rho_s = eN_D$, while it is 0 outside the space-charge region. The electric field is maximal at the interface and then decreases linearly until the end of the

8. Semiconductor junctions

space-charge region. The magnitude of the field along the space charge region is given by

$$|\xi(x)| = \frac{qN_D}{\epsilon_0\epsilon_s}(W-x), \tag{8.46}$$

where ϵ_s is the dielectric constant of the semiconductor. The voltage drop across the space charge region is given by the area under the electric field curve,

$$V_{\text{bi}} - V = \frac{\xi(0)W}{2} = \frac{qN_D W^2}{2\epsilon_0\epsilon_s}, \tag{8.47}$$

where the applied V is $V = V_F$ for forward bias and $V = -V_R$ for reverse bias. From Eq. (8.47) we find W to be

$$W = \sqrt{\frac{2\epsilon_0\epsilon_s(V_{\text{bi}} - V)}{qN_D}}. \tag{8.48}$$

8.3.1 The Schottky barrier

As we mentioned at the beginning of this section, we distinguish between rectifying and ohmic metal-semiconductor junctions. Rectifying junctions are also called *Schottky barriers*, because the German physicist Walter Schottky first described the physics of these junctions in 1938.

A metal-semiconductor junction is rectifying if the barrier height is large, i.e. $\phi_{Bn} \ll k_B T$ or $\phi_{Bp} \ll k_B T$. In contrast to p-n junctions, where current transport is mainly due to transport of the minority carriers, for Schottky barriers, the transport of the majority carriers is mainly responsible for the current transport. The dominant transport mechanism is *thermionic emission* of the majority carriers over the barrier into the metal. As we have seen in Figure 6.5, electrons in the conduction band have a droplet-shaped distribution with respect to the energy $n(E)$. The fraction of electrons with energies above a certain barrier energy E_B is given by

$$n_{th} = \int_{E_B}^{\infty} E(n)\,dE. \tag{8.49}$$

For a junction between a metal and an n-type semiconductor we find

$$n_{th} = N_C \exp\left(-\frac{q\Phi_{Bn}}{k_B T}\right). \tag{8.50}$$

Figure 8.15 depicts the energy-dependent electron distribution and the resulting currents for different applied voltages. If no voltage is applied, as shown in Figure 8.15 (a), the currents due to thermionic emission are equivalent in both directions, so no net current is flowing. However, when a forward bias is applied, the conduction band is shifted upwards with respect to the metal and the barrier decreases. As the barrier decreases, the density of the thermionic electrons n_{th} is increasing. Therefore, the thermionic current from the semiconductor into the metal is larger than that flowing the other way. In contrast, if a reverse bias is applied, n_{th} in the semiconductor decreases with respect to the value in the metal. As a consequence, a net current is flowing from the metal into the semiconductor.

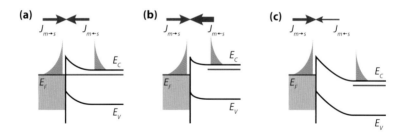

Figure 8.15: The energy distributions of the electrons, the thermionic electrons, and the resulting currents in a metal-semiconductor junction with *n*-type substrate under (a) no external bias; (b) *forward bias*; and (c) reverse bias.

The general current–voltage characteristic of a metal-semiconductor contact under thermionic emission is given by

$$J(V) = J_s \left(e^{qV/k_BT} - 1\right), \tag{8.51}$$

where the *saturation current density* J_s is given by

$$J_s = A^* T^2 e^{-q\phi_{Bn}/k_BT}. \tag{8.52}$$

The constant A^* is the *Richardson constant for thermionic emission*, which is given by

$$A^* \equiv \frac{4\pi q m_n^* k^2}{h^3}, \tag{8.53}$$

with the effective electron mass m_n^*. A detailed derivation can be found in [24, 31]. Also transport of minority carriers occurs, but it is orders of magnitude smaller than the majority carrier transport. Hence, it can be neglected for our application.

8.3.2 Ohmic contacts

If the resistance of the metal-semiconductor junction is negligible with respect to the bulk resistance of the semiconductor, the junction is called *ohmic*. In such a junction, the voltage drop is small and hardly affects the device performance.

As a figure of merit, the *specific contact resistance* R_C can be used. It is defined as

$$R_C := \left(\frac{\partial J}{\partial V}\right)^{-1}_{V=0}. \tag{8.54}$$

If the semiconductor has a low doping concentration, R_C is given by

$$R_C = \frac{k}{eA^*T} \exp\left(\frac{q\phi_{Bn}}{k_BT}\right). \tag{8.55}$$

Hence, for reducing the contact resistance, the barrier height should be kept low.

The situation looks different if a junction is formed between a metal and a *highly-doped* semiconductor. The width of the depletion region W decreases if doping is increased, as

8. Semiconductor junctions

we can see from Eq. (8.48). As a result, quantum-mechanic *tunnelling* of the charge carriers through the barrier becomes more probable. In this case, tunnelling can be the major current transport mechanism. As a consequence, R_C is also determined by tunnelling. We find [31]

$$R_C \propto \exp\left(\frac{4\sqrt{m_n^* \epsilon_s \phi_{Bn}}}{\sqrt{N_D}\hbar}\right). \tag{8.56}$$

Hence, if the doping concentration is increased, the contact resistance decreases.

In solar cells, the contact resistance at a metal-semiconductor junction should be as small as possible. Reducing the barrier height is often not possible as the materials that determine the barrier cannot be optimized for the barrier height as they fulfil other purposes. However, close to the metal-semiconductor junction a highly-doped region can be introduced that enables tunnelling and hence reduces the contact resistance as seen in *Eq.* (8.56).

8.4 Exercises

8.1 The *p*-type region of a silicon *p-n* junction is doped with 10^{16} boron atoms per cubic centimetre, and the *n*-type region is doped with 10^{18} phosphorus atoms per cubic centimetre. Assume a step *p-n* junction and that all doping atoms are ionized. The intrinsic carrier concentration in silicon at 300 K is 1.5×10^{10} cm^{-3}.

(a) What are the electron and hole concentrations in the *p*- and *n*-type regions at thermal equilibrium?

(b) Calculate the built-in voltage V_{bi} at 300 K.

(c) Calculate the width of the depletion region at 300 K (the relative permittivity of Si is $\epsilon_r = 11.7$).

8.2 A crystalline silicon solar cell generates a photocurrent density of $J_{ph} = 35$ mA/cm^2 at $T = 300$ K. The saturation current density is $J_0 = 1.95 \times 10^{-10}$ mA/cm^2. Assuming that the solar cell behaves as an ideal *p-n* junction, calculate the open circuit voltage V_{oc}.

8.3 Which of the following statements is true?

(a) The open circuit voltage decreases when the lifetime of the minority charge carriers increases.

(b) The open circuit voltage increases when the irradiance increases.

(c) The open circuit voltage increases when the intrinsic density of charge carriers increases.

(d) The open circuit voltage does not depend on the doping.

8.4 Figure 8.16 (a) shows the band diagrams of two materials with different bandgaps that are going to form a *P-n* heterojunction. Which one of the four options shown in Figure 8.16 (b) represents the actual band diagram of the *P-n* junction?

8.5 The width of the space charge region in a *p-n* junction is reduced by applying a voltage bias across the *p-n* junction. Which statement is correct?

(a) The voltage is a forward bias and the diffusion of the majority charge carriers becomes more dominant.

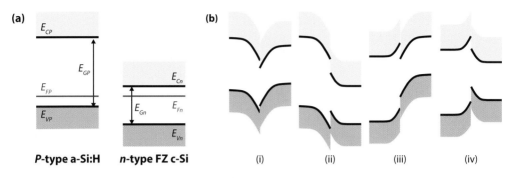

Figure 8.16

(b) The voltage is a reverse bias and the diffusion of the majority charge carriers becomes more dominant.

(c) The voltage is a forward bias and the drift of the minority charge carriers becomes more dominant.

(d) The voltage is a reverse bias and the drift of the minority charge carriers becomes more dominant.

8.6 Consider a crystalline silicon *p-n* junction solar cell with an area of 1×1 cm^2, and thickness of 180 μm. The doping of the *p* and *n* regions of the solar cell are $N_A = 5 \times 10^{16}$ cm^{-3} and $N_D = 9 \times 10^{18}$ cm^{-3}, respectively. The quality of the doped crystalline silicon regions is expressed by the lifetime and diffusion length of minority carriers; in *p*-type $\tau_n = 150$ μs and $D_n = 18$ cm^2/s, respectively, and in *n*-type $\tau_p = 70$ μs and $D_p = 6$ cm^2/s, respectively. Under standard test conditions (AM 1.5G) the solar cell generates a photocurrent of $I_L = 0.041$A. Assume that all the doping atoms are ionized and the solar cell is an ideal Shockley diode. Assume room temperature (300 K) and thermal equilibrium condition.

(a) Calculate the minority carrier concentrations of the *p*– and *n*-type regions, and also determine the position of the Fermi level with respect to the conduction band in the *p*-type and *n*-type quasi-neutral regions.

(b) Calculate the built-in voltage of the *p-n* junction.

(c) From the formula of built-in voltage, it is obvious that higher doping can result in higher built-in voltage, which means solar cell can have higher potential V_{oc}. However, in real solar cell manufacturing, the *p-n* junction is not heavily doped at both sides. Could you explain the reason or the disadvantages of heavy doping at both sides?

(d) Calculate the width of the solar cell's *p-n* junction depletion region, and then compare it to the thickness of the wafer. Express the fraction of the depletion region as a percentage of the total thickness of the wafer. Explain the reason why the width depletion region can be much smaller than the thickness of wafer.

(e) Draw the band diagram of the *p-n* junction by using the Fermi levels calculated in point (a).

8.7 Consider a *p*– doped c-Si wafer with an intrinsic carrier concentration at room temperature of $n_i = 1.1 \times 10^{10}$ cm^{-3} and a boron doping density of $N_A = 3 \times 10^{16}$ cm^{-3}

(a) What are the electron and hole carrier concentrations? What is the position of the Fermi level with respect to the conduction band?

(b) The *p*-layer is exposed to light. The absorption of light generates an additional hole and carrier concentration of $\Delta_p = \Delta_n = 1 \times 10^{14}$ cm^{-3}. What are the minority and majority carrier concentrations under illumination?

(c) The diffusion coefficient of electrons in silicon is $D = 36$ cm^{2}s^{-1}. In a silicon material the electron density drops linearly over a distance of 2 µm from $n = 2.7 \times 10^{16}$ cm^{-3} down to $n = 1 \times 10^{15}$ cm^{-3}. What is the electron diffusion current density induced by such a density gradient? What is the direction of the diffusion?

(d) Assume the same material as in (c). The electron mobility in silicon is 1350 cm^2 (Vs)$^{-1}$. What voltage over the gradient zone of 2 µm would be required to compensate the electron diffusion flux with an electron drift flux?

(e) Assume room temperature and mobility of electrons in a *p*-type c-Si wafer to be $\mu_n \approx 1250$ cm^{2}V^{-1}s^{-1}, which corresponds to doping of $N_A = 10^{14}$ cm^{-3}, and $\tau_n = 10^{-6}$ s, calculate the electron diffusion length.

(f) A drift current density of $J_{drift} = 120$ A/cm^2 is required in *p*-type c-Si (hole mobility $\mu_p \approx 480$ cm^2 V^{-1}s^{-1}) with an applied electric field of E = 20 V/cm. What doping concentration is required to achieve this current?

(g) Excess electrons in concentrations of 10^{15} cm^{-3} have been generated in *p*-type c-Si. The excess carrier lifetime is 10 µs. The generation stops at time $t = 0$. Calculate the excess electron concentration for:
(i) $t = 0$; (ii) $t = 1$ µs; (iii) $t = 4$ µs.

(h) Using the parameters from previous question, calculate the recombination rate of the excess electrons for:
i) $t = 0$; ii) $t = 1$ µs, iii) $t = 4$ µs.

8.8 Consider a light source that emits light only in the wavelength range 310 nm to 620 nm with a constant spectral irradiance $I = 3.00$ W/(m^2 nm).

(a) What is the total intensity of the light source, expressed in W/m^2? What is the total photon flux of the light source, expressed in photons/(m^2 s)?

(b) Find the optimum bandgap E_g for a solar cell that is illuminated by the above mentioned light source. Assume that this solar cell has a single junction and except for spectral mismatch losses, does not suffer any losses.

For questions (c), (d), (e) and (f) assume that a solar cell material with $E_g = 1.0$ eV is used.

(c) What would the voltage of this solar cell be?

(d) What would the current density of this solar cell be?

(e) What would the power conversion efficiency of this solar cell be?

(f) Sketch a graph of the spectral irradiance of the light source versus wavelength. For every wavelength indicate the part of the energy converted into electricity. Determine the solar cell's power conversion efficiency from this graph and compare it to the value found in question (e).

(g) Now consider that solar cell materials with higher bandgaps are available as well: 1.5eV, 2.0 eV, 2.5 eV, 3.0 eV, 3.5 eV and 4.0 eV. Calculate the solar cell efficiency for the above mentioned bandgap values and identify which material would give the highest efficiency.

Table 8.1: Bandgap and open circuit voltage of single junction solar cells based on various semiconductor materials.

Technology	E_G [eV]	V_{oc} [mV]	W [eV]	W/E_G
Ge	0.67	194		
CIGS	1.1	689		
c-Si (heterojunction)	1.12	729		
GaAs	1.42	1000		
CdTe	1.45	845		
a-Si:H	1.76	950		
AlGaInP	2.10	1560		

(h) Give three additional losses that were not included in the calculation.

8.9 Consider a Schottky barrier between Cu (work function of 4.65 V) and n-type Si with an electron affinity of 4.01 eV and doping density, $N_d = 1 \times 10^{16}$ cm^{-3}.

(a) Sketch the band diagram of the Schottky barrier under a forward bias of $V_a = 0.25$ V. Give the value of the potential barrier for electrons in the conduction band.

(b) Calculate the depletion-region width at 0 V bias and at a bias of $V_a = 0.25$ V.

8.10 Consider an n-p junction solar cell with external parameters J_{sc}, V_{oc}, and FF. The thickness of the n-type region is much lower than that of the p-type region. This solar cell is uniformly illuminated with generation rate, G_0, leading to an excess carrier concentration of δn and δp.

(a) Show that in open circuit condition the open circuit voltage is equal to:

$$V_{oc} = \frac{k_B T}{q} \ln\left(\frac{np}{n_i^2}\right). \quad (8.57)$$

(b) The bandgap voltage offset W is a useful parameter to study the solar cell quality. Basically W is a measure of how close electron and hole quasi-Fermi levels are to the conduction and valence band edges, respectively. It is defined as:

$$W = \frac{E_G}{q} - V_{oc}.$$

Show that W can be expressed as:

$$W = \frac{k_B T}{q} \ln\left(\frac{N_C N_V}{np}\right). \quad (8.58)$$

(c) The performance of solar cells under concentrated light (irradiance larger than 1 sun) is generally better in reference to standard test conditions. How is this reflected in Eqs. (8.57) and Eqs. (8.58)?

(d) Calculate the W and ratio W/E_{gap} of the various solar cells shown in Table 8.1. Which technologies have, in absolute and relative terms, the largest fraction of band-gap energy available at the solar cell terminals?

8. Semiconductor junctions

Table 8.2

	$Al_{0.4}Ga_{0.6}As$	GaAs
Bandgap (eV)	1.92	1.42
Electron affinity (V)	3.63	4.07
Conduction band density of states, N_C (cm^{-3})	5.0×10^{17}	4.7×10^{17}
Valence band density of states, N_V (cm^{-3})	8.0×10^{18}	7.0×10^{18}
Relative permittivity	11.6	12.9

8.11 An N-p heterojunction with doping concentrations $N_{dN} = 2 \times 10^{17}$ cm^{-3} and $N_{ap} = 6 \times 10^{17}$ cm^{-3}, consists of the semiconductor materials listed in Table 8.2.

(a) Draw the band diagram for this junction.

(b) How much does the spike in the conduction band extend above the conduction band of the quasi-neutral p-type region?

9
Solar Cell Parameters and Equivalent Circuit

9.1 External solar cell parameters

The main parameters that are used to characterize the performance of solar cells are the *peak power* P_{max}, the *short circuit current density* J_{sc}, the *open circuit voltage* V_{oc}, and the *fill factor FF*. These parameters are determined from the illuminated *J-V* characteristic as illustrated in Figure 8.10. The *conversion efficiency* η can be determined from these parameters.

9.1.1 Standard test conditions

For a reliable measurement of the *J-V* characteristics, it is vital to perform the measurements under *standard test conditions* (STC). This means that the total irradiance on the solar cell that should be measured is equal to 1000 W/m². Further, the spectrum should resemble the AM1.5 spectrum that we discussed in Section 5.5. Additionally, the temperature of the solar cell should be kept constant at 25 °C. As we will see in Section 20.3, the performance of a solar cell strongly depends on the temperature.

9.1.2 Short circuit current density

The *short circuit current* I_{sc} is the current that flows through the external circuit when the electrodes of the solar cell are short circuited. The short circuit current of a solar cell depends on the photon flux incident on the solar cell, which is determined by the spectrum of the incident light. For standard solar cell measurements, the spectrum is standardized to the AM1.5 spectrum. The I_{sc} depends on the area of the solar cell. In order to remove the dependence of the solar cell area on I_{sc}, the *short-circuit current density* is often used to describe the maximum current delivered by a solar cell. The maximum current that the solar cell can deliver strongly depends on the optical properties of the solar cell, such as absorption in the absorber layer and reflection.

In the ideal case, J_{sc} is equal to J_{ph}, which can easily be derived from Eq. (8.33). J_{ph} can be approximated by Eq. (8.34), which shows that in the case of an ideal diode (for example no surface recombination) and uniform generation, the critical material parameters that determine J_{ph} are the diffusion lengths of minority carriers. Crystalline silicon solar cells can deliver under an AM1.5 spectrum a maximum possible current density of 46 mA/cm². In laboratory c-Si solar cells the measured J_{sc} is above 42 mA/cm², while commercial solar cells have a J_{sc} exceeding 35 mA/cm².

9.1.3 Open circuit voltage

The *open-circuit voltage* is the voltage at which no current flows through the external circuit. It is the maximum voltage that a solar cell can deliver. V_{oc} corresponds to the forward bias voltage, at which the dark current density compensates the photocurrent density. V_{oc} depends on the photo generated current density and can be calculated from Eq. (8.33) assuming that the net current is zero,

$$V_{oc} = \frac{k_B T}{q} \ln\left(\frac{J_{ph}}{J_0} + 1\right) \approx \frac{k_B T}{q} \ln\left(\frac{J_{ph}}{J_0}\right), \quad (9.1)$$

where the approximation is justified because of $J_{ph} \gg J_0$

Equation (9.1) shows that V_{oc} depends on the saturation current density of the solar cell and the photo generated current. While J_{ph} typically has a small variation, the key effect is the saturation current, since this may vary by orders of magnitude. The saturation current density, J_0, depends on the recombination in the solar cell. Therefore, V_{oc} is a measure of the amount of recombination in the device. Laboratory crystalline silicon solar cells have a V_{oc} of up to 720 mV under the standard AM1.5 conditions, while commercial solar cells typically have V_{oc} exceeding 600 mV.

9.1.4 Fill factor

The fill factor is the ratio between the maximum power ($P_{max} = J_{mpp} V_{mpp}$) generated by a solar cell and the product of V_{oc} with J_{sc} (see Figure 8.10),

$$FF = \frac{J_{mpp} V_{mpp}}{J_{sc} V_{oc}}. \quad (9.2)$$

The subscript "mpp" in Eq. (9.2) denotes the *maximum power point* (MPP) of the solar cell, i.e. the point on the *J-V* characteristic of the solar cell, at which the solar cell has the maximal power output. To optimize the operation of PV systems, it is very important, to operate the solar cells (or PV modules) at the MPP. This is ensured with *maximum power point tracking* (MPPT), which is discussed in great detail in Section 19.1.

9. Solar cell parameters and equivalent circuit

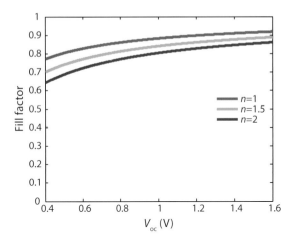

Figure 9.1: The FF as a function of V_{oc} for a solar cell with ideal diode behaviour.

Assuming that the solar cell behaves as an ideal diode, the fill factor can be expressed as a function of open circuit voltage V_{oc} [35],

$$FF = \frac{v_{oc} - \ln(v_{oc} + 0.72)}{v_{oc} + 1}, \quad (9.3)$$

where

$$v_{oc} = V_{oc} \frac{q}{k_B T} \quad (9.4)$$

is a normalized voltage. Equation (9.3) is a good approximation of the ideal value of FF for $v_{oc} > 10$. The FF as a function of V_{oc} is illustrated in Figure 9.1. This figure shows that FF does not change drastically with a change in V_{oc}. For a solar cell with a particular absorber, large variations in V_{oc} are not common. For example, at standard illumination conditions, the difference between the maximum open circuit voltage measured for a silicon laboratory device and a typical commercial solar cell is about 120 mV, giving a maximal FF of 0.85 and 0.83, respectively. However, the variation in maximum FF can be significant for solar cells made from different materials. For example, a GaAs solar cell may have an FF approaching 0.89.

However, in practical solar cells the dark diode current Eq. (8.23) does not obey the Boltzmann approximation. The non-ideal diode is approximated by introducing an *ideality factor n*, into the Boltzmann factor,

$$\exp \frac{qV_a}{nk_B T}. \quad (9.5)$$

Figure 9.1 also demonstrates the importance of the diode ideality factor when introduced into the normalized voltage in Eq. (9.3). The ideality factor is a measure of the junction quality and the type of recombination in a solar cell. For the ideal junction where the recombination is represented by the recombination of the minority carriers in the quasi-neutral regions, the n is equal to 1. However, when other recombination mechanisms occur, the n can have a value of 2. A high n value not only lowers the FF, but since it signals a high recombination, it leads to a low V_{oc}. Eq. (9.3) describes a maximum achievable FF. In practice the FF is often lower due to the presence of parasitic resistive losses.

9.1.5 Conversion efficiency

The *conversion efficiency* is calculated as the ratio between the maximal generated power and the incident power. As mentioned above, solar cells are measured under the STC, where the incident light is described by the AM1.5 spectrum and has an irradiance of $I_{in} = 1000 \text{ W/m}^2$,

$$\eta = \frac{P_{max}}{I_{in}} = \frac{J_{MPP} V_{MPP}}{I_{in}} = \frac{J_{sc} V_{oc} FF}{I_{in}}. \tag{9.6}$$

Typical external parameters of a crystalline silicon solar cell as shown are; $J_{sc} \approx 35$ mA/cm^2, V_{oc} up to 0.65 V and FF in the range 0.75 to 0.80. The conversion efficiency lies in the range of 17 to 18%.

Example

A crystalline silicon solar cell generates a photo current density of $J_{ph} = 35$ mA/cm^2. The wafer is doped with 10^{17} acceptor atoms per cubic centimetre and the emitter layer is formed with a uniform concentration of 10^{19} donors per cubic centimetre. The minority-carrier diffusion length in the p-type region and n-type region is 500×10^{-6} m and 10×10^{-6} m, respectively. Further, the intrinsic carrier concentration in silicon at 300 K is 1.5×10^{10} cm^{-3}, the mobility of electrons in the p-type region is $\mu_n = 1000$ cm^{2}V^{-1}s^{-1} and holes in the n-type region is $\mu_p = 100$ cm^{2}V^{-1}s^{-1}. Assume that the solar cell behaves as an ideal diode. Calculate the built-in voltage, the open circuit voltage and the conversion efficiency of the cell.

$J_{ph} = 350$ Am^{-2}.
$N_A = 10^{17}$ cm^{-3} = 10^{23} m^{-3}.
$N_D = 10^{19}$ cm^{-3} = 10^{25} m^{-3}.
$L_N = 500 \times 10^{-6}$ m.
$L_P = 10 \times 10^{-6}$ m.
$D_N = (k_B T/q)\mu_n = 0.0258$ V $\times 1000 \times 10^{-4}$ cm^{2}V^{-1}s^{-1} = 2.58×10^{-3} m^{2}s^{-1}.
$D_P = (k_B T/q)\mu_p = 0.0258$ V $\times\ 100 \times 10^{-4}$ cm^{2}V^{-1}s^{-1} = 2.58×10^{-4} m^{2}s^{-1}.

Using Eq. (8.16) we calculate the built-in voltage of the cell,

$$\psi_0 = \frac{k_B T}{q} \ln\left(\frac{N_A N_D}{n_i^2}\right) = 0.0258 \text{ V} \times \ln\left[\frac{10^{23} 10^{25}}{(1.5 \times 10^{16})^2}\right] = 0.93 \text{ V}.$$

According to the assumption that the solar cell behaves as an ideal diode, the Shockley equation describing the J-V characteristic is applicable. Using Eq. (8.25) we determine the saturation-current density,

$$J_0 = q n_i^2 \left(\frac{D_N}{L_N N_A} + \frac{D_P}{L_P N_D} \right) = 1.602 \times 10^{-19} \,\text{C} \, (1.5 \times 10^{16})^2 \,\text{m}^{-6}$$

$$\times \left(\frac{2.58 \times 10^{-3} \,\text{m}^2\text{s}^{-1}}{500 \times 10^{-6} \,\text{m} \, 10^{23} \,\text{m}^{-3}} + \frac{2.58 \times 10^{-4} \,\text{m}^2\text{s}^{-1}}{100 \times 10^{-6} \,\text{m} \, 10^{25} \,\text{m}^{-3}} \right)$$

$$= 1.95 \times 10^{-9} \, \frac{\text{C}}{\text{m}^2\text{s}} = 1.95 \times 10^{-9} \, \frac{\text{A}}{\text{m}^2}.$$

Using Eq. (9.1) we determine the open circuit voltage,

$$V_{oc} = \frac{k_B T}{q} \ln \left(\frac{J_{ph}}{J_0} + 1 \right) = 0.0258 \,\text{V} \, \ln \left(\frac{350 \,\text{Am}^{-2}}{1.95 \cdot 10^{-9} \,\text{Am}^{-2}} + 1 \right) = 0.67 \,\text{V}.$$

The fill factor of the cell can be calculated from Eq. (9.3). First, we normalize V_{oc},

$$v_{oc} = V_{oc} \Big/ \frac{k_B T}{q} = \frac{0.67 \,\text{V}}{0.0258 \,\text{V}} = 26.8.$$

Hence,

$$FF = \frac{v_{oc} - \ln(v_{oc} + 0.72)}{v_{oc} + 1} = 0.84.$$

Finally, the conversion efficiency is determined using Eq. (9.6),

$$\eta = \frac{J_{sc} V_{oc} FF}{P_{in}} = \frac{350 \,\text{Am}^{-2} \, 0.67 \,\text{V} \, 0.84}{1000 \,\text{Wm}^{-2}} = 0.197 = 19.7\%.$$

9.2 The external quantum efficiency

The external quantum efficiency EQE(λ) is the fraction of photons incident on the solar cell that create electron-hole pairs in the absorber which are successfully collected. It is wavelength dependent and is usually measured by illuminating the solar cell with monochromatic light of wavelength λ and measuring the photocurrent I_{ph} through the solar cell. The external quantum efficiency is then determined as

$$\text{EQE}(\lambda) = \frac{I_{ph}(\lambda)}{q \Psi_{ph,\lambda}}, \qquad (9.7)$$

where q is the elementary charge and $\Psi_{ph,\lambda}$ is the spectral photon flow incident on the solar cell. Since I_{ph} is dependent on the bias voltage, the bias voltage must be fixed during measurement. The photon flow is usually determined by measuring the EQE of a calibrated photo diode under the same light source.

The shape of this EQE curve is determined by optical and electrical losses such as parasitic absorption and recombination losses, respectively, which can make the analysis

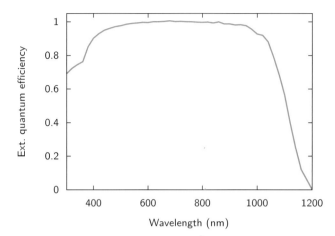

Figure 9.2: The external quantum efficiency of a high quality crystalline silicon-based solar cell.

complex. Figure 9.2 illustrates a typical EQE for a high quality crystalline silicon- based solar cell. In such a solar cell the minority-carrier diffusion length in the crystalline silicon substrate is very long and surface recombination is virtually suppressed. In this case we can identify the major optical loss mechanisms in the EQE for such a solar cell: For short wavelengths only a small fraction of the light is converted into electron-hole pairs. Most photons are already absorbed in the layers that the light traverses prior to the absorber layer; this is called parasitic absorption.

For long wavelengths, the penetration depth, which we defined in Section 4.4, exceeds the optical thickness of the absorber. Then the absorber itself becomes transparent so that most of the light leaves the solar cell before it can be absorbed. We can see that for this type of solar cells the EQE is close to 1 for a broad wavelength band. Hence, in this band almost all absorbed photons are converted into electron-hole pairs that can leave the solar cell.

For solar cells in which the minority-carrier diffusion length is shorter than the wafer thickness and/or surface recombination is not suppressed, the EQE curve will be affected. In essence the EQE curve will drop to lower values reflecting recombination losses in the device.

When a bias voltage of 0 V is applied, the measured photocurrent density equals the short circuit current density. In the case of *p-i-n* solar cells, when applying a sufficiently large reverse bias voltage, it can be assured that nearly all photo generated charge carriers in the intrinsic layer are collected. Thus, this measurement can be used to study the optical effectiveness of the design, i.e. light management and parasitic absorption in inactive layers, such as the transparent conducting oxide TCO layer, doped layers and the back reflector.

Measuring the EQE

EQE spectra are measured using an EQE-setup that is also called a *spectral response setup*. For this measurement, a wavelength selective light source, a calibrated light detector and a current meter are usually required. The light source generally used is a *xenon gas discharge lamp* that has a very broad spectrum covering all the wavelengths important for the solar cell performance. With the help of filters and monochromators a very narrow wavelength band of photon energies can be selected that can then be incident on the solar cell.

As already seen in Eq. (9.7), EQE(λ) is proportional to the current divided by the photon flow. While the current can be easily determined using an Ampere meter, the photon flow must be determined indirectly. This is done by performing a measurement with a calibrated photodetector (or solar cell), of which the EQE is known. Via this measurement we find

$$\Psi_{\text{ph},\lambda} = \frac{I_{\text{ph}}^{\text{ref}}(\lambda)}{q\,\text{EQE}^{\text{ref}}(\lambda)}. \quad (9.8)$$

By combining Eqs. (9.7) and (9.8) we therefore obtain

$$\text{EQE}(\lambda) = \text{EQE}^{\text{ref}}(\lambda)\frac{I_{\text{ph}}(\lambda)}{I_{\text{ph}}^{\text{ref}}(\lambda)}. \quad (9.9)$$

Hence, the EQE can be determined by performing two current measurements. Of course it is very important that the light source is sufficiently stable during the whole measurement as we assume that the photon flow in the reference measurement and the actual measurement is unchanged.

If we perform the EQE measurement under short circuit conditions, the measurement can be used to determine the *short circuit current density* J_{sc}. Determining J_{sc} via the EQE has the advantage that it is independent of the spectral shape of the light source used, in contrast to determining the J_{sc} via a J-V measurement. Secondly, on lab scale the real contact area of solar cells is not accurately determined during J-V measurements. When using shading masks, the EQE measurement is independent of the contact area. Hence, for accurately measuring the short circuit current density, it is not sufficient to rely on J-V measurements only, but a spectral response setup has to be used.

For determining J_{sc} we combine the photon flow at a certain wavelength with the EQE at this wavelength, leading to the flow of electrons leaving the solar cell at this wavelength. J_{sc} then is obtained by integrating across all the relevant wavelengths,

$$J_{\text{sc}} = -q \int_{\lambda_1}^{\lambda_2} \text{EQE}(\lambda)\Phi_{\text{ph},\lambda}^{\text{AM1.5}}\,d\lambda, \quad (9.10)$$

with the spectral photon flux $\Phi_{\text{ph},\lambda}$. For crystalline silicon, the important range would be from 300 to 1200 nm.

9.3 The equivalent circuit

The J-V characteristic of an illuminated solar cell that behaves as the ideal diode is given by Eq. (8.33),

$$J(V) = J_{\text{rec}}(V) - J_{\text{gen}}(V) - J_{\text{ph}} = J_0\left[\exp\left(\frac{qV}{k_B T}\right) - 1\right] - J_{\text{ph}}.$$

This behaviour can be described by a simple equivalent circuit, illustrated in Figure 9.3 (a), in which a diode and a current source are connected in parallel. The diode is formed by a

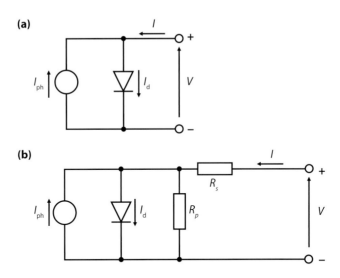

Figure 9.3: The equivalent circuit of (a) an ideal solar cell; and (b) a solar cell with series resistance R_s and shunt resistance R_p.

p-n junction. The first term in Eq. (8.33) describes the dark diode current density while the second term describes the photo generated current density. In practice the FF is influenced by a series resistance R_s, and a shunt resistance R_p. The influence of these parameters on the J-V characteristic of the solar cell can be studied using the equivalent circuit presented in Figure 9.3 (b). The J-V characteristic of the one-diode equivalent circuit with the series resistance and the shunt resistance is given by

$$J = J_0 \left\{ \exp\left[\frac{q(V - AJR_s)}{k_B T}\right] - 1 \right\} + \frac{V - AJR_s}{R_p} - J_{ph}, \tag{9.11}$$

where A is the area of the solar cell. The effect of R_s and R_p on the J-V characteristic is illustrated in Fig. 9.4.

In real solar cells the FF is influenced by additional recombination occurring in the p-n junction. This non-ideal diode is often represented in the equivalent circuit by two diodes, an ideal one with an ideality factor equal to unity and a non-ideal diode with an ideality factor greater than one. The equivalent circuit of a real solar cell is presented in Figure 9.5. The J-V characteristic of the two-diode equivalent circuit is given by

$$J = J_{01} \left\{ \exp\left[\frac{q(V - AJR_s)}{n_1 k_B T}\right] - 1 \right\} + J_{02} \left\{ \exp\left[\frac{q(V - AJR_s)}{n_2 k_B T}\right] - 1 \right\} + \frac{V - AJR_s}{R_p} - J_{ph}, \tag{9.12}$$

where J_{01} and J_{02} are the saturation current densities of the two diodes, respectively. n_1 and n_2 are the ideality factors of the two diodes.

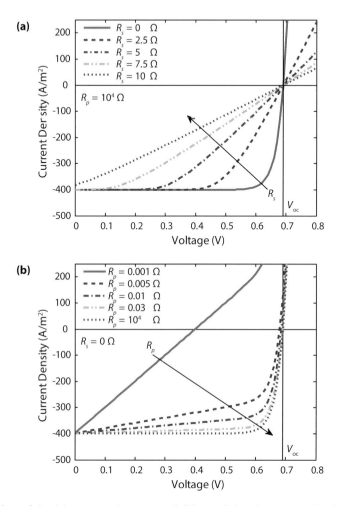

Figure 9.4: Effect of the (a) series resistance; and (b) parallel resistance on the *J-V* characteristic of a solar cell.

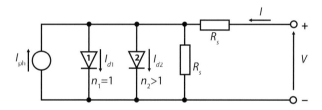

Figure 9.5: The equivalent circuit of a solar cell based on the two-diode model.

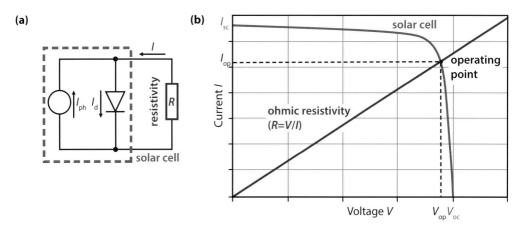

Figure 9.6: (a) Illustrating an ideal solar cell connected to an resistivity R. (b) The operating point of this simple system is at the intersection of the I-V curves of the solar cell and R.

Operation of a solar cell

At this point it is good to discuss how a solar cell actually operates. For this purpose, we combine the ideal solar cell from Figure 9.3 (a) with an ohmic resistivity R, as illustrated in Figure 9.6 (a).

Figure 9.6 (b) shows the I-V characteristics of a solar cell under illumination. Note, that we are looking at a solar cell with an area A that is connected to a resistor. Therefore, we are not looking at the current density J but at the current $I = AJ$ in this case. Further, Figure 9.6 (b) shows the I-V curve of the ohmic resistivity, which is simply a straight line because $R = V/I$.

The *operating point* of this very simple PV system is the point at which the two I-V curves cross. At this point, the solar cell produces a power $P_{op} = I_{op}V_{op}$. This power is dissipated as heat in the resistivity.

9.4 Exercises

9.1 What is the fill factor of a solar cell having maximum power $P_{max} = 1.5$ W? The open circuit voltage and the short circuit current of the cell are $V_{oc} = 0.6$ V and $J_{sc} = 2.8$ A, respectively.

9.2 Calculate the efficiency of a silicon-based solar cell having short circuit current density of $J_{sc} = 42.2$ mA/cm^2, open circuit voltage $V_{oc} = 706$ mV and fill factor FF = 0.828. Remember that the irradiation at standard test conditions is 100 mW/cm^2.

9.3 A solar cell with dimensions 12 cm × 12 cm is illuminated at standard test conditions. The cell has the following external parameters: $V_{oc} = 0.8$ V, $I_{sc} = 3$ A, $V_{MPP} = 0.75$ V, $I_{MPP} = 2.5$ A. Calculate the fill factor and efficiency of this solar cell.

9.4 John has used a solar simulator setup to measure the relationship between the voltage and the current of a small photovoltaic module (40 cm long and 40 cm wide). The measurement setup maintains the standard measurement conditions: the temperature is controlled to 25 °C, the incident spectrum is the AM1.5 spectrum with an incident power density of

9. Solar cell parameters and equivalent circuit

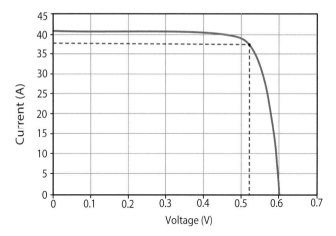

Figure 9.7

1000 W/m². The result is illustrated in Figure 9.7. John determined that the maximum power he could get out of this module is 19.5 W.

(a) Determine the short-circuit current density.

(b) Calculate the fill factor of the module.

(c) What is the efficiency of the module?

(d) John decides to connect two of these modules with a cable in series. This results in an additional 2 mΩ series resistance loss. What is the new fill factor? (Hint: use the voltage drop at the maximum power point.)

9.5 Figure 9.8 shows an ideal and a non-ideal J-V curve.

(a) Based on this plot and assuming STC, give approximately the following numerical values:

  i. The fill factor of the ideal cell.
  ii. The fill factor of the non-ideal cell.
  iii. The efficiency of the ideal J-V curve.
  iv. The efficiency of the non-ideal J-V curve.

(b) Which of the following statements is true?

  i. The inverse of the slope of the non-ideal curve at $V = V_{oc}$ is approximately equal to the shunt resistance.
  ii. An increase in shunt resistance will result in a decrease in fill factor in reference to the non-ideal case.
  iii. A large current in the shunt branch will increase the efficiency of the solar cell.
  iv. The inverse of the slope of the non-ideal curve at $V = V_{oc}$ is approximately equal to the series resistance.

9.6 Which of the following statements about the external quantum efficiency (EQE) is false?

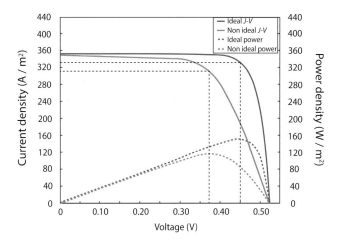

Figure 9.8

(a) The EQE of a solar cell depends both on the absorption of light and the collection of charge carriers.

(b) The EQE of a crystalline silicon solar cell cannot be less than 1.

(c) The EQE of a crystalline silicon solar cell cannot be greater than 1.

(d) Photons with energy lower than the bandgap energy contribute to the increase of the EQE of the solar cell.

9.7 The characterization of a solar cell under standard test conditions (STC) reveals that the efficiency is 18.6%. What would the efficiency of the solar cell be, if the EQE was halved and all the other external parameters remained the same?

9.8 Figure 9.9 shows the external quantum efficiency (EQE) of a solar cell as a function of the wavelength under short circuit condition ($V = 0$ V).

(a) What is the bandgap (in eV) of the absorber layer of the solar cell?

(b) Calculate the short circuit current density J_{sc} of the solar cell for the wavelength range of 300–1300 nm, if the incident (total) photon flux for this wavelength range is $\Phi_{ph} = 4.05 \times 10^{21}$ m^{-2}s^{-1}.

9.9 Which of the sketches shown in Figure 9.10 represents the equivalent circuit of an ideal solar cell under illumination?

9.10 You are given a PV module of 40×80 cm^2 and resistors with a resistance of 1 Ω, 2 Ω, 5 Ω and 10 Ω. Your goal is to run the current from the PV module through one of the resistors and use it as a heating element. The I-V characteristics of the PV module are given in Table 9.1. Assume standard test conditions (module temperature of 25 °C and irradiance of 1000 W/m^2).

(a) Sketch the I-V curve of the PV module and of each of the resistors in a single graph.

(b) Use the sketch to approximate the power generated when 1 Ω, 2 Ω, 5 Ω or 10 Ω resistors are used.

9. Solar cell parameters and equivalent circuit

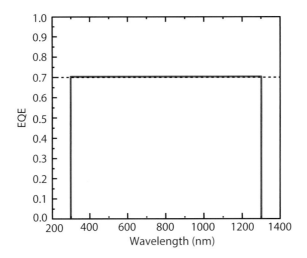

Figure 9.9

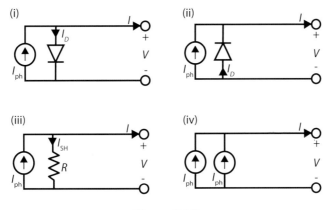

Figure 9.10

Table 9.1

V [V]	I [A]
0	4.2
5	4.1
10	4.0
15	3.8
20	2.1
22	0.0

(c) What is the power conversion efficiency of this module approximately?

9.11 700 nm monochromatic light with an irradiance of 20 mW/cm^2 is incident on a 100 cm^2 solar cell. External quantum efficiency of the solar cell at the wavelength of 700 nm is 80%. What is the maximum photo generated current achievable with this solar cell?

9.12 Consider a crystalline silicon p-n junction solar cell with an area $A = 10^{-4}$ m^2 and thickness of 1.8×10^{-4} m. The doping of the p- and n- type regions of the solar cell is $N_A = 4 \times 10^{22}$ m^{-3} and $N_D = 2 \times 10^{25}$ m^{-3} respectively. The quality of the crystalline silicon regions is expressed by the lifetime and mobility of minority carriers which in p-type is $\tau_n = 1.5 \times 10^{-4}$ s and $\mu_n = 0.07$ m^{2}V^{-1}s^{-1} respectively, and in n-type is $\tau_p = 7.0 \times 10^{-5}$ s and $\mu_p = 0.023$ m^{2}V^{-1}s^{-1} respectively. Under standard test conditions the solar cell generates a photocurrent of $I_{ph} = 0.041$ A. Assume that all the doping atoms are ionized, the solar cell is an ideal diode and the short circuit current density equals the photocurrent density. Also assume room temperature (300 K) and thermal equilibrium conditions.

(a) Calculate the open circuit voltage V_{oc}, short circuit current density J_{sc}, and fill factor FF.

(b) Calculate the minority-carriers concentrations in $p-$ and n-type regions.

(c) Calculate the positions of the Fermi levels (E_f - E_c) inside p and n-type regions.

(d) Calculate the built-in voltage across the p-n junction.

(e) Draw the band diagram of the solar cell under illumination at the short circuit and the open circuit condition, showing the Fermi levels and built-in voltage.

9.13 Consider a Si n^+-p junction solar cell with external parameters J_{sc}, V_{oc}, and FF. The thickness of the n^+-type region is much smaller than that of the p-type region. This solar cell is uniformly illuminated with generation rate, G_L, leading to an excess carrier concentration Δn and Δp. If the solar cell is in open circuit condition, show that in open circuit condition,

$$np = n_i^2 e^{qV_{oc}/k_B T}.$$

10
Losses and efficiency limits

In the previous chapters we have learned the basic physical principles of solar cells. In this chapter we will bring the different building blocks together and analyse how efficient a solar cell can be in theory. After discussing different efficiency limits and the major loss mechanisms, we will finalise this chapter with the formulation of *three design rules* that should always be kept in mind when designing solar cells.

It is very important to understand why a solar cell cannot convert 100% of the incident light into electricity. Different efficiency limits can be formulated, each taking different effects into account.

10.1 The thermodynamic limit

The most general efficiency limit is the *thermodynamic efficiency limit*. In this limit, the photovoltaic device is seen as a thermodynamic *heat engine*, as illustrated in Figure 10.1. Such a heat engine operates between two heat reservoirs; a hot one with temperature T_H and a cold one with temperature T_C. For the heat engine, three energy flows are relevant. First, the heat flow $\dot{Q}_H$ from the hot reservoir to the engine. Secondly, the *work* $\dot{W}$ that is performed by the engine and thirdly, the heat flowing from the engine to the cold reservoir that serves as a *heat sink*, $\dot{Q}_C$. Clearly, the third energy flow is a loss and consequently, the efficiency of the heat engine is given by

$$\eta = \frac{\dot{W}}{\dot{Q}_H}. \tag{10.1}$$

The *second law of thermodynamics* teaches us that the entropy of an independent system never decreases. It only increases or stays the same. While the heat flows $\dot{Q}_H$ and $\dot{Q}_C$

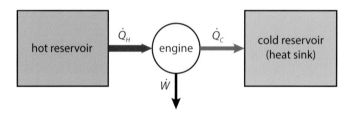

Figure 10.1: Illustrating the major heat flows in a generic heat engine.

carry entropy, the performed work W is an entropy-free form of energy. Thermodynamics teaches us that there is an efficiency limit for the transformation of heat into entropy-free energy. An (ideal) engine that has this maximal efficiency is called a *Carnot engine* and its efficiency is given by

$$\eta_{\text{Carnot}} = 1 - \frac{T_C}{T_H}. \tag{10.2}$$

For a Carnot engine, the entropy does not increase. Note that all the temperatures must be given in a temperature scale where the absolute zero takes the value 0 e.g., the Kelvin scale. From Eq. (10.2) we can already see two important trends that are basically true for every heat engine, such as steam engines or combustion engines. The efficiency increases if the higher temperature T_H is increased and/or the lower temperature T_C is decreased.

Let us now look at a solar cell that we imagine as a heat engine operating between an *absorber* of temperature T_A (this is our hot reservoir) and a cold reservoir, which is given by the surroundings and that we assume to be of temperature $T_C = 300$ K. What this heat engine actually does is convert the energy stored in the heat of the absorber into entropy-less chemical energy that is stored in the electron-hole pairs. Here, we assume that the transformation of chemical energy into electrical energy happens lossless, i.e. with an efficiency of 1. Hence, the efficiency of this thermodynamic heat engine is given by

$$\eta_{\text{TD}} = 1 - \frac{T_C}{T_A}. \tag{10.3}$$

The absorber will be heated as it absorbs sunlight. As we look at the ideal situation, we assume the absorber to be a blackbody that absorbs all incident radiation. Further, we assume the Sun to be a blackbody of temperature $T_S = 6000$ K. As we have seen in Chapter 5, the solar irradiance incident onto the absorber is given by

$$I_e^S = \sigma T_S^4 \Omega_{\text{inc}}, \tag{10.4}$$

where Ω_{inc} is the solid angle covered by the incident sunlight. As the absorber is a blackbody of temperature T_A it will also emit radiation. The emittance of the absorber is given by

$$E_e^A = \sigma T_A^4 \Omega_{\text{emit}}. \tag{10.5}$$

Ω_{emit} is the solid angle into which the absorber can emit.

The efficiency of the absorption process is given by

$$\eta_A = \frac{I_e^S - E_e^A}{I_e^S} = 1 - \frac{E_e^A}{I_e^S} = 1 - \frac{\Omega_{\text{emit}}}{\Omega_{\text{inc}}} \frac{T_A^4}{T_S^4}. \tag{10.6}$$

10. Losses and efficiency limits

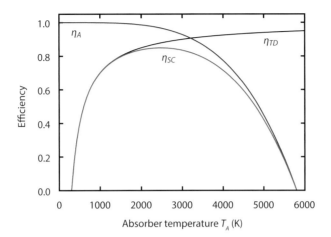

Figure 10.2: The absorber efficiency η_A, the thermodynamic efficiency η_{TD} and the combined solar cell efficiency η_{SC} under full concentration for a solar temperature of 5,800 K and an ambient temperature of 300 K.

The absorber efficiency can be increased by increasing Ω_{inc}, which can be achieved by *concentrating* the incident sunlight. Under *maximal concentration* sunlight will be incident onto the absorber from all angles of the hemisphere, i.e. $\Omega_{\text{inc}}^{\max} = 2\pi$. We assume the absorber to be open towards the surroundings and hence the Sun will be on the top side. Its bottom side is connected to the heat engine such that radiative loss only can happen via the top side. Therefore, also $\Omega_{\text{emit}} = 2\pi$. Hence, the maximal absorber efficiency is achieved under maximal concentration and it is given by

$$\eta_A^{\max} = 1 - \frac{T_A^4}{T_S^4}. \tag{10.7}$$

Note that η_A is greater when T_A is low, while the efficiency of the heat engine η_{TD} is greater when T_A is high.

For the total efficiency of the ideal solar cell we combine Eq. (10.3) with Eq. (10.7) and obtain

$$\eta_{SC} = \left(1 - \frac{T_A^4}{T_S^4}\right)\left(1 - \frac{T_C}{T_A}\right). \tag{10.8}$$

Figure 10.2 shows the absorber efficiency, the thermodynamic efficiency and the solar cell efficiency. We see that the solar cell efficiency reaches its maximum of about 85% for an absorber temperature of 2,480 K. Please note that the solar cell model presented in this section does not resemble a real solar cell; is only intended to discuss the physical limit of converting solar radiation into electricity. Several much more detailed studies on the thermodynamic limit have been performed. We want to refer the interested reader to works by Würfel [25] and Markvart et al. [36–38].

10.2 The Shockley-Queisser limit

We now will take a look at the theoretical limit for single-junction solar cells. This limit is usually referred to as the *Shockley–Queisser* (SQ) limit, as they where the first ones to formulate this limit based purely on physical assumptions and without using empirically determined constants [27]. We will derive the SQ limit in a two-step approach. First, we will discuss the losses due to *spectral mismatch*. Secondly, we will also take into account that the solar cell will have a temperature different from 0 K which means that it emits electromagnetic radiation according to Planck's law. Just like William B. Shockley (1910–1989) and Hans-Joachim Queisser (1931–), we will do this with the *detailed balance* approach.

10.2.1 Spectral mismatch

There are two principal losses that strongly reduce the energy conversion efficiency of single-junction solar cells. As discussed in Chapter 8, an important part of a solar cell is the absorber layer, in which the photons of the incident radiation are efficiently absorbed resulting in a creation of electron-hole pairs. In most cases, the absorber layer is formed by a semiconductor material, which we characterise by its bandgap energy E_G. In principle, only photons with energy higher than the bandgap energy of the absorber can generate electron-hole pairs. Since the electrons and holes tend to occupy energy levels at the bottom of the conduction band and the top of the valence band, respectively, the extra energy that the electron-hole pairs receive from the photons is released as heat into the semiconductor lattice in the *thermalization* process. Photons with energy lower than the bandgap energy of the absorber are in principle not absorbed and cannot generate electron-hole pairs. Therefore these photons are not involved in the energy conversion process. The *non-absorption* of photons carrying less energy than the semiconductor band gap and the *excess energy* of photons, larger than the bandgap, are the two main losses in the energy conversion process using solar cells. Both of these losses are thus related to the spectral mismatch between the energy distribution of photons in the solar spectrum and the bandgap of a semiconductor material.

Shockley and Queisser call the efficiency that is obtained when taking the spectral mismatch losses into account the *ultimate efficiency*, given according to the hypothesis that 'each photon with energy greater than $h\nu_G$ produces one electronic charge q at a voltage of $V_G = h\nu_G/e$' [27].

Let us now determine the fraction of energy of the incident radiation spectrum that is absorbed by a single-junction solar cell. If we denote λ_G as the wavelength of photons that corresponds to the bandgap energy of the absorber of the solar cell, only the photons with $\lambda \leq \lambda_G$ are absorbed. The fraction p_{abs} of the incident power that is absorbed by a solar cell and used for energy conversion can be expressed as

$$p_{\text{abs}} = \frac{\int_0^{\lambda_G} \frac{hc}{\lambda} \Phi_{\text{ph},\lambda} \, d\lambda}{\int_0^{\infty} \frac{hc}{\lambda} \Phi_{\text{ph},\lambda} \, d\lambda}, \tag{10.9}$$

where $\Phi_{\text{ph},\lambda}$ is the spectral photon flux of the incident light as defined in Chapter 5. The fraction of the absorbed photon energy exceeding the bandgap energy is lost because of thermalization. The fraction of the absorbed energy that the solar cell can deliver as useful

10. Losses and efficiency limits

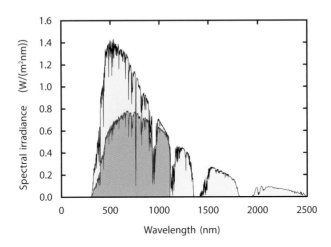

Figure 10.3: The fraction of the AM1.5 spectrum that can be converted into a usable energy by a crystalline silicon solar cell with $E_G = 1.12$ eV.

energy is then given by

$$p_{\text{use}} = \frac{E_G \int_0^{\lambda_G} \Phi_{\text{ph},\lambda} \, d\lambda}{\int_0^{\lambda_G} \frac{hc}{\lambda} \Phi_{\text{ph},\lambda} \, d\lambda}. \quad (10.10)$$

By combining Eqs. (10.9) and (10.10), we can determine the *ultimate conversion efficiency*,

$$\eta_{\text{ult}} = p_{\text{abs}} p_{\text{use}} = \frac{E_G \int_0^{\lambda_G} \Phi_{\text{ph},\lambda} \, d\lambda}{\int_0^{\infty} \frac{hc}{\lambda} \Phi_{\text{ph},\lambda} \, d\lambda}. \quad (10.11)$$

Figure 10.3 illustrates the fraction of the AM1.5 spectrum that can be converted into a usable energy by a crystalline silicon solar cell. Figure 10.4 shows the ultimate conversion efficiency with respect to the absorber bandgap for three different solar radiation spectra: blackbody radiation at 6,000 K, AM0 and AM1.5. The figure demonstrates that in the case of a crystalline silicon solar cell ($E_G = 1.12$ eV) the losses due to spectral mismatch account for almost 50%. It also shows that an absorber material for a single junction solar cell has an optimal bandgap of 1.1 eV and 1.0 eV for the AM0 and AM1.5 spectra, respectively. Note that the maximum conversion efficiency for the AM1.5 spectrum is higher than that for AM0, while the AM0 spectrum has a higher overall power density. This is because the AM1.5 spectrum has a lower power density in parts of the spectrum that are not contributing to the energy conversion process, as can be seen in Fig. 10.3. The dips in the AM1.5 spectrum also result in the irregular shape of the conversion efficiency as a function of the bandgap.

10.2.2 Detailed balance limit of the efficiency

Similar to *Shockley and Queisser* we will now formulate the *detailed balance limit of the efficiency*. But before we start we will briefly discuss the reason that the ultimate efficiency

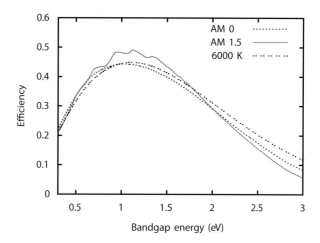

Figure 10.4: The ultimate conversion efficiency for the blackbody spectrum at 6,000 K, the AM0 and AM1.5 solar radiation spectra, limited only by the spectral mismatch as a function of the bandgap of a semiconductor absorber in single junction solar cells.

formulated above is not physically meaningful for solar cells with temperatures higher than 0 K.

Let us estimate that the solar cell is embedded in an environment of ambient temperature of 300 K and that the solar cell temperature is also 300 K. As the solar cell will be in thermal equilibrium with its surroundings, it will absorb thermal radiation according to the ambient temperature *and* it will also emit the same amount of radiation. Therefore recombination of electron-hole pairs will be present in the semiconductor leading to a recombination current density different from zero. As we have seen in Eq. (9.1), the open circuit voltage will be reduced with increasing recombination current, which is an efficiency loss.

For deriving the detailed balance limit we first recall the definition of the efficiency from Eq. (9.6),

$$\eta = \frac{J_{\text{ph}} V_{\text{oc}} FF}{P_{\text{in}}}. \tag{10.12}$$

For calculating η_{ult} we made the assumption that '*each photon with energy greater than $h\nu_G$ produces one electronic charge q at a voltage of $V_G = h\nu_G/q$*'. Under the same assumption, we obtain for the *short circuit current density*

$$J_{\text{ph}}(E_G) = -q \int_0^{\lambda_G} \Phi_{\text{ph},\lambda} \, d\lambda \tag{10.13}$$

with $\lambda_G = hc/E_G$. Note that here we implicitly assumed that the photo generated current density J_{ph} is equivalent to the short circuit current density. This approximation is valid as the recombination current originating from thermal emission is orders of magnitude lower than the photo generated current. By combining Eqs. (10.11) and (10.13) we find

$$J_{\text{ph}} = -\frac{q}{E_G} P_{\text{in}} \eta_{\text{ult}} = -\frac{P_{\text{in}} \eta_{\text{ult}}}{V_G}. \tag{10.14}$$

10. Losses and efficiency limits

Let us now define the *bandgap utilization efficiency* η_V that is given by

$$\eta_V = \frac{V_{oc}}{V_G} \tag{10.15}$$

and tells us the fraction of the bandgap that can be used as open circuit voltage (Shockley and Queisser use the letter v for this efficiency). We now combine Eqs. (10.12), (10.14) and (10.15) and find

$$\eta = \eta_{\text{ult}} \eta_V FF. \tag{10.16}$$

For determining the efficiency in the detailed balance limit, we first must determine the bandgap utilization efficiency and the fill factor. Let us start with η_V.

According to Eq. (9.1), the open circuit voltage will be reduced with increasing recombination current density, which is an efficiency loss. It is given as

$$V_{oc} = \frac{k_B T}{q} \ln\left(\frac{J_{\text{ph}}}{J_0} + 1\right). \tag{10.17}$$

The only unknown in this equation is the dark current density J_0. We assume the solar cell to be in *thermal equilibrium* with its surroundings at an ambient temperature of $T_a = 300$ K. Further, we assume that the solar cell absorbs and emits as a blackbody for wavelengths shorter than the bandgap wavelength of the solar cell absorber. For wavelengths longer than the bandgap we assume the solar cell to be completely transparent, thus to neither absorb nor emit. This is the same assumption that we already used for the absorption of sunlight.

Using the equation for the *blackbody radiance* $L_{e\lambda}^{BB}$ as given in Eq. (5.18a) we find for the radiative recombination current density

$$\begin{aligned} J_0(E_G) &= -2q \int_0^{\lambda_G} \int_{2\pi} L_{e\lambda}^{BB}(\lambda; T_a) \cos\theta \, d\Omega \, d\lambda \\ &= -2q\pi \int_0^{\lambda_G} \frac{2hc^2}{\lambda^5} \left[\exp\left(\frac{hc}{\lambda k_B T_a}\right) - 1\right]^{-1} d\lambda, \end{aligned} \tag{10.18}$$

where the factor 2 arises from the fact that we assume the solar cell to emit thermal radiation from both its front and back sides.

Combining Eq. (10.15) with (10.17) we find

$$\eta_V(E_G) = k_B T/E_G \ln\left[\frac{J_{\text{ph}}(E_G)}{J_0(E_G)} + 1\right]. \tag{10.19}$$

Figure 10.5 shows the bandgap utilization efficiency for three different spectra of the incident sunlight. For a bandgap of 1.12 eV this efficiency is about $\eta_V \approx 77\%$.

For the fill factor we take the empirical but very accurate approximation

$$FF = \frac{v_{oc} - \ln(v_{oc} + 0.72)}{v_{oc} + 1} \tag{10.20}$$

with $v_{oc} = qV_{oc}/k_B T$. We already discussed this approximation in Eq. (9.3).

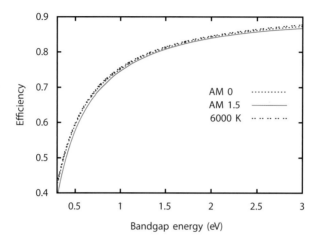

Figure 10.5: The bandgap utilization efficiency η_V for the blackbody spectrum at 6,000 K, and the AM0 and AM1.5 solar spectra.

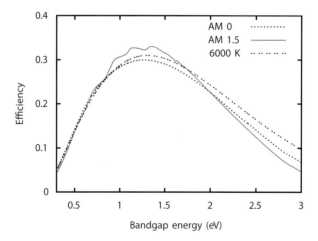

Figure 10.6: The Shockley–Queisser efficiency limit for the blackbody spectrum at 6,000 K, and the AM0 and AM1.5 solar spectra.

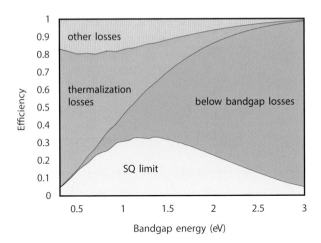

Figure 10.7: The major loss mechanisms in the Shockley–Queisser limit. For this calculation the AM1.5 spectrum was used as incident light.

Figure 10.6 finally shows the Shockley–Queisser efficiency limit for three different spectra of the incident light. For the AM1.5 spectrum the limit is about 33.1% at 1.34 eV. For AM0 it is 30.1% at 1.26 eV.

The major loss mechanisms that are taken into account in the Shockley–Queisser limit are illustrated in Fig. 10.7. The major losses are non-absorbed photons below the bandgap and thermalised energy of photons above the bandgap. The other losses are due to the voltage loss because of thermal radiation and the fill factor being different from 100%.

10.2.3 Efficiency limit for silicon solar cells

It is very important to note that the Shockley–Queisser (SQ) limit is not directly applicable to solar cells made from crystalline silicon. The reason for this is that silicon is a so-called *indirect bandgap* semiconductor as we will discuss in detail in Chapter 12. This means that Auger recombination, which is a non-radiative recombination mechanism, is dominant. For the derivation of the SQ limit we assume that only radiative recombination is present. Clearly, this assumption cannot be valid for crystalline silicon solar cells. Several attempts to calculate the efficiency limit while taking radiative recombination mechanisms into account were performed in the past. A study from 2013 by Richter et al. derives an efficiency limit of 29.43% for silicon solar cells.

As the Shockley–Queisser limit only considers radiative recombination, it is most valid for direct bandgap materials such as GaAs. Because of its direct bandgap, radiative recombination is the limiting recombination mechanism for GaAs.

10.3 Additional losses

The Shockley–Queisser limit is a very idealised model. For example all optical losses are neglected. Now we will discuss several loss mechanisms that have to be taken into account

10.3.1 Optical losses

For the Shockley-Queisser limit, we only took the bandgap energy E_G into account for deriving the efficiency limit. However, the real performance is also strongly influenced by the optical properties given as the complex refractive index $\tilde{n} = n - ik$, which is a function of the wavelength.

As we already discussed in Section 4.3, part of the light is reflected and the other part is transmitted when light arrives on an interface between two media. The interface is therefore characterised by the wavelength-dependent reflectivity $R(\lambda)$ and transmittance $T(\lambda)$. All the reflections and transmissions at the different interfaces in the solar cell result in a total reflectance between the solar cell and the surrounding air. Hence, a part of the incident energy that can be converted into a usable energy by the solar cell is lost by reflection. We shall denote the total *effective* reflectivity in the wavelength range of interest as R^*.

As we will discuss in more detail in Chapter 12, in most c-Si solar cells thin metal strips are placed on the front side of the solar cell that serve as front electrodes. The metal-covered area does not allow the light to enter the solar cell because it reflects or slightly absorbs the incident light. The area that is covered by the electrode effectively decreases the active area of the solar cell. We denote the total area of the cell as A_{tot} and the cell area that is not covered by the electrode as A_f, the fraction of the active area of the cell is determined by the ratio

$$C_f = \frac{A_f}{A_{\text{tot}}}, \qquad (10.21)$$

which is called the active area coverage factor C_f. The resulting loss is called the *shading loss*. The design of the front electrode is of great importance since it should minimise losses due to the series resistance of the front electrode, i.e. should be designed with sufficient cross-section. The optimal design of the front electrode is therefore a trade-off between a high coverage factor and a sufficiently low series resistance of the front electrode.

When light penetrates into a material, it will be partially absorbed as it propagates through the material. The absorption of light in the material depends on its absorption coefficient and the layer thickness, as we have seen in Section 4.4. In general, light is absorbed in all layers of the solar cell. All the absorption in layers other than the absorber layer is loss. It is called the *parasitic absorption*. Further, due to the limited thickness of the absorber layer, not all the light entering the absorber layer is absorbed. Incomplete absorption in the absorber due to its limited thickness is an additional loss that lowers the energy conversion efficiency. The incomplete absorption loss can be described by the internal optical quantum efficiency IQE_{op}, which is defined as the probability of a photon being absorbed in the absorber material. Since there is a chance that a highly energetic photon can generate more than one electron-hole pair, we also define the *quantum efficiency for carrier generation* η_G which represents the number of electron-hole pairs generated by one absorbed photon. Usually η_G is assumed to be unity.

10.3.2 Solar cell collection losses

Not all charge carriers that are generated in a solar cell are collected at the electrodes. The photo generated carriers are the excess carriers with respect to the thermal equilibrium and are subjected to the recombination. The carriers recombine in the bulk, at the interfaces, and/or at the surfaces of the junction, as discussed in Chapter 7. The recombination is determined by the electronic properties of the materials that form the junction, such as density of states within the bandgap because of trap states. The trap density N_T strongly influences the minority-carrier lifetimes.

The contributions of both the electronic and optical properties of the solar cell materials to the photovoltaic performance are taken into account in the absolute external quantum efficiency that we already defined in Chapter 9. The EQE can be approximated by

$$\text{EQE}(\lambda) = (1 - R^*)\text{IQE}_{\text{op}}(\lambda)\eta_G(\lambda)\text{IQE}_{\text{el}}(\lambda), \tag{10.22}$$

where IQE_{el} denotes the electrical quantum efficiency and is defined as the probability that a photo generated carrier is collected.

When we take the shading losses and the EQE into account, we find the short circuit current density to be

$$J_{\text{sc}} = J_{\text{ph}}(1 - R^*)\text{IQE}_{\text{op}}^*\eta_G^*\text{IQE}_{\text{el}}^*C_f, \tag{10.23}$$

where the * denotes averages across the relevant wavelength range and J_{ph} is as defined in Eq. (10.13).

10.3.3 Additional limiting factors

We have seen in Section 10.2 that the V_{oc} of a solar cell depends on the saturation current J_0 and the photo generated current J_{sc} of the solar cell. The saturation current density depends on the recombination in the solar cell. Recombination cannot be avoided and depends on the doping of different regions (n-type and p-type) of the junction and the electronic quality of materials forming the junction. The doping levels and the recombination determine the bandgap utilization efficiency η_V that we already defined in Section 10.2.

For determining the Shockley–Queisser limit we assumed for the FF that the solar cell behaves as an ideal diode. In a real solar cell, however, the FF is lower than the ideal value because of the following reasons:

- The voltage drop due to the *series resistance* R_s of a solar cell, which is introduced by the resistance of the main current path through which the photo generated carriers arrive to the external circuit. The contributions to the series resistance come from the bulk resistance of the junction, the contact resistance between the junction and electrodes, and the resistance of the electrodes themselves.

- The voltage drop due to *leakage currents*, which is characterised by the shunt resistance R_p of a solar cell. The leakage current is caused by the current through local defects in the junction or due to shunts at the edges of solar cells.

- The *recombination* in a non-ideal solar cell results in a decrease of the FF.

10.3.4 Conversion efficiency

Again, we start with the expression for the efficiency

$$\eta = \frac{J_{sc} V_{oc} FF}{P_{in}}. \tag{10.24}$$

Using Eqs. (10.23) and (10.14) we find

$$\begin{aligned}\eta &= \frac{\eta_{ult} V_{oc} FF}{V_G}(1-R^*)\text{IQE}^*_{op}\eta^*_G \text{IQE}^*_{el} C_f \\ &= p_{abs} p_{use}(1-R^*)\text{IQE}^*_{op}\eta^*_G \text{IQE}^*_{el} C_f \eta_V FF,\end{aligned} \tag{10.25}$$

where we used Eqs. (10.9) and (10.10). By filling in the definitions for p_{abs}, p_{use}, η_V and C_f, we obtain [39].

$$\eta = \underbrace{\frac{\int_0^{\lambda_G} \frac{hc}{\lambda}\Phi_{ph,\lambda}\,d\lambda}{\int_0^{\infty} \frac{hc}{\lambda}\Phi_{ph,\lambda}\,d\lambda}}_{1} \underbrace{\frac{E_G \int_0^{\lambda_G}\Phi_{ph,\lambda}\,d\lambda}{\int_0^{\lambda_G} \frac{hc}{\lambda}\Phi_{ph,\lambda}\,d\lambda}}_{2} \underbrace{(1-R^*)}_{3} \underbrace{\text{IQE}^*_{op}\eta^*_G}_{4} \underbrace{\text{IQE}^*_{el}}_{5} \underbrace{\frac{A_f}{A_{tot}}}_{6} \underbrace{\frac{eV_{oc}}{E_G}}_{7} \underbrace{FF}_{8}. \tag{10.26}$$

This describes the conversion efficiency of a solar cell in terms of components that represent particular losses in energy conversion:

1. due to non-absorption of long wavelengths;
2. due to thermalization of the excess energy of photons;
3. due to the total reflection;
4. by incomplete absorption due to the finite thickness;
5. due to recombination;
6. by metal electrode coverage, shading losses;
7. due to voltage factor;
8. due to fill factor.

10.4 Design rules for solar cells

Now, as we have extensively discussed all the factors that limit the efficiency of a solar cell we are able to distil three *design rules*. We will use these design rules in Part III when we discuss different PV technologies. The three design rules are:

1. utilization of the band gap energy,
2. spectral utilization,
3. light management.

We will now take a closer look at each of these design rules.

10. Losses and efficiency limits

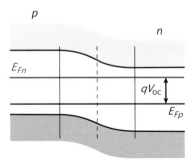

Figure 10.8: The open circuit voltage of a solar cell is determined by the splitting of the quasi-Fermi levels.

10.4.1 Bandgap utilization

As we have seen earlier in this chapter, the open circuit voltage V_{oc} is always below the voltage V_G corresponding to the bandgap. This loss is characterised with the bandgap utilization efficiency η_V. The open circuit voltage is determined by the extent to which the quasi-Fermi levels are able to split, which is limited by the charge carrier-recombination mechanisms. We discussed the different recombination mechanisms already in detail in Chapter 7. Here, we briefly summarise them and discuss how they affect the bandgap utilization.

Figure 10.8 shows a *p-n* junction, with the *p*-doped region on the left and the *n*-doped region on the right. Further, the quasi-Fermi levels are depicted. The extent of splitting between the quasi-Fermi levels determines the open circuit voltage. The open circuit voltage, for example expressed in Eq. (10.17) can also be expressed in terms of the generation rate G_L, the life time τ_0 of the minority charge carriers and the intrinsic density of the charge carriers n_i in the semiconductor material. Under the assumption that G_L is spatially homogeneous across the junction, we find

$$V_{oc} \approx \frac{2k_B T}{q} \ln\left(\frac{G_L \tau_0}{n_i}\right). \tag{10.27}$$

The derivation of this equation is outside the scope of this book.

Let us now take a closer look at Eq. (10.27). If we increase the irradiance, or in other words, the generation rate of charge carriers, the open circuit voltage is increased. This is a welcome effect which is utilised in concentrator photovoltaics (CPV), which we will discuss in Section 15.8. Secondly, we see that the lifetime τ_0 plays an important role. The larger the lifetime of the minority charge carrier, the larger the open circuit voltage can be. Or in other words, the longer the lifetime, the larger the possible splitting between the quasi-Fermi levels and the larger the fraction of the bandgap energy that can be utilised.

The lifetime of the minority charge carrier is determined by the recombination rate. As discussed in Chapter 7, we have to consider three different recombination mechanisms: radiative, Shockley–Read–Hall, and Auger recombination. While radiative and Auger recombination depend on the semiconductor itself, SRH recombination is proportional to the density of traps or impurities in the semiconductor. In the three recombination mechanisms, energy and momentum are transferred from charge carriers to phonons or photons.

The efficiency of the different recombination processes depends on the nature of the bandgap of the semiconductor material used. We distinguish between *direct* and *indirect* bandgap semiconductors. Crystalline silicon is an indirect bandgap material. The radiative recombination in an indirect bandgap material is inefficient and recombination will be dominated by the Auger mechanism. For direct bandgap materials such as GaAs under moderate illumination conditions, radiative recombination will be the dominant loss mechanism of charge carriers. For very high illumination conditions, Auger recombination starts to play a role as well.

To summarize, we find that in the defect-rich solar cells, the open circuit voltage is limited by the SRH recombination. In low-defect solar cells based on indirect bandgap materials, the open circuit voltage is limited by Auger recombination. In low-defect solar cells based on direct bandgap materials, the open circuit voltage is limited by radiative recombination.

Besides bandgap utilization, it is also important to discuss the relationship between the maximum thickness for the absorber layer of a solar cell and the dominant recombination mechanism. As we have seen in Chapter 7, the recombination mechanism also affects the diffusion length of the minority charge carrier. The diffusion length L_n of minority electrons is given by

$$L_n = \sqrt{D_n \tau_n}, \tag{10.28}$$

where D_n is the diffusion coefficient and τ is the lifetime of the minority charge carrier. Similarly, we can formulate the diffusion length for minority holes, L_p.

It is important to realise that ideally the thickness of the absorber layer should not exceed the diffusion length. In order to understand this requirement, we consider photons that penetrate far into the absorber layer before being absorbed and generating charge carriers. The charge carriers generated deep in the absorber need to diffuse to the *p-n* junction or the back contact for separation and collection. If the distance these charge carriers need to diffuse exceeds the diffusion length, these excited charge carriers will likely recombine before arriving at the *p-n* junction or back contact. This means that a substantial fraction of the charge carriers generated at a distance greater than the diffusion length from the *p-n* junction or the back contact will not be collected and hence is lost. Only a fraction of the generated carrier density smaller than $1/e$ is collected, where e is the base of the natural logarithm. If the charge carriers are generated within the diffusion length of the *p-n* junction or back contact, the likelihood for collection is much greater. This means that the minority carrier diffusion length limits the maximum thickness of the solar cell.

Mathematically speaking, the influence of the absorber thickness on V_{oc} is given by

$$V_{oc} \approx C + \frac{k_B T}{q} \ln\left(\frac{L}{W}\right), \tag{10.29}$$

when the surface recombination velocity is $S = 0$. L is the diffusion length in the absorber, just as above, and W is the thickness (or width) of the absorber. A derivation of this equation is given in Appendix C.

To summarise this section, the open circuit voltage is limited by the dominant recombination mechanism. The dominance of radiative, Auger or Shockley–Reed–Hall recombination depends on the type of semiconductor materials used in the solar cell and the

10. Losses and efficiency limits

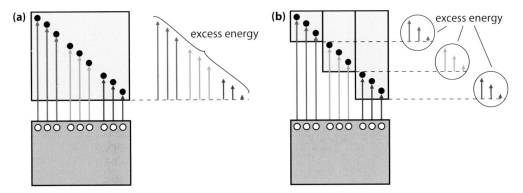

Figure 10.9: Illustrating the lost excess energy in (a) a single-juntion; and (b) a multi-junction solar cell.

illumination conditions. We will discuss several different cases on Part III on PV technology.

10.4.2 Spectral utilization

The spectral utilization is mainly determined by the choice of materials from which the solar cell is made. As we have seen in Section 10.2, and mainly Eq. (10.13), the photocurrent density is determined by the bandgap of the material. For a bandgap of 0.62 eV corresponding to a wavelength of 2,000 nm, we could theoretically generate a short circuit current density of 62 mA/cm^2. If we consider c-Si, having a band gap of 1.12 eV (1107 nm), we arrive at a theoretical current density of 44 mA/cm^2.

The optimal bandgap for single-junction solar cells is determined by the Shockley–Queisser limit, as illustrated in Figure 10.6. For single-junction solar cells, semiconductor materials such as silicon, gallium arsenide and cadmium telluride have a band gap close to the optimum.

In Part III we will discuss various concepts that are allowed to surpass the Shockley–Queisser limit. Here, we will briefly discuss the concept of *multi-junction solar cells*. In these devices, solar cells with different bandgaps are stacked on top of each other. As illustrated in Figure 10.9, the excess energy can be reduced significantly, and the spectral utilization will improve.

10.4.3 Light management

The third and last design rule that we discuss is *light management*. In an ideal solar cell, all light that is incident on the solar cell should be absorbed in the absorber layer. As we have discussed in Section 4.4, the intensity of light decreases exponentially as it travels through an absorptive medium. This is described by the Lambert–Beer law that we formulated in Eq. (4.25),

$$I(d) = I_0 \exp(-\alpha d). \tag{10.30}$$

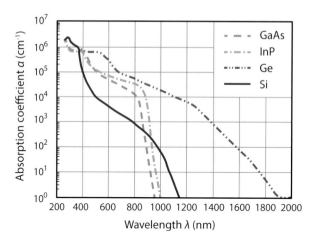

Figure 10.10: Absorption coefficients of different semiconductors.

From the Lambert–Beer law it follows that at the side at which the light is entering the film, more light is absorbed relative to the back side. The total fraction of the incident light absorbed in the material is equal to the light intensity entering the absorber layer minus the intensity transmitted through the absorber layer,

$$I^{\text{abs}}(d) = I_0[1 - \exp(-\alpha d)]. \tag{10.31}$$

Ideally, we would like a solar cell to absorb 100% of the incident light. Such an absorber is called *optically thick* and has a transmissivity very close to 0. As we can see from the Lambert–Beer law, this can be achieved either by absorbers with a large thickness d or with very large absorption coefficients α.

Figure 10.10 shows the absorption coefficients for four different semiconductor materials: germanium (Ge), silicon (Si), gallium arsenide (GaAs) and indium phosphide (InP). We notice that Ge has the lowest bandgap. It starts to absorb at long wavelengths, which corresponds to a low photon energy. GaAs has the highest bandgap, as it starts to absorb light at the smallest wavelength, or highest photon energy. In addition, if we focus on the visible spectral part from 300 nm to 700 nm, we see that the absorption coefficients of InP and GaAs are significantly higher than for Si. This is related to the fact that InP and GaAs are direct bandgap materials as discussed earlier. Materials with an indirect bandgap have smaller absorption coefficients. Only in the very blue and ultraviolet parts below 400 nm, Si has a direct bandgap transition. Silicon is a relatively poor absorber. Therefore for the same fraction of light, thicker absorber layers are required in comparison to GaAs.

In general, for all semiconductor materials the absorption coefficient in the blue is orders of magnitude higher than in the red. Therefore the penetration depth of blue light into the absorber layer is rather small. In crystalline silicon, the blue light is already fully absorbed within a few nanometres. The red light requires an absorption path length of 60 μm to be fully absorbed. The infrared light is hardly absorbed, and after an optical path length of 100 μm only about 10% of the light intensity is absorbed.

As the absorption of photons generates excited charge carriers, the wavelength dependence of the absorption coefficient determines the local generation profile of the charge

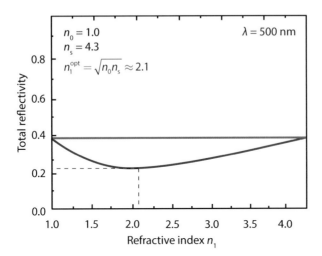

Figure 10.11: Illustrating the effect of an interlayer with refractive index n_1 between n_0 and n_s, on the reflectivity.

carriers. At the front side where the light enters the absorbing film, the generation of charge carriers is significantly higher than at the back. It follows that the EQE values measured in the blue correspond to charge carriers generated close to the front of the solar cell, whereas the EQE in the red part represents charge carriers generated throughout the entire absorber layer.

Further, it is important to reduce the optical loss mechanisms such as *shading losses*, *reflection*, and *parasitic absorption* that we already discussed in Section 10.3.1.

For reducing the reflection, anti-reflective coatings (ARC) can be used. Light that is impinging upon a surface between two media with different refractive indices will always be partly reflected and partly transmitted. In order to reduce losses, it is important to minimise these reflective losses.

The first method is based on a clever utilization of the Fresnel equations that we introduced in Eqs. (4.12) and (4.13). For understanding how this can work, we will first take a look at interfaces with silicon, the most commonly used material for solar cells. Let us consider light of 500 nm wavelength falling onto an air–silicon interface perpendicularly. At 500 nm, the refractive index of air is $n_0 = 1$ and that of silicon is $n_s = 4.3$. With the Fresnel equations we hence find that the optical losses due to reflection are significant at 38.8%.

The reflection can be significantly reduced by introducing an interlayer with a refractive index n_1 with a value in between that of n_0 and n_s. If no multiple reflection or interference is taken into account, it can easily be shown that the reflectivity becomes minimal if

$$n_1 = \sqrt{n_0 n_s}. \tag{10.32}$$

This is also seen in Figure 10.11, where n_1 takes all the values in between n_0 and n_s. In this example, including a single interlayer can reduce the reflection at the interface from 38.8% down to 22.9%. If more than one interlayer is used, the reflection can be reduced even further. This technique is called *refractive index grading*.

In another approach, constructive and destructive interference of light is utilised. In Chapter 4 we already discussed that light can be considered as an electromagnetic wave. Waves have the interesting property that they can interfere with each other, they can be *superimposed*. For understanding this we look at two waves A and B that have the same wavelength and cover the same portion of space,

$$A(x, t) = A_0 e^{ikx - i\omega t}, \tag{10.33a}$$

$$B(x, t) = B_0 e^{ikx - i\omega t + i\phi}. \tag{10.33b}$$

The letter ϕ denotes the *phase shift* between A and B and is very important. We can superimpose the two waves by simply adding them

$$\begin{aligned} C(x, t) &= A(x, t) + B(x, t) \\ &= A_0 e^{ikx - i\omega t} + B_0 e^{ikx - i\omega t + i\phi} \\ &= A_0 e^{ikx - i\omega t} + B_0 e^{ikx - i\omega t} e^{i\phi} \\ &= \left(A_0 + B_0 e^{i\phi} \right) e^{ikx - i\omega t}. \end{aligned} \tag{10.34}$$

The amplitude of the superimposed wave is thus

$$C_0 = A_0 + B_0 e^{i\phi}. \tag{10.35}$$

Depending on the phase shift, the superimposed wave will be stronger or weaker than A and B. If ϕ is *in phase*, i.e. a multiple of 2π, i.e. $0, 2\pi, 4\pi, \ldots$, we will have maximal amplification of waves,

$$C_0 = A_0 + B_0 \times 1 = A_0 + B_0. \tag{10.36}$$

This situation is called *constructive interference*.

But if ϕ is *in antiphase*, i.e. from the set $\pi, 3\pi, 5\pi, \ldots$ we have maximal attenuation of the waves, or *destructive interference*,

$$C_0 = A_0 + B_0 \cdot (-1) = A_0 - B_0. \tag{10.37}$$

Based on this principle, we can design an anti-reflection coating, as illustrated in Figure 10.12 (a). The wave reflected from the material-air interface and the wave reflected from the air-material interface are in antiphase. As a result the total amplitude of the electric field of the outgoing wave is smaller and hence the total irradiance coupled out of the system is smaller as well.

Let us now look at a layer in a solar cell with thickness d and a refractive index n_1. It can easily be shown that the two waves reflected from the front and back interface of this layer are in antiphase, when the product $n_1 d$ is equal to the wavelength divided by four,

$$n_1 d = \frac{\lambda_0}{4}. \tag{10.38}$$

Using an anti-reflection coating based on interference demands that the typical length scale of the interlayer thickness must be in the order of the wavelength.

The last approach that we discuss for realising anti-reflective coating is to use *textured interfaces*. Here, we consider the case where the typical length scales of the surface features

10. Losses and efficiency limits

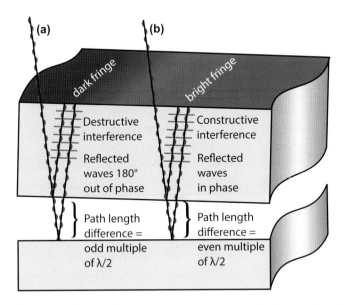

Figure 10.12: Illustrating the working principle of an anti-reflective coating based on interference [40].

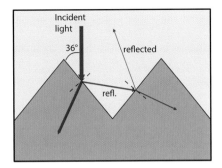

Figure 10.13: Illustrating the effect of texturing.

are larger than the typical wavelength of light. In this case, which is also called the *geometrical limit*, the reflection and transmission of the light rays are fully determined by the Fresnel equations (Eqs. (4.12) and (4.13)) and Snell's law (Eq. (4.11)).

The texturing helps to enhance the coupling of light into the layer. For example, light that is perpendicularly incident maybe reflected from one part of the textured surface to a second, somewhere else on the interfaces, as illustrated in Figure 10.13. Here, another fraction of the light will be transmitted into the layer and effectively less light will be reflected, when compared to a flat interface in-between the same materials.

In summary, we have discussed three types of anti-reflective coatings: Rayleigh films with intermediate refractive indexes, anti-reflection coatings based on destructive interference and enhanced in-coupling of light due to scattering at textured interfaces.

At the end of this chapter, we want to discuss something that is very important for thin-film solar cells. In order to reduce both production costs and bulk recombination, it is desirable to have the absorber layer as thin as possible. On the other hand, it should be *optically thick* in order to absorb as much light as possible. In principle, the light should be reflected back and forth inside the absorber until everything is absorbed. However, at every internal reflection part of the light is transmitted out of the film.

If the light could travel through the layer at an angle greater than the critical angles of the front and back interfaces of the absorber, it could stay there until everything is absorbed without any loss. Unfortunately, it is very difficult if not impossible to design such a solar cell.

Note that textured interfaces do not help in this case. For the same reason that more light can be coupled into the absorber because of texturing, more light is coupled out of the absorber!

10.5 Exercises

10.1 Consider two solar cells: solar cell A has a bandgap energy of 1 eV and solar cell B has a bandgap energy of 1.7 eV. From all kind of possible losses, assume only optical losses. Which of the following statements is true?

(a) Non-absorption losses are higher in solar cell A than in solar cell B.

(b) Heating losses are higher in solar cell A than in solar cell B.

(c) According to the Shockley–Queisser analysis, radiative recombination is more important in solar cell B than in solar cell A.

(d) According to the Shockley–Queisser analysis, solar cell A can have an efficiency up to 50%.

10.2 Consider an incoming monochromatic light beam of wavelength $\lambda = 800$ nm incident on a c-Si layer. How thick should the layer be in order to absorb 90% of the incoming light? Assume the absorption coefficient to be $\alpha(800\text{ nm}) = 10^3$ cm^{-1}.

10.3 The typical thickness of the absorber layer of a c-Si solar cell is around 300 μm. Assume that the absorption coefficient for infrared light is $\alpha(1,100\text{ nm}) = 10$ cm^{-1}. How much of the light with $\lambda = 1100$ nm is not absorbed by the absorber layer? Give the answer as a percentage of the incident intensity I_0.

10. Losses and efficiency limits

10.4 Imagine that you are in the laboratory and can decide the thickness of the Si layer for your solar cell. You want to optimise the solar cell performance for a wavelength of $\lambda = 1,000$ nm, for which the absorption coefficient is $\alpha(1,000\text{nm}) = 10^2$ cm^{-1}. For silicon, the minority carrier diffusivity is around $D = 27$ cm^2/s and the minority-carrier lifetime is around $\tau = 15$ μs. Which thickness would you choose? (Hint: use the Lambert–Beer law (Eq. (4.25)).)

(a) $d_{Si} = 100$ μm.

(b) $d_{Si} = 180$ μm.

(c) $d_{Si} = 300$ μm.

10.5 In Figure 10.10 we have seen the absorption coefficient as a function of the wavelength for several semiconductor materials. Let us consider monochromatic light of photons with energy $E_{ph} = 1.55$ eV that is incident in a film with thickness d. If we ignore possible reflection losses at the rear and front interfaces of the film, what thickness is required to achieve a light absorption of 90% for the four materials?

10.6 A solar cell is illuminated by a light source that has an output power of 6 watts and emits monochromatic light at a wavelength of 620 nm.

(a) How many photons per second are emitted by the light source?

(b) Assume that the solar cell converts all photons into electron-hole pairs and there are no recombination losses. What current would the solar cell generate?

(c) Assume that the electrons and holes do not lose any energy on their way to the contacts. What voltage would the solar cell generate?

(d) What would be the power conversion efficiency of the solar cell if it generated the current and voltage calculated above? Can this be realised in practice?

Part III

PV technology

11
A short history of solar cells

We will start our discussion on PV technology with a brief summary of the *history of solar energy*. In the seventh century BCE, humans were already using magnifying glasses to concentrate sunlight and hence to make fire. Later, the ancient Greeks and Romans used concentrating mirrors for the same purpose.

In the 18th century the Swiss physicist Horace-Bénédict de Saussure built heat traps, which are a kind of miniature greenhouse. He constructed hot boxes, consisting of a glass box within another bigger glass box, with a total number of up to five boxes. When exposed to direct solar irradiation, the temperature in the innermost box could rise up to 108°C; warm enough to boil water and cook food. These boxes can be considered as the World's first solar collectors.

In 1839, the French physicist Alexandre-Edmond Becquerel discovered the photovoltaic effect at the age of only 19 years. He observed this effect in an electrolytic cell, which was made out of two platinum electrodes, placed in an *electrolyte*. An electrolyte is an electrically conducting solution; Becquerel used silver chloride dissolved in an acidic solution. He observed that the current of the cell was enhanced when his setup was irradiated with sunlight. A photograph is shown in Figure 11.1.

In the 1860s and 1870s, the French inventor Augustin Mouchot developed solar powered steam engines using the World's first *parabolic trough* solar collector, that we will discuss in Chapter 22. Mouchot's motivation was his believe that the coal resources were limited. At that time, coal was *the* energy source for driving steam engines. However, as coal became cheaper, the French government decided that solar energy was too expensive and stopped funding Mouchot's research.

In 1876, the British natural philosopher William Grylls Adams together with his student Richard Evans Day demonstrated the photovoltaic effect in a junction based on platinum and the semiconductor selenium, however with a very poor performance. Seven years later, the American inventor Charles Fritts managed to make a PV device based on a gold-

Figure 11.1: Alexandre-Edmond Becquerel [41].

Figure 11.2: Daryl M. Chapin, Calvin S. Fuller, and Gerald L. Pearson, the developers of the first modern solar cell [42].

selenium junction. The energy conversion efficiency of that device was 1%.

In 1887, the German physicist Heinrich Hertz discovered the photoelectric effect, already briefly mentioned in Chapter 3. In this effect, electrons are emitted from a material that has absorbed light with a frequency exceeding a material-dependent threshold frequency. In 1905 Albert Einstein published a paper in which he explained the photoelectric effect by assuming that light energy was being carried with quantized packages of energy [23], which we nowadays call *photons*.

In 1918 the Polish chemist Jan Czochralski invented a method to grow high quality crystalline materials. Nowadays this technique is very important for growing monocrystalline silicon used for high quality silicon solar cells that we will study in detail in Chapter 12. The development of the c-Si technology started in the second half of the 20th century.

In 1953, the American chemist Dan Trivich was the first one to perform theoretical calculations on the solar cell performance for materials with different bandgaps.

The real development of solar cells as we know them today, started at the *Bell Laboratories* in the United States. In 1954, their scientists Daryl M. Chapin, Calvin S. Fuller, and

Gerald L. Pearson, made a silicon-based solar cell with an efficiency of about 6% [43]. Figure 11.2 shows them in their laboratory. In the same year, D. C. Reynolds et al. reported on the photovoltaic effect for cadmium sulphide (CdS), a II-VI semiconductor [44].

In the mid- and late 1950s several companies and laboratories started to develop silicon-based solar cells in order to power satellites orbiting the Earth. Among these were RCA Corporation, Hoffman Electronics Corporation but also the United States Army Signal Corps. In those days, research on PV technology was mainly driven by supplying space applications with energy. For example, the American satellite *Vanguard 1*, which was launched by the US Navy in 1958, was powered by solar cells from Hoffman Electronics. It was the fourth artificial Earth satellite and the first one to be powered with solar cells. It was operating until 1964 and still is orbiting Earth. In 1962 Bell Telephone Laboratories launched the first solar powered telecommunications satellite and in 1966 NASA launched the first Orbiting Astronomical Observatory, which was powered by a 1 kW photovoltaic solar array.

In 1968, the Italian scientist Giovanni Francia built the first concentrated solar power plant near Genoa, Italy. The plant was able to produce 1 MW with superheated steam at 100 bar and 500 °C.

In 1970, the Soviet physicist Zhores Alferov developed solar cells based on a gallium arsenide heterojunction. This was the first solar cell based on III-V semiconductor materials which we will discuss in Section 13.2. In 1976, Dave E. Carlson and Chris R. Wronski developed the first thin-film photovoltaic devices based on amorphous silicon at RCA Laboratories. We will discuss this technology in Section 13.3. In 1978, the Japanese companies SHARP and Tokyo Electronic Application Laboratory brought the first solar powered calculators on the market.

Because of the 1970s oil crisis, which led to sharply rising oil prices, the public interest in photovoltaic technology for terrestrial application increased in that decade. In that time, PV technology moved from a niche technology for space application to a technology applicable for terrestrial applications. In the late 1970s and 1980s many companies started to develop PV modules and systems for terrestrial applications. Solar cells are still very important for space applications as seen in Figure 11.3, which shows a solar panel array on the International Space Station (ISS).

In 1980 the first thin-film solar cells based on a copper-sulphide/cadmium-sulphide junction were demonstrated with a conversion efficiency above 10% at the University of Delaware. In 1985, crystalline silicon solar cells with efficiencies above 20% were demonstrated at the University of New South Wales in Australia.

From 1984 through 1991 the world's largest solar thermal energy generating facility in the world was built in the Mojave Desert in California. It consists of nine plants with a combined capacity of 354 megawatts.

In 1991 the first high efficiency dye-sensitized solar cell was published by the École polytechnique fédérale de Lausanne in Switzerland by Michael Grätzel and co-workers. The dye-sensitized solar cell is a kind of photo electrochemical system, in which a semiconductor material based on molecular sensitizers is placed between a photoanode and an electrolyte. We will introduce this technology in Section 13.6.

In 1994, the US *National Renewable Energy Laboratory* in Golden, Colorado, demonstrated a concentrator solar cell based on III-V semiconductor materials. Their cell, based on an indium-gallium-phosphide/gallium-arsenide tandem junction, exceeded the 30% conver-

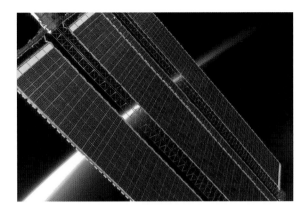

Figure 11.3: A solar panel array on the International Space Station (ISS) [45].

sion limit.

In 1999, the total global installed photovoltaic power passed 1 GW$_p$. Starting from about 2000, environmental and economic issues started to become more and more important in the public discussion, which renewed the public interest in solar energy. Since 2000, the PV market therefore transformed from a regional market to a global market, as discussed in Chapter 2. Germany took the lead with a progressive feed-in tariff policy, leading to a large national solar market and industry [19].

Since about 2008, the Chinese government has been heavily investing in their PV industry. As a result, China has been the dominant PV module manufacturer for several years now. In 2012 the worldwide solar energy capacity surpassed the magic barrier of 100 GW$_p$ [17]. Between 1999 and 2012, the installed PV capacity has hence grown by a factor of 100. In other words, during these last 13 years, the average annual growth of the installed PV capacity was about 40%.

12
Crystalline silicon solar cells

As we already discussed in Chapter 6, most semiconductor materials have a *crystalline lattice* structure. As a starting point for our discussion on crystalline silicon PV technology, we will take a closer look at some properties of the crystal lattice.

12.1 Crystalline silicon

In such a lattice, the atoms are arranged in a certain pattern that repeats itself. The lattice thus has *long-range order* and *symmetry*. However, the pattern is not the same in every direction. In other words, if we make large cuts through the lattice, the various planes would not look the same.

Crystalline silicon has a density of 2.3290 g/cm^3 and a *diamond cubic crystal structure* with a lattice constant of 543.07 pm, as illustrated in Fig. 12.1. Figure 12.2 shows two different sections through a crystalline silicon lattice, which originally consisted of $3 \times 3 \times 3\times$ unit cells. The first surface shown in Figure 12.2 (a) is the *100 surface*, whose surface normal is in the 100 direction. At a 100 surface, each Si atom has two back bonds and two valence electrons pointing to the front. The second surface, shown in Figure 12.2 (b), is the *111 surface*. Here, every Si atom has three back bonds and one valence electron pointing towards the plane normal.

To understand the importance of these directions, we look at the *electronic band dispersion diagram* for silicon, shown in Figure 12.3. On the vertical axis, the energy position of the valence and conduction bands is shown. The horizontal axis shows the crystal momentum, i.e. the momentum of the charge carriers. The white area represents the energy levels in the forbidden bandgap. The bandgap of silicon is determined by the lowest energy point of the conduction band at X, which corresponds to the 100 direction, and the highest energy value of the valence band, at Γ. The *band gap energy* is the difference between

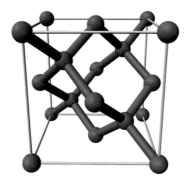

Figure 12.1: A unit cell of a diamond cubic crystal [30].

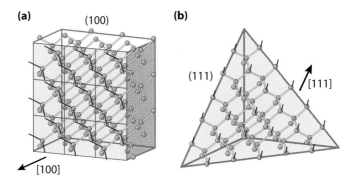

Figure 12.2: The (a) 100; and (b) 111 surfaces of a silicon crystal.

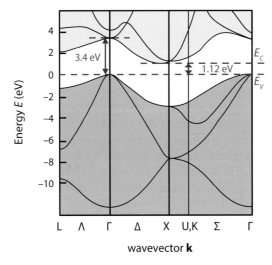

Figure 12.3: The band diagram of crystalline silicon.

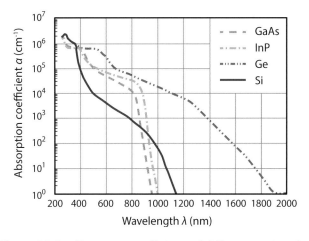

Figure 12.4: Absorption coefficients of different semiconductors.

those two levels and is equal to 1.12 eV, or 1107 nm, when expressed in wavelengths. 1,107 nm is in the infrared part of the spectrum of light. This bandgap is an *indirect bandgap*, because the charge carriers must change in energy *and* momentum to be excited from the valence to the conduction band. As we can see, crystalline silicon has a direct transition as well. This transition has an energy of 3.4 eV, which is equivalent to a wavelength of 364 nm, which is in the blue spectral part.

Because of the required change in momentum, for an indirect bandgap material it is less likely that a photon with an energy exceeding the bandgap can excite the electron, with respect to a direct bandgap material like gallium arsenide (GaAs) or indium phosphide (InP). Consequently, the absorption coefficient of crystalline silicon is significantly lower than that of direct bandgap materials, as we can see in Figure 12.4. While in the visible part of the spectrum c-Si absorbs less than the GaAs and InP, below 364 nm it absorbs just as much as GaAs and InP, because there silicon has a direct band-to-band transition as well. Germanium (Ge), also indicated in the figure, is an indirect bandgap material, just like silicon. The bandgap of Ge is 0.67 eV, which means it already starts to absorb light at wavelengths shorter than 1,850 nm. In the visible part of the spectrum, germanium has some direct transitions as well.

Let us now take another look at the design rules for solar cells that we introduced in Section 10.4. First, we look at *spectral utilization*. The c-Si band gap of 1.12 eV means that in theory we can generate a maximum short circuit current density of 45 mA per square centimetre. Let us now consider the second design rule, i.e. *light management*. First, we look at a wavelength around 800 nm, where c-Si has an absorption coefficient of 1,000 cm^{-1}. Using the Lambert–Beer law [see Eq. (4.25)], we can easily calculate that absorbing 90% of the incident light at 800 nm requires an absorption path length of 23 μm. Secondly, we look at 970 nm wavelength, where c-Si has an absorption coefficient of 100 cm^{-1}. Hence, an absorption path length of 230 μm is required to absorb 90% of the light. 230 μm is a typical thickness for silicon wafers. This calculation demonstrates that light management techniques become important for crystalline silicon absorber layers above a wavelength of

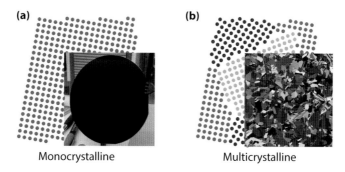

Figure 12.5: Illustrating (a) a monocrystalline; and (b) a multicrystalline silicon wafer.

about 900 nm.

Let us now consider the design rule of *bandgap utilization*, which is determined by the recombination losses. As silicon is an indirect bandgap material, only Auger recombination and Shockley–Read–Hall (SRH) recombination will determine the open circuit voltage, while radiative recombination can be neglected. Considering SRH recombination, the recombination rate of the charge carriers is related to the electrons trapped at defect states. When looking at the defect density in the bulk of silicon, we can differentiate between two major types of silicon wafers: *monocrystalline silicon* and *multicrystalline silicon*, which is also called *polycrystalline silicon*.

Monocrystalline silicon, also known as single-crystalline silicon, is a crystalline solid in which the crystal lattice is continuous and unbroken without any grain boundaries over the entire bulk, up to the edges. In contrast, polycrystalline silicon, often simply abbreviated as polysilicon, is a material that consists of many small crystalline grains, with random orientations. Between these grains are grain boundaries. Figure 12.5 shows two pictures of monocrystalline and multicrystalline wafers. While a monocrystalline silicon wafer has one uniform colour, in multicrystalline silicon, the various grains are clearly visible to the human eye. At the grain boundaries we find lattice mismatches, resulting in many defects at these boundaries. As a consequence, the charge-carrier lifetime for polycrystalline silicon is shorter than for monocrystalline silicon, because of the SRH recombination. The more grain boundaries in the material, the shorter the lifetime of the charge carriers. Hence, the grain size plays an important role in the recombination rate.

Figure 12.6 shows the relationship between the open circuit voltage and the average grain size for various solar cells developed around the world, based on multicrystalline wafers [46]. The larger the grain size, the longer the charge-carrier lifetimes and the larger the bandgap utilization and hence the open circuit voltage will be. On the right-hand side of the graph the open circuit voltages of various solar cells, based on monocrystalline wafers, are shown. As monocrystalline silicon has no grain boundary, much larger open circuit voltages can be obtained.

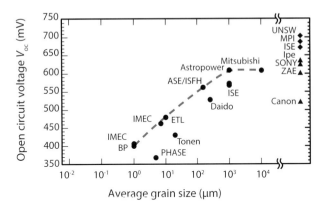

Figure 12.6: The relationship between the open circuit voltage and the average grain size (reprinted from Thin Solid Films, vol. 403–404, R. Bergmann and J. Werner, The future of crystalline silicon films on foreign substrates, pages 162–169, Copyright (2002), with permission from Elsevier) [46].

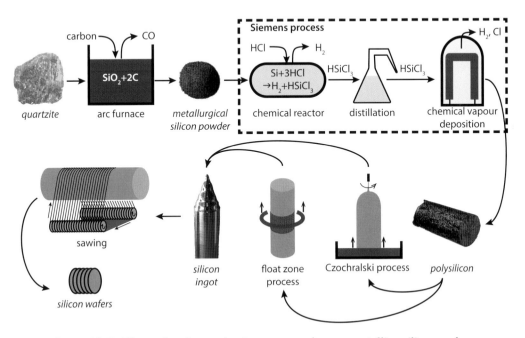

Figure 12.7: Illustrating the production process of monocrystalline silicon wafers.

12.2 Production of silicon wafers

After the initial considerations on designing c-Si solar cells, we will now discuss how monocrystalline and multicrystalline silicon wafers can be produced. In Figure 12.7 we illustrate the production process of monocrystalline silicon wafers.

The lowest quality of silicon is the so-called metallurgical silicon, which is made from *quartzite*. Quartzite is a rock consisting of almost pure silicon dioxide (SiO_2). To produce silicon the quartzite is melted in a submerged-electrode arc furnace by heating it up to around 1,900°C, as illustrated in Figure 12.7. Then, the molten quartzite is mixed with carbon. As a carbon source, a mixture of coal, coke and wood chips is used. The carbon then starts reacting with the SiO_2. Since the reactions are rather complex, we will not discuss them in detail here. The overall reaction however can be written as

$$SiO_2 + 2\,C \rightarrow Si + 2\,CO. \tag{12.1}$$

As a result, carbon monoxide (CO) is formed, which will leave the furnace in the gas phase. In this way, the quartzite is purified from the silicon. After the reactions are finished, the molten silicon that was created during the process is drawn off the furnace and solidified. The purity of metallurgic silicon, shown as a powder in Figure 12.7, is around 98% to 99%.

About 70% of the metallurgical silicon produced worldwide is used in the aluminium casting industry for making aluminium silicon alloys, which are used in automotive engine blocks. Around 30% is used for making a variety of chemical products like silicones. Only around 1% of metallurgical silicon is used as a raw product for making electronic-grade silicon.

The silicon material with the next higher level of purity is called polysilicon. It is made from a powder of metallurgical silicon in the *Siemens process*. In the process, the metallurgical silicon is brought into a reactor and exposed to hydrogen chloride (HCl) at elevated temperatures in the presence of a catalyst. The silicon reacts with the hydrogen chloride,

$$Si + 3\,HCl \rightarrow H_2 + HSiCl_3, \tag{12.2}$$

leading to the creation of trichlorosilane ($HSiCl_3$). This is a molecule that contains one silicon atom, three chlorine atoms and one hydrogen atom. Then, the trichlorosilane gas is cooled and liquified. Using *distillation*, impurities with boiling points higher or lower than $HSiCl_3$ are removed. The purified trichlorosilane is evaporated again in another reactor and mixed with hydrogen gas. There, the trichlorosilane is decomposed at hot rods of highly purified Si, which are at a high temperature between around 850 °C and 1,050 °C. The Si atoms are deposited on the rod whereas the chlorine and hydrogen atoms are desorbed from the rod surface back into the gas phase. As a result a pure silicon material is grown. This method of depositing silicon on the rod is one example of *chemical vapour deposition* (CVD). As the exhaust gas still contains chlorosilanes and hydrogen, these gases are recycled and used again: chlorosilane is liquified, distilled and reused. The hydrogen is cleaned and thereafter recycled back in to the reactor. The Siemens process consumes a lot of energy.

Polysilicon granules can also be produced using *Fluid Bed Reactors* (not shown in Figure 12.7). This process is operated at lower temperatures and consumes much less energy.

Polycrystalline silicon can have a purity as high as 99.9999%, or in other words, only one out of one million atoms is different from Si.

The last approach we briefly mention is that of *upgraded metallurgical silicon* (not shown in Figure 12.7). In this process metallurgical silicon is chemically refined by blowing gases through the silicon melt, removing the impurities. Although processing is cheap, the silicon purity is lower than that achieved with the Siemens or the fluid bed reactor approaches.

Now we introduce two methods that are used industrially for making monocrystalline silicon *ingots*, i.e. large cylinders of silicon that consist of one crystal only. This means that inside the ingot no grain boundaries are present. Such a monocrystalline ingot and both methods are sketched in Figure 12.7.

The first method we discuss is based on the *Czochralski process* that was discovered by the Polish scientist Jan Czochralski in 1916. In this method, highly purified silicon is melted in a crucible at typical temperature of 1,500°C. Boron or phosphorus can be added for making *p*-doped or *n*-doped silicon, respectively. A seed crystal that is mounted on a rotating shaft is dipped into the molten silicon. The orientation of this seed crystal is well defined; it is either 100 or 111 oriented. The melt solidifies at the seed crystal and adopts the orientation of the crystal. The crystal is rotating and is pulled upwards slowly, allowing the formation of a large, single-crystal cylindrical column from the melt – the ingot. To conduct the process successfully, temperature gradients, the rate of pulling the shaft upwards, and the rotational speed must be well controlled. Due to improved process control throughout the years, nowadays ingots of diameters of 200 mm or even 300 mm with lengths of up to two metres can be fabricated. To prevent the incorporation of impurities, the Czochralski process takes place in an inert atmosphere, like argon gas. The crucible is made from quartz, which partly dissolves in the melt as well. Consequently, monocrystalline silicon made with the Czochralski method has a relatively high oxygen level.

The second method to make monocrystalline silicon is the *float zone* process, which allows the fabrication of ingots with extremely low densities of impurities like oxygen and carbon. As a source material, a polycrystalline rod made with the Siemens process is used. The end of the rod is heated up and melted using an induction coil operating at radio frequency (RF). The molten part is then brought into contact with the seed crystals, where it solidifies again and adopts the orientation of the seed crystal. Again 100 or 111 orientations are being used. As the molten zone is moved along the polysilicon rod, the single-crystal ingot is grown as well. Many impurities remain in and move along with the molten zone. Nowadays, during the process nitrogen is intentionally added in order to improve the control over microdefects and the mechanical strength of the wafers. One advantage of the float zone technique is that the molten silicon is not in contact with other materials like quartz, as is the case when using the Czochralski method. In the float zone process the molten silicon is only in contact with the inert gas such as argon. The silicon can be doped by adding doping gasses like diborane (B_2H_6) and phosphine (PH_3) to the inert gas to get *p*-doped and *n*-doped silicon, respectively. The diameter of float-zone processed ingots is generally not larger than 150 mm, as the size is limited by surface tensions during the growth.

Both monocrystalline silicon ingots and *multicrystalline silicon ingots*, which consist of many small crystalline grains, can be fabricated (not shown in Figure 12.7). This can be made by melting highly purified silicon in a crucible and pouring the molten silicon into a cubic-shaped *growth crucible*. There, the molten silicon solidifies into a multicrystalline ingot in a process called *silicon casting*. If both melting and solidification is done in the same crucible it is referred to as *directional solidification*. Such multicrystalline ingots can have a front surface area of up to 70×70 cm^2 and a height of up to 25 cm.

Now, as we know how to produce monocrystalline and multicrystalline ingots we will discuss how to make *wafers* out of them. The process that is used to make the wafers is *sawing*, as illustrated in Figure 12.7. Logically, sawing will damage the surface of the wafers. Therefore, this processing step is followed by a polishing step. The biggest disadvantage of sawing is that a significant fraction of the silicon is lost as *kerf loss*, which is usually determined by the thickness of the wire or saw used for sawing. Generally, it is in the order of 100 μm. As typical wafers used in modern solar cells have thicknesses in the order of 150 μm up to 200 μm, the kerf loss is very significant.

A completely different process for making silicon wafers is the *silicon ribbon* technique (not shown in Figure 12.7). As this technique does not include any sawing steps, no kerf loss occurs. In the silicon ribbon technique a string is used that is resistant against high temperatures. This string is pulled up from a silicon melt. The silicon solidifies on the string and hence a sheet of crystalline silicon is pulled out of the melt. Then, the ribbon is cut into wafers. Before the wafers can be processed further in order to make solar cells, some surface treatments are required. The electronic quality of ribbon silicon is not as high as that of monocrystalline silicon.

To summarize, we discussed how to make metallurgical silicon out of quartzite and how to fabricate polysilicon. We have seen that monocrystalline ingots are made using either the Czochralski or the float-zone process, while multicrystalline ingots are made using a casting method. Wafers are fabricated by sawing these ingots. A method that does not have any kerf losses is the ribbon silicon approach.

12.3 Designing c-Si solar cells

In this section, we briefly discuss the operating principles of c-Si solar cells. In particular, we discuss several technical aspects that play an important role in the collection of the light, excitation of charge carriers, and the reduction of optical losses.

In Chapter 8 we discussed how an illuminated p-n junction can operate as a solar cell. In the illustrations used there, both the p-doped and n-doped regions have the same thickness. This is not the case in real c-Si devices. For example, the most conventional type of c-Si solar cells is built from a p-type silicon wafer, as sketched in Figure 12.8. However, the n-type layer on the top of the p-wafer is much thinner than the wafer; it typically has a thickness of around 0.3 μm. Often, this layer is called the *emitter* layer. As mentioned before, the whole wafer typically has thicknesses between 100 and 300 μm.

For monochromatic light the generation profile shows exponential decay (and hence a straight line on a logarithmic scale as in Figure 12.9) because of the Lambert–Beer law [Eq. (4.25)]. Figure 12.9 shows the generation profile of silicon that is illuminated under the AM1.5 spectrum. This generation profile does not show such a behaviour because of the wavelength-dependent absorption coefficient (see e.g. Figure 12.4). The largest fraction of the light is absorbed close to the front surface of the solar cell. In the first 10 μm by far the most charge carriers are generated. By making the front emitter layer very thin, a large

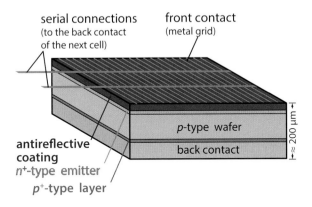

Figure 12.8: Scheme of a modern crystalline silicon cell.

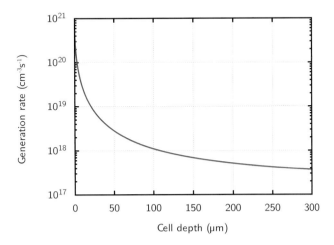

Figure 12.9: The generation profile in crystalline silicon illuminated under AM1.5. Note the logarithmic scale on the y-axis.

fraction of the light-excited charge carriers generated by the incoming light are created within the diffusion length of the *p-n* junction.

Now we will discuss the collection of charge carriers in a crystalline silicon solar cell. The crucial components that play a role in charge collection are the emitter layer, the metal front contacts and the back contact. First, the emitter layer: The minority charge carriers, which are excited by the light, are separated at the *p-n* junction; the minority electrons in the *p*-layer drift to the *n*-layer, where they have to be collected. Since the silicon *n*-emitter is not sufficiently conductive we have to use the much more conductive metal contacts, which are placed on top of the emitter layer. Very often, the metal contacts are made of cheap aluminium.

This means that the electrons have to diffuse laterally through the emitter layer to the electric front contact to be collected. What factors are important for good transport of the electrons to the contact? One is that the lifetime of the charge carriers needs to be high. A high lifetime guarantees large open circuit voltages or, in other words, the optimal utilization of the bandgap energy. For increasing the lifetime, recombination losses must be reduced as much as possible.

Recombination not only reduces the V_{oc}, it also limits the collected current. As mentioned earlier, in silicon, two recombination mechanisms are present: Shockley–Read–Hall recombination and Auger recombination. First, let us take a look at Shockley-Read-Hall recombination at the surface, that we introduced in Section 7.5. A bare c-Si surface contains many defects, because the surface silicon atoms have some valence electrons that cannot make molecular orbitals due to the absence of neighbouring atoms. These valence orbitals containing only one electron at the surface act like defects. They are also called *dangling bonds*. At the dangling bonds, the charge carriers can recombine through the SRH process. The probability and speed at which charge carriers can recombine is usually expressed in terms of the *surface recombination velocity*. Since a large fraction of the charge carriers are generated close to the front surface, a high surface recombination velocity at the emitter front surface will lead to significant charge carrier losses and consequently lower short circuit current densities. In high quality monocrystalline silicon wafers, for example, no defect-rich boundaries are present in the bulk. Thus, the lifetime of charge carriers is limited by the recombination processes at the wafer surface.

In order to reduce the surface recombination two approaches are used. First, the defect concentration on the surface is reduced by depositing a thin layer of a different material on top of the surface. This material partially restores the bonding environment of the silicon atoms. In addition, the material must be an insulator; it must force the electrons to remain in and move through the emitter layer. Typical materials used for these *chemical passivation* layers are silicon oxides (SiO_x) and silicon nitrides (Si_xN_y).[1] A silicon oxide layer can be formed by heating up the silicon surface in an oxygen-rich atmosphere, leading to the oxidation of the surface silicon atoms. Si_3N_4 can be deposited using plasma-enhanced chemical vapour deposition (PE-CVD) that we will discuss in more detail in Chapters 13 and 14.

A second approach for reducing the surface recombination velocity is to reduce the minority charge carrier density near the surface. As the surface recombination velocity is limited by the minority charge carrier's density, it is beneficial to have the minority charge

---

[1]Stoichiometric silicon oxide and silicon nitride have the chemical formula SiO_2 and Si_3N_4, respectively. As the layers used for passivation are not stoichiometric, we indicate the elementary fractions with x and y.

carrier density at the surface as low as possible. By increasing the doping of the emitter layer, the density of the minority charge carriers can be reduced, which results in lower surface recombination velocities. However, this is in competition with the diffusion length of the minority charge carriers. The blue part of the solar spectrum leads to the generation of many charge carriers very close to the surface, i.e. in the emitter layer. For utilizing these light-excited minority charge carriers, the diffusion length of the holes has to be large enough to reach the depletion zone at the *p-n* junction. However, increasing the doping levels leads to a decreasing diffusion length of the minority holes in the emitter. Therefore, too high doping levels or too thick emitter layers would result in a poor blue response or – in other words – low external quantum efficiency (EQE, see Chapter 9) values in the blue part of the spectrum. Such an emitter layer could be called a *'dead layer'* as the light-excited minority charge carriers cannot be collected.

Next, we take a closer look at the metal-emitter interface. Because electrons must be easily conducted from the emitter to the metal, insulating passivation layers such as SiO_x or Si_xN_y cannot be used. Therefore, the metal-semiconductor interface has more defects and hence an undesirably high interface recombination velocity. Additionally, a metal-semiconductor junction induces a barrier for the majority charge carriers, as we have seen in Section 8.3. Hence, this barrier will give rise to a higher contact resistance. Again high doping levels can reduce the recombination velocity at the metal-semiconductor interface and also reduce the contact resistance. In order to minimize the recombination at the interface defects as much as possible, the area of the metal-semiconductor interface must be minimized and the emitter directly below the interface should be heavily doped, which is indicated with n^{++}. The sides of the metal contacts are buried in the insulating passivation layer. The area below the contact has been heavily doped. These two approaches reduce the recombination and collection losses at the metal contact.

The solar cell shown in Figure 12.8 has a classic metal grid pattern on top. We see two paths for the electrons in the middle of top surface of the solar cell. They are called *busbars*. The small stripes going from the busbars to the edges of the solar cell are called the *fingers*. Let now R be the resistance of such a finger. If L, W, and H are the length, width and height of the fingers, respectively, and ρ is the resistivity of the metal, R is given by

$$R = \rho \frac{L}{WH}. \tag{12.3}$$

This equation shows that the longer the path length for an electron, the larger the resistance the electrons experience. Further, the smaller the cross-section ($W \cdot H$) of the finger, the larger the resistance will be. Note that the resistance of the contacts will act as a series resistance in the equivalent electric circuit. Larger series resistance will result in lower fill factors of the solar cell as discussed in Chapter 9. Hence, electrically large cross-sections of the fingers are desirable.

To arrive at a metallic contact, electrons in the emitter have to travel laterally through the emitter. Because of the resistivity of the *n*-type silicon, the charge carriers in the emitter layer also experience a resistance. It can be shown that the power loss due to the resistivity of the emitter layer scales with the spacing between two fingers to the power 3.

As the metal contacts are at the front surface, they act as unwelcome shading objects, or in other words light incident on the metallic front contact area cannot be absorbed in the PV-active layers. Therefore, the contact area should be kept as small as possible, which is in

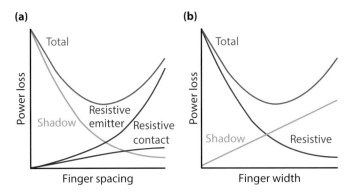

Figure 12.10: Power loss due to (a) spacing; and (b) widths of the metallic fingers on top of a c-Si solar cell.

competition with the fact that the finger cross-section should be maximized. So basically, the finger height should be as high as possible while the finger width, you would like to have as small as possible to comply with these requirements.

We can see that several effects compete with each other. Figure 12.10 (a) shows the relationship between power losses and finger spacing. With increasing finger spacing the power losses of the solar cell decrease because of less shading. On the other hand, the losses due to the increased resistivity in the emitter layer increase. Hence, there is an optimal spacing distance at which the power loss is minimal. A similar plot can be made for power loss versus finger width, shown in Figure 12.10 (b). The larger the W, the larger the shading losses will be. But with increasing W the resistance decreases. Again, here an optimum exists at which the power losses are minimal. We see that optimizing the front contact pattern is a complex interplay between the finger width and spacing.

For designing the back contact we find similar issues. Just as the electrons are to be collected in the front n-type layer, the holes are to be collected at the back contact. Electrons are the only charge carriers that exist in metal. Therefore the holes have to recombine with the electrons at the back semiconductor-metal interface. If the distance between the p-n interface and the back contact is smaller than the typical diffusion length of the minority electrons, the minority electrons can be lost at the defects of the back contact interface because of SRH recombination.

Several methods can be used to reduce this loss. First, the area between the metal contact and the semiconductor can be reduced, just as for the front contact. To do this, point contacts can be used, while the rest of the rear surface is passivated by an insulating passivation layer, similar to the one already discussed for the emitter front surface. The recombination loss of electrons at the back contact can be further reduced by introducing a *back surface field*. A highly p-doped region is placed above the point contacts which is indicated by p^+.

To understand how the back surface field works, we take a look at the band diagram shown in Figure 12.11. The interface between the normally doped p-region and the highly doped p^+-region acts like an n-p junction. Here, this junction acts as a barrier that prevents minority electrons in the p-region from diffusing to the back surface. Further, the space-

12. Crystalline silicon solar cells

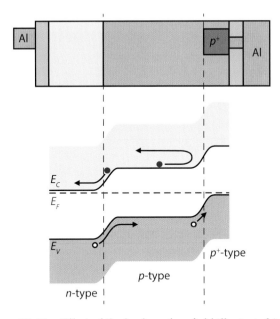

Figure 12.11: Effect of the back surface field illustrated in a band diagram.

charged field behaves like a passivation layer for the defects at the back contact interface and allows higher minority carrier densities in the *p*-doped bulk.

After this thorough discussion on managing the charge carriers, we now take a closer look at managing the photons in a crystalline silicon solar cell. Several optical loss mechanisms must be addressed. These are shading, reflection losses, parasitic absorption losses in the non-PV-active layers, and transmission through the back of the solar cell. As already mentioned, *shading losses* are caused by the metallic front contact grid.

Secondly, *reflection* from the front surface is an important loss mechanism. We briefly mention two approaches to design *anti-reflective coatings* (ARC) for reducing these losses. First, reflection can be minimized based on the *Rayleigh film* principle: by putting a film with a refractive index smaller than that of silicon wafer between the cell and the wafer, losses can be reduced. The optimal value for the refractive index of the intermediate layer equals the square root of the product of refractive indexes of the two other media,

$$n_{\text{opt}} = \sqrt{n_{\text{air}} n_{\text{Si}}}. \tag{12.4}$$

At a wavelength of 500 nm, the optimal refractive index for a layer in-between air and silicon is 2.1. Note that in practice a solar cell is encapsulated under a glass or polymer plate, which will have a beneficial effect on the refractive index grading as well, reducing the reflection losses further. Secondly, using the concept of destructive interference, the thickness and refractive index of an anti-reflection coating can be chosen such that in a certain wavelength range the reflection is minimized. This happens when the light reflected from the air-ARC interface is in antiphase with the light reflected from the ARC-Si interface, as discussed in Chapter 4. We find that the thickness of such a layer should be *a quarter of the*

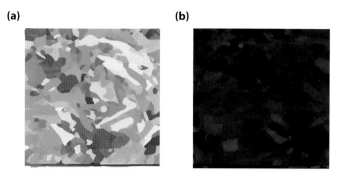

Figure 12.12: A multicrystalline silicon wafer (a) without; and (b) with an anti-reflective coating from silicon nitride.

wavelength in the layer,

$$d_{\text{ARC}} = \frac{\lambda_0}{4n}, \tag{12.5}$$

where λ_0 denotes the wavelength *in vacuo* and n is the refractive index of the anti-reflective layer. For a refractive index of 2.1, a layer thickness of 60 nm would lead to destructive interference at 500 nm. As mentioned earlier, a typical material for passivation is silicon nitride. Figure 12.12 (a) shows a multicrystalline wafer without any ARC. It appears silverish, which means that it is highly reflective. In Figure 12.12 (b) a similar wafer is shown after it was passivated with Si_xN_y. We see that it has a dark blue appearance, hence its reflection is much lower. Interestingly enough, the refractive index of Si_xN_y at 500 nm is in the range of 2 to 2.2, close to the optimum mentioned earlier. The blue appearance indicates that reflection in the blue is stronger than at other wavelengths.

The reflection losses can also be minimized by texturing the wafer surface. Light that is reflected at the textured surface can be reflected at angles such that it is incident somewhere else on the surface, where it can still be coupled into the silicon. Additionally, the scattering at textured surfaces will couple the light under angles different from the interface normal to the wafer. Therefore, the average path length of the light in the absorber will be increased, which leads to stronger absorption of the light. An example of a typical pyramid-textured c-Si wafer that was processed in the DIMES Technology Centre in Delft is shown in Figure 12.13 (a). The process to create such textures is discussed in the Section 12.4. With a different etching approach, it is even possible to make silicon appear completely black, as shown in Figure 12.13 (b). This is called *black silicon*.

12.4 Fabricating c-Si solar cells

After the discussion on the different aspects that are important when designing c-Si solar cells, we now will discuss how such solar cells can actually be fabricated. Figure 12.14 shows a schematic flowchart for the fabrication of c-Si solar cells.

The first step to fabricate a solar cell is to clean a silicon wafer, which is produced as discussed in Section 12.2. This cleaning happens in an acidic or alkaline wet etch process and is also important as it removes damage in the wafer due to the sawing process when

12. Crystalline silicon solar cells

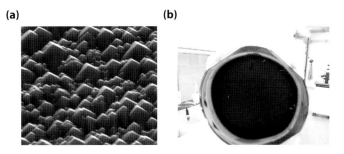

Figure 12.13: (a) A *textured* multicrystalline Si wafer; and (b) a wafer with a *black silicon* surface.

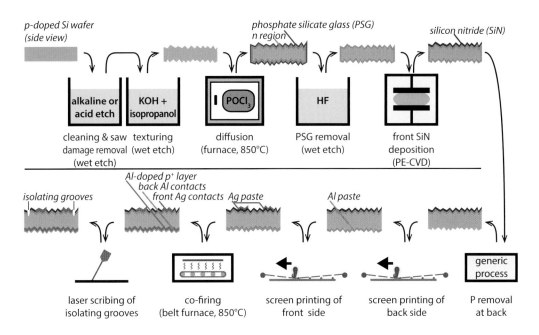

Figure 12.14: Schematic flowchart for making crystalline silicon solar cells.

wafers are cut out of an ingot. As we have seen in the Section 12.3, most c-Si solar cells are textured in order to increase the absorption of the incident light and to reduce reflection losses. The textures usually are created during a wet etching process using a mixture of *potassium hydroxide* (KOH) and *2-propanol* (isopropanol, C_3H_7OH). Interestingly enough, 100 surfaces are etched much faster than 111 surfaces. Such behaviour is called *anisotropic etching*. Thus, when a 100 wafer is etched, the 100 surface will be etched away, while the (*slanted*) 111 surface facets will remain. Therefore, we obtain a wafer with a pyramid structure consisting of 111 facets only.

Next, the emitter layer has to be created, for example during a *solid state diffusion* process. In this process, the wafers are placed in a furnace at around 850 °C. In the furnace, a phosphorus-containing chemical is present, which acts as a source for the P atoms that are used as n-dopants. Often, *phosphoryl chloride* ($POCl_3$) is used as a P source. At these high temperatures, the P atoms react with the surface, and they are mobile in the silicon crystal. According to Fick's law, they can diffuse into the wafer,

$$J(x) = -D\frac{\mathrm{d}n}{\mathrm{d}x}, \qquad (12.6)$$

where the particle flux J (in $cm^{-2}s^{-1}$) is proportional to the negative gradient of the particle density n (in cm^{-3}). D (in $cm^2 s^{-1}$) is the diffusion constant. The diffusion process has to be controlled such that the dopants penetrate into the solid to establish the desired emitter thickness. During the diffusion process not only is the emitter created, but also a thin layer of *phosphosilicate glass* (PSG) – a mixture of P_2O_5 and SiO_2. This PSG layer has to be removed, which usually happens in a wet etching process using *hydrofluoric acid* (HF).

Now, a *silicon nitride* layer is deposited onto the emitter, which acts as anti-reflective coating and passivating layer, as we have seen in Section 12.3. Different processes can be used for the deposition of this layer, such as *plasma-enhanced chemical vapour deposition* (PE-CVD) that we discuss in more detail in Chapter 14.

During the diffusion process an n-doped layer was not only created at the front side, but also on the back side. This layer at the back has to be removed, otherwise we would have an n-p-n device that would not be a working solar cell. The removal can be done using many different processes, for example wet etching. Alternatively, sometimes a protective layer is deposited onto the back side prior to the emitter diffusion. This layer would prevent the creation of an n-doped layer at the back.

Now the p-n junction in principle is finished; we only need to make the electrical contacts at the front and the rear that allow us to connect the solar cell with an electric circuit or with other solar cells in a PV module. In industry these contacts are usually fabricated using screen printing processes followed by firing. These processes are explained in detail in Chapter 14. Different screen printing pastes are used, which are usually based on aluminium or silver as a metal. Ag paste is for the front contact commonly printed on top of the passivation layer. As we have seen in Section 12.3, the design of the front metal grid needs careful optimization between resistive and shading losses. In contrast the rear side is totally covered with Al.

To make real contacts out of the screen printing pastes, the cell is put in a belt furnace at 850 °C. This process is called *firing*; if both the top and front contacts are fired at the same time, we call it *co-firing*. As a result of the co-firing process the front side Ag paste etches away the underlying SiN layer, forming a direct contact with the emitter.

During the co-firing process Al atoms diffuse into the wafer at the rear side and act as a *p*-type dopant, forming a p^+ layer at the rear of the device. Because of this layer, a back surface field (BSF) is created, which enhances the performance of the solar cell as we have seen in Section 12.3. However, during this process a eutectic layer is also created at the rear side, which leads to high parasitic absorption and hence low internal reflection of light that penetrates through the entire solar cell to the back. Usually, also Ag busbars are screen-printed on the back side for the interconnections with other cells.

The last step is to create isolating grooves at the border of the solar cells. This for example can be done with laser scribing. Isolating grooves are very important to prevent leakage currents along the sides of the solar cells. Such leaking currents can be very detrimental to the solar cell performance.

12.5 High-efficiency concepts

In the last section of this chapter we discuss three examples of high-efficiency concepts based on crystalline silicon technology. As already discussed, different types of silicon wafers with different qualities can be used. Naturally, to achieve the highest efficiencies, the bulk recombination must be as small as possible. Therefore the high efficiency concepts are based on monocrystalline wafers.

12.5.1 The PERL concept

The first high efficiency concept was developed in the late 1980s and the early 1990s at the *University of New South Wales* in the group led by Martin Green. Figure 12.15 shows an illustration of the PERL concept, which uses a *p*-type float zone silicon wafer. With this concept, conversion efficiencies of 25% were achieved [47]. The abbreviation PERL stands for *Passivated Emitter Rear Locally diffused*. This abbreviation indicates two important concepts that have been integrated into this technology: first, the optical losses of the PERL solar cell at the front side are minimized using three techniques:

1. The top surface of the solar cell is textured with inverted pyramid structures. This microscopic texture allows a fraction of the reflected light to be incident on the front surface for a second time, which enhances the total amount of light coupled into the solar cell.

2. The inverted pyramid structure is covered with a double-layer anti-reflective coating (ARC), which results in an extremely low top surface reflection. Often a double layer coating of magnesium fluoride (MgF_2) and zinc sulfide (ZnS) is used as an anti-reflection coating.

3. The contact area at the front side has to be as small as possible, to reduce the shading losses. In the PERL concept the very thin and fine metal fingers are processed using photolithography technology.

Secondly, the emitter layer is well designed. As discussed in the previous sections, the emitter should be highly doped underneath the contacts, which in the PERL concept is achieved by heavily phosphorus-diffused regions. The rest of the emitter is moderately doped, or in other words lightly diffused, to preserve an excellent blue response. The

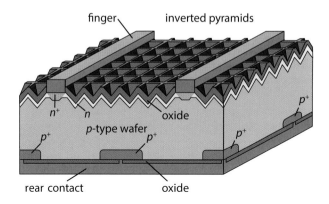

Figure 12.15: Illustrating the structure of a PERL solar cell.

emitter is passivated with a silicon oxide layer on top of the emitter to suppress surface recombination as much as possible. With the PERL concept, the surface recombination velocity could be reduced so far that open circuit voltages with values of above 700 mV could be obtained.

At the rear surface of the solar cell, point contacts have been used in combination with thermal oxide passivation layers, which passivate the non-contacted area, and hence reduce the undesirable surface recombination. A highly doped boron region, created by local boron diffusion, operates as a local back surface field to limit the recombination of the minority electrons at the metal back contact.

The PERL concept includes some expensive processing steps. Therefore, the Chinese company *Suntech* developed, in collaboration with University of New South Wales, a more commercially viable crystalline silicon cell technology, which was inspired by the PERL design.

12.5.2 The interdigitated back contact solar cell

A second successful cell concept is the *interdigitated back contact* (IBC) solar cell. The main idea of the IBC concept is to have no shading losses at the front metal contact grid at all. All the contacts responsible for collecting charge carriers at the n- and p-sides are positioned at the back of the crystalline wafer solar cell. A sketch of such a solar cell is shown in Figure 12.16 (a).

An advantage of the IBC concept is that monocrystalline float-zone n-type wafers can be used. This is interesting because n-type wafers have some useful advantages with respect to p-type wafers. First, the n-type wafers do not suffer from light induced degradation. In p-type wafers both boron and oxygen are present, which under light exposure start to make complexes that act like defects. This light-induced degradation causes a reduction of the power output of 2–3% after the first few weeks of installation. No such effect is present in n-type wafers. The second advantage is that n-type silicon is not that sensitive to impurities such as iron. As a result, less effort has to be made to fabricate high quality n-type silicon, and thus high quality n-type silicon can be processed cheaper than p-type

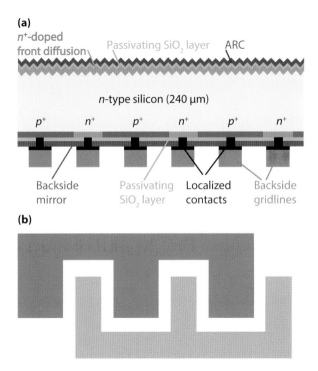

Figure 12.16: (a) Illustrating the structure of an IBC solar cell and (b) the contacts on its back.

silicon. On the other hand, p-doped wafers have the advantage that the boron doping is more homogeneously distributed across the wafer as this is possible for n-type wafers. This means that within one n-type wafer the electronic properties can vary, which lowers the yield of solar cell production based on n-type monocrystalline wafers.

Although IBC cells are made from n-type wafers, they lack one large p-n junction. Instead, IBC cells have many localized junctions. The holes are separated at a junction between the p^+ silicon and the n-type silicon, whereas the electrons are collected using n^+-type silicon. The semiconductor-metal interfaces are kept as small as possible in order to reduce the undesired recombination at this defect-rich interface. Another advantage is that the cross-section of the metal fingers can be made much larger, because they are at the back and therefore do not cause any shading losses. Thus, resistive losses at the metallic contacts can be reduced. Since both electric contacts are on the back side, it contains two metal grids, as illustrated in Figure 12.16 (b). The passivation layer should made from a low refractive index material such that it operates like a backside mirror. It will reflect the light above 900 nm, which is not absorbed during the first pass back into the absorber layer. Thus, this layer enhances the absorption path length.

At the front side of the IBC cell, losses of light-excited charge carriers due to surface recombination are suppressed by a *front surface field* similar to the back surface field discussed earlier. This field is created with a highly doped n^+ region at the front of the surface. Thus, an n^+-n junction is created that acts like an n-p junction. It will act as a barrier that prevents the light-excited minority holes in the n-region from diffusing towards the front surface. The front surface field behaves like a passivation for the defects at the front interface and allows higher levels for the hole minority density in the n-doped bulk.

Reflective losses at the front side are reduced in a similar way as for PERL solar cells: deposition of double-layered anti-reflection coatings and texturing of the front surface.

The IBC concept is commercialized by the US company *SunPower Corp.*, who have achieved high solar cell efficiencies of 24.2%.

12.5.3 Heterojunction solar cells

The third high efficiency concept is the *silicon heterojunction* (SHJ) solar cell. Before we discuss the technological details, we briefly recall the principles of heterojunctions that were discussed in Section 8.2.

Homojunctions, which are present in all the c-Si solar cell types discussed so far in this chapter, are fabricated by different doping types within the same semiconductor material. Hence, the bandgap in the p- and n-regions is the same. A junction consisting of a p-doped semiconductor material and an n-doped semiconductor made from another material is called a *heterojunction*. In SHJ cells, the heterojunction is formed in-between two different silicon-based semiconductor materials. On the one hand, we use an n-type float zone monocrystalline silicon wafer. The other material is hydrogenated amorphous silicon (a-Si:H), which we will discuss in more detail in Section 13.3. a-Si:H is a silicon material in which the atoms are not ordered in a crystalline lattice but in a disordered lattice. For the moment we must only keep in mind that a-Si:H has a bandgap of around 1.7 eV which is considerably higher than that of c-Si (1.12 eV).

In Figure 12.17 a band diagram of a heterojunction between n-doped crystalline silicon and p-doped amorphous silicon in the dark and thermal equilibrium is sketched. We see

12. Crystalline silicon solar cells

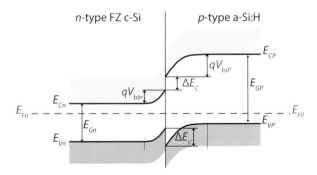

Figure 12.17: Illustrating the band diagram of a heterojunction.

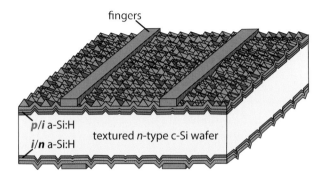

Figure 12.18: Illustrating the structure of a silicon heterojunction solar cell.

that next to the induced field, because of the space-charge region, some local energy steps are introduced. These steps are caused by the two different bandgaps for the *p* and *n* regions. The valence band is positioned higher in the *p*-type amorphous silicon than in the *n*-type crystalline silicon. This will allow the minority charge carriers in the *n*-type c-Si, the holes, to drift to the *p*-type silicon. However, the holes experience a small barrier. While they could not travel across such a barrier in classical mechanics, quantum mechanics allows them to *tunnel* across this barrier.

Let us now take a closer look at SHJ solar cells. The SHJ concept was developed by the Japanese company Sanyo, currently a part of the Japanese *Panasonic Corp.*, who called it *heterojunction with intrinsic thin layer* (HIT). As we can see in Figure 12.18, the SHJ cell configuration has two junctions: the junction at the front side is formed using a thin layer of only five nanometers of intrinsic amorphous silicon which is indicated by the colour red. A thin layer of *p*-doped amorphous silicon is deposited on top and indicated with the colour blue. The heterojunction forces the holes to drift to the *p*-layer. At the rear surface a similar junction is made. First, a thin layer of intrinsic amorphous silicon is deposited on the wafer surface, indicated by red. On top of the intrinsic layer an *n*-doped amorphous silicon is deposited, indicated by yellow.

As discussed earlier, for high quality wafers, like this *n*-type float zone monocrystalline silicon wafer, the recombination of charge carriers at the surface determines the charge-carrier lifetime. The advantage of the SHJ concept is the amorphous silicon layer, which acts as a very good passivation layer. With this approach the highest possible charge carrier lifetimes are accomplished. As a consequence, c-Si wafer-based heterojunction solar cells achieve the highest open circuit voltages among the different crystalline silicon technologies. The current record cell[2] has an open circuit voltage of 0.74 V and an efficiency of 25.6% [48].

How do the charge carriers travel to the contacts? The conductive properties of the *p*-doped amorphous silicon are relatively poor. While in homojunction solar cells the lateral diffusion to the contacts takes place in the emitter layer, in a SHJ solar cell this occurs through a transparent conducting oxide (TCO) material such as indium tin oxide (ITO), which is deposited on top of the *p*-doped layer. TCOs are discussed in more detail in Section 13.1.

The same contacting scheme is applied at the *n*-type back side. This means that this solar cell can be used in a bifacial configuration: it can collect light from the front, and scattered and diffused light falling on the back side of the solar cell. Another important advantage of the SHJ technology is that the amorphous silicon layers are deposited using the cheap plasma-enhanced chemical vapour deposition (PE-CVD) technology at low temperatures, not higher than 200 °C. Therefore, making the front and back surface fields in SHJ solar cells is very cheap.

In this chapter we have discussed many aspects of crystalline silicon solar cell technology. Before the solar cells can be installed, they must be packed as a *PV module*. How such modules are made is discussed in Chapter 15.

12.6 Exercises

12.1 Which of the following statements is true about mono- and multicrystalline silicon?

  (a) SRH recombinations are more important in multicrystalline than in monocrystalline silicon.

  (b) Higher open circuit voltages are obtained with multicrystalline than with monocrystalline silicon.

  (c) The open circuit voltage increases as the grain size of the multicrystalline silicon absorber layer decreases.

  (d) The lifetime of minority-charge carriers is higher in multicrystalline than in monocrystalline silicon.

12.2 What is the correct order, from lower purity to higher purity, of the following types of silicon?

  (a) Metallurgical silicon / monocrystalline silicon / polysilicon.

  (b) Monocrystalline silicon / polysilicon / metallurgical silicon.

  (c) Metallurgical silicon / polysilicon / monocrystalline silicon.

  (d) Polysilicon / metallurgical silicon / monocrystalline silicon.

12.3 Which of the following processes is used in order to fabricate monocrystalline silicon? (More than one correct answer possible.)

---

[2] As of October 2014.

(a) The Czochralski process.
(b) The fluid bed reactor process.
(c) The float zone process.
(d) The Siemens process.

12.4 Which process involves dipping a seed crystal into a molten silicon liquid and pulling it upwards at a specific rotational speed?

(a) Multicrystalline ingot casting.
(b) Float zone pulling.
(c) Czochralski casting.
(d) Silicon wire casting.

12.5 Which of the following processes is used in order to fabricate polycrystalline silicon? (More than one correct answer possible.)

(a) The Czochralski process.
(b) The fluid bed reactor process.
(c) The float zone process.
(d) The Siemens process.

12.6 Why should the front n-layer of a c-Si solar cell be thinner than the p-layer?

(a) The solar cell would actually have the same performance if the n-layer were thicker than the p-layer, as long as a p-n junction is created between these two.
(b) Because the diffusion length of the photogenerated holes is larger in the n-layer than in the p-layer.
(c) Because n-doping is required to absorb the incoming light.
(d) Because most of the photons are absorbed close to the front surface of the solar cell and the thickness should be smaller than the hole diffusion length in the emitter.

12.7 Silicon wafers can be made p-doped by diffusing boron into the wafer. Imagine that boron is diffused into a wafer with no previous boron in it at a temperature of 1,100°C for five hours. The diffusion can be described by Fick's law (Eq. (12.6)). If the concentration of boron at the surface is 10^{18} cm^{-3}, calculate the depth below the surface at which the boron concentration is 10^{17} cm^{-3}. The boron diffusion flux is $J = 3 \times 10^9$ cm^{-2}s^{-1} and the diffusion coefficient for boron diffusing in silicon at 1,100°C is $D = 1.5 \times 10^{-12}$ cm^{2}s^{-1}.

12.8 What kinds of losses are related to the design of the top contact of a solar cell?

(a) Shading losses.
(b) Resistive losses in the top contact.
(c) Resistive losses in the emitter.
(d) All of the above.

12.9 Which of the following statements is false?

(a) Increasing the length of the finger increases the resistive losses.
(b) Increasing the finger spacing decreases the power losses due to shadowing.

(c) Increasing the finger width increases the power losses due to shadowing.

(d) Increasing the finger width increases the resistive losses.

12.10 Consider a c-Si solar cell with fingers having a resistance of $R = 0.1\,\Omega$. What would be the finger resistance if the finger width was doubled and the finger height was set to one-third of its initial value?

12.11 We have discussed that a bare crystalline silicon surface contains many defects which act as SRH recombination centres. How can the surface recombination at the air/n-silicon interface be reduced? (More than one correct answer possible.)

(a) By decreasing the doping of the n-layer.

(b) By increasing the doping of the n-layer.

(c) By depositing a thin insulating layer on top of the n-layer.

(d) By depositing a thin conductive layer on top of the n-layer.

12.12 In state-of-the-art crystalline silicon solar cell technology, which is based on p-type silicon, a back surface field (BSF) is implemented. This BSF is implemented, because in this way …

(a) …the barrier for hole collection at the back contact is reduced, increasing the surface recombination velocity at the back contact.

(b) …the barrier for electron collection at the back contact is reduced, decreasing the surface recombination velocity at the back contact.

(c) …a barrier is created for hole collection at the back contact, increasing the surface recombination velocity at the back contact.

(d) …a barrier is created for electron collection at the back contact, decreasing the surface recombination velocity at the back contact.

12.13 Which of the following concepts are used in the PERL (passivated emitter rear locally diffused) solar cell? (More than one correct answer possible.)

(a) Inverted pyramids.

(b) Multi-junctions.

(c) Bifacial configuration.

(d) Photolithographic fabrication of front metal grid.

12.14 What is *not* true about the thermal oxide layer?

(a) A thermal oxide layer can serve as a passivation layer.

(b) A thermal oxide layer acts as an insulator and forces the electrons to flow in the emitter.

(c) It creates a back surface field at the back side of the solar cell and thus it reduces the surface recombination velocity.

(d) The oxide layer can be thermally grown.

13
Thin-film solar cells

In Chapter 12 we discussed the PV technology based on c-Si wafers, which currently is by far the dominant PV technology. It is very likely that it will stay dominant for a long time.

In this chapter we will look at an alternative family of technologies, namely *thin-film technologies*, also referred to as the *second generation* PV technology. These solar cells are made from films that are much thinner than the wafers that form the base for first generation PV. According to Chopra et al. [50], 'a *thin film* is a film that is created *ab initio* by the random nucleation process of individually condensing/reacting atomic/ionic/molecular species on a substrate. The structural, chemical, metallurgical and physical properties of such a material are strongly dependent on a large number of deposition parameters and may also be thickness dependent.'

Thin-film solar cells were expected to become much cheaper than first generation solar cells. However, due to the current price decline in wafer-based solar cells, thin-film solar cells have not yet become economically viable.[1] In general thin-film cells have a lower efficiency than c-Si solar cells, with GaAs being an exception to this rule of thumb [47]. In contrast to wafer-based silicon solar cells, which are self-supporting, thin-film solar cells require a carrier that gives them mechanical stability. Usual carrier materials are glass, stainless steel or polymer foils. It is thus possible to produce flexible thin-film solar cells.

In thin-film solar cells the active semiconductor layers are sandwiched between a transparent conductive oxide (TCO) layer and the electric back contact. Often a back reflector is introduced at the back of the cell in order to minimize transmissive solar cell losses. As we will see in this chapter, many different semiconductors are used for thin-film solar cells. Figure 13.1 shows the abundance of elements in the Earth's crust. Some semiconductors require very rare elements such as *indium* (In), *selenium* (Se), or *tellurium* (Te). For terawatt scale photovoltaics, solar cells should be based on abundant elements only.

---

[1] According to the *PHOTON module price index*, the price for wafer-based modules has decreased around 40% within one year as of 25 May 2012 [51].

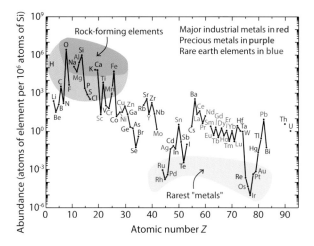

Figure 13.1: The abundance of elements (as atomic fraction) in the Earth's upper crust. Data from Fact Sheet 087-02 courtesy of the US Geological Survey [49].

Thin-film PV technologies can easily fill a book on their own, see for example the book edited by Poortmans and Arkhipov [52]. In this chapter we can only give a general introduction into the different thin-film technologies. We will focus on the working principles of the various devices, the current status and the future challenges of the various technologies. But before we start with this discussion, we will begin with a short introduction on *transparent conducting oxides* (TCOs).

13.1 Transparent conducting oxides

Due to the paramount importance of the *transparent conducting oxide* (TCO) layer for the solar cell performance we briefly discuss its main properties. The TCO layer acts as electric front contact of the solar cell. Furthermore, it guides the incident light to the active layers. It should therefore be both highly conductive and highly transparent in the active wavelength range. The first resistance measurements on thin films of what we nowadays call TCOs were published by Bädeker in 1907 [53].

Typical TCO layers are made from fluor-doped tin oxide (SnO_2:F), aluminium-doped zinc oxide (ZnO:Al), boron-doped zinc oxide (ZnO:B), hydrogen-doped (hydrogenated) indium oxide (In_2O_3:H) [54], and indium tin oxide, which is a mixture of about 90% indium oxide (In_2O_3) and 10% tin oxide (SnO_2). These films are processed using either sputtering (ZnO:Al), low pressure chemical vapour deposition (LPCVD, used for example for ZnO:B), metal organic chemical vapour deposition (MO CVD), or atmospheric-pressure chemical vapour deposition (AP CVD, used for example for SnO_2:F).

Figure 13.2 shows the transmission, reflection and absorption spectra of a flat ZnO:Al layer. Following Kluth, we divide this spectrum into three parts [55]. For short wavelengths, the transmission is very low due to the high absorption of light with energies higher than the bandgap. For longer wavelengths, with photon energies below the bandgap, the trans-

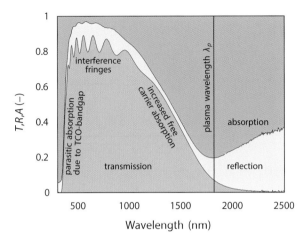

Figure 13.2: Transmission, reflection and absorption of a ZnO:Al layer ($d = 880$ nm).

mission is very high. Here we see interference fringes that can be used to determine the film thickness. After a broad highly transmissive wavelength band, the absorption increases again. This absorption is called free carrier absorption and can be explained with the Drude model of metals that was developed by Paul Drude in 1900 [56, 57]. In this model, the frequency-dependent electric permittivity is given by

$$\epsilon(\omega) = [n(\omega) - i\kappa(\omega)]^2$$
$$= 1 + \chi(\omega) = 1 - \frac{\omega_p^2}{\omega^2 + i\frac{\omega}{\tau}}, \quad (13.1)$$

where $\chi(\omega)$ is the dielectric susceptibility, τ is the relaxation time,[2] $n - i\kappa$ is the complex refractive index and ω_p denotes the *plasma frequency* that is given by

$$\omega_p = \frac{N_f q^2}{\epsilon_0 m_e^{*2}}. \quad (13.2)$$

Here, N_f is the density of free charge carriers, q is the elementary charge, ϵ_0 is the permittivity of vacuum and m_e^* is the effective electron mass in the TCO layer.

The real and imaginary parts of the susceptibility are given by

$$(\Re\chi)(\omega) = -\omega_p^2 \frac{\tau^2}{\omega^2\tau^2 + 1}, \quad (13.3a)$$
$$\text{and} \quad (13.3b)$$
$$(\Im\chi)(\omega) = \omega_p^2 \frac{\tau/\omega}{\omega^2\tau^2 + 1}. \quad (13.3c)$$

If $\omega\tau \gg 1$, ϵ can be simplified to

$$\epsilon(\omega) \approx 1 - \frac{\omega_p^2}{\omega^2}, \quad (13.4)$$

---

[2] The relaxation time denotes the average time between two collisions, *i.e.* two abrupt changes of velocity, of the electrons.

while the imaginary part is negligible.

If this approximation is valid around ω_p, the material is transparent for $\omega > \omega_p$ ($\epsilon > 0$). For $\omega < \omega_p$, ϵ becomes negative, i.e. the refractive index is purely imaginary and the material therefore has a reflectivity of 1. In this case, the material changes dramatically from transparent to reflective, as ω_p is crossed.

If the approximation $\omega\tau \gg 1$ is not valid, $\Im\chi$ cannot be neglected. The imaginary part will increase with decreasing frequency, i.e. with increasing wavelengths the absorption increases. For wavelengths longer than the plasma wavelength the material becomes more reflective, which we also see in Figure 13.2. For application in solar cells, the TCO should be highly transparent in the active region of the absorber. Therefore the plasma wavelength should at least be longer than the bandgap wavelength of the absorber. On the other hand the plasma frequency is proportional to the free carrier density N_f. A longer plasma wavelength therefore corresponds to a lower N_f. Finding an optimum between high transparency and high carrier densities is an important issue in designing TCOs for solar cell applications.

Even though the Drude model gives a good approximation of the free carrier related phenomena in TCOs, this model is often too simple. Therefore several authors used extended Drude models with more parameters [58–60].

Of all the TCO materials currently available, the trade-off between transparency and conductivity is best for indium tin oxide [61]. However, indium is a rare Earth element with a very low abundance of 0.05 ppm in the Earth's crust, similar to the abundance of silver (0.07 ppm) and mercury (0.04 ppm) [62], which makes it less preferable for cheap large-scale PV applications. Therefore other TCO materials are thoroughly investigated and used in industry. Among them are aluminium-doped zinc oxide, boron-doped zinc oxide and fluorine-doped tin oxide. The abundances of the used elements are: aluminium: 7.96%, zinc: 65 ppm, boron: 11 ppm, fluorine: 525 ppm, and tin: 2.3 ppm [62].

Often, TCOs in thin-film solar cells are nanotextured. These textures scatter the incident light and hence help to prolong the average path length of the photons. As a consequence, absorption in the absorber layer can be increased leading to an increased photocurrent density and hence efficiency. Some examples for nanotextured TCO layers are discussed in Appendix D.

13.2 The III-V PV technology

Of all the thin-film technologies, the III-V (spoken 'three-five') technology results in the highest conversion efficiencies under both AM1.5 standard test conditions and concentrated Sun conditions. Hence, it is mainly used in space applications and concentrator photovoltaics (CPV), which we discuss in Section 15.8.

As some concepts use a crystalline germanium or a GaAs wafer as substrate, it might not be considered as a real thin-film PV technology, in contrast to thin-film silicon, CdTe, CIGS or organic PV, which we will discuss later in this chapter. However, the III-V based absorber layers themselves can be considered as thin compared to the thickness of crystalline silicon wafers.

The III-V materials are based on the elements with *three valence electrons* like aluminium (Al), gallium (Ga) or indium (In) and elements with *five valence electrons* like phosphorus (P)

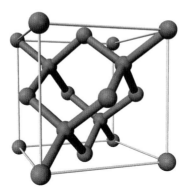

Figure 13.3: A unit cell of a gallium arsenide crystal in the zincblende structure [63].

or arsenic (As). Various different semiconductor materials such as *gallium arsenide* (GaAs), *gallium phosphide* (GaP), *indium phosphide* (InP), *indium arsenide* (InAs), and more complex alloys like GaInAs, GaInP, AlGaInAs and AlGaInP have been explored.

We will now take a closer look at GaAs, which is the most common III-V semiconductor for solar cells. GaAs has a *zincblende crystal structure*, illustrated in Figure 13.3. It is very similar to the diamond cubic crystal structure with the difference that it has atoms of alternating elements at its lattice sites. In contrast to c-Si, where every Si atom has four neighbours of the same kind, in GaAs, every Ga has four As neighbours, while every As atom has four Ga neighbours. When compared to silicon, GaAs has a slightly larger lattice constant of 565.35 pm, and is significantly denser than silicon with a density of 5.3176 g/cm^3 (Si: 543.07 pm, 2.3290 g/cm^3). Note that both Ga and As are roughly twice as dense as silicon. In contrast to silicon, which is highly abundant, the abundance of gallium in the Earth's crust is only about 14 ppm [64]. GaAs therefore is a very expensive material. Arsenic is highly toxic; it is strongly suggested that GaAs is carcinogenic for humans [65].

Figure 13.4 shows the electronic band dispersion diagram of gallium arsenide. As mentioned before, GaAs is a *direct bandgap* material, i.e., the highest energy level in the valence band is vertically aligned with the lowest energy level in the conduction band. Hence, only transfer of energy is required to excite an electron from the valence to the conduction band, but no transfer of momentum is required.

The bandgap of GaAs is 1.424 eV [24]. Here we look again at the absorption coefficient versus the wavelength. Consequently, as you can see in Figure 12.4, the absorption coefficient of GaAs is significantly larger than that of silicon. The same is true for InP, another III-V material that also is shown in Fig. 12.4. Because of the high absorption coefficient, the same amount of light can be absorbed in a film more than one order of magnitude as thin when compared to silicon. Another advantage of direct III-V semiconductor materials is their sharp bandgap. Above E_g, the absorption coefficient increases quickly.

Let us now take a look at the utilization of the bandgap energy. Since GaAs is a direct bandgap material, *radiative recombination processes* become important. On the other hand, Shockley–Read–Hall recombination can be kept at a low level because III-V films can be deposited by *epitaxy processes* that result in high purity films.

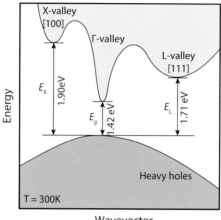

Figure 13.4: The electronic band dispersion diagram of gallium arsenide.

13.2.1 Multi-junction cells

III-V PV devices can reach very high efficiencies because they are often based on the multi-junction concept, which means that more than one bandgap is used. As discussed in Chapter 10, the maximum theoretical efficiency of single-junction cells is described by the Shockley-Queisser limit. A large fraction of the energy of the energetic photons are lost as heat, while photons with energies below the bandgap are lost as they are not absorbed.

For example, if we use a low bandgap material, a large fraction of the energy carried by the photons is not used. However, if we use more bandgaps, the same amount of photons can be used but less energy is wasted as heat. Thus, large parts of the solar spectrum and large parts of the energy in the solar spectrum can be utilized at the same time, if more than one *p-n* junctions are used.

In Figure 13.5 a typical III-V triple junction cell is shown. As substrate, a germanium (Ge) wafer is used. From this wafer, the *bottom cell* is created. Germanium has a bandgap of 0.67 eV. The *middle cell* is based on GaAs with a gap of about 1.4 eV. The *top cell* is based on GaInP with a bandgap in the order of 1.86 eV.

Let us now take a closer look at how a multi-junction solar cell works. Light will enter the device from the top. As the spectral part with the most energetic photons like blue light has the smallest penetration depth in materials, the junction with the highest bandgap always acts as the top cell. On the other hand, as the near infrared light outside the visible spectrum has the largest penetration depth, the bottom cell is the cell with the lowest bandgap.

Figure 13.6 shows the *J-V* curve of the three single *p-n* junctions. We observe *p-n* junction 1 has the highest open circuit voltage and the lowest short circuit current density, which means that this *p-n* junction has the highest bandgap. In contrast, *p-n* junction 3 has a low open circuit voltage and a high current density, consequently it has the lowest bandgap. *p-n* junction 2 has a bandgap in between. Hence, if we are designing a triple-junction cell out of these three junctions, junction 1 will act as the top cell, junction 2 will act as the middle cell and junction 3 will act as the bottom cell.

13. Thin-film solar cells

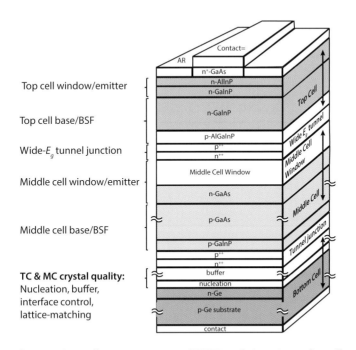

Figure 13.5: Illustrating a typical III-V triple junction solar cell.

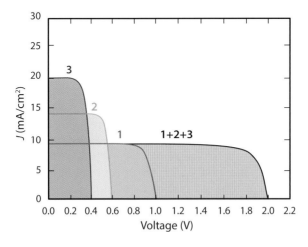

Figure 13.6: The J-V curves of the three junctions used in the III-V triple junction cell and J-V curves of the III-V triple junction cell (in black).

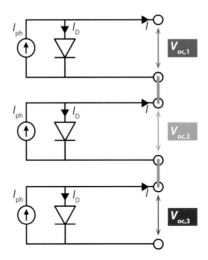

Figure 13.7: The equivalent circuit of the three junctions connected in series.

To understand how the *J-V* curve of the triple junction looks, we take a look at the equivalent circuit. Every *p-n* junction in the multi-junction cell can be represented by the circuit of a single-junction cell, as discussed in Chapter 9. As the three junctions are stacked onto each other, they are connected to each other *in series*, as illustrated in Figure 13.7. In a series connection, the voltages of the individual cell add up in the triple junction cell. Further, the current density in a series connection is equal over the entire solar cell, hence the current density is determined by the *p-n* junction generating the lowest current. The resulting *J-V* curve is also illustrated in Figure 13.6. We see that the voltages add up and the current is determined by the cell delivering the lowest current.

Figure 13.8 (a) shows a typical band diagram of such a triple junction. The top cell with high bandgap is shown at the left-hand side and the bottom cell is at the right-hand side. However, this band diagram does not represent reality. If we were to place three *p-n* junctions in series, for example the *p*-layer of the top cell and the *n*-layer of the middle cell would form a *p-n* junction in the reverse direction than the *p-n* junctions of the three single-junction solar cells. These reverse junctions would significantly lower the voltage of the total triple junction.

To prevent the creation of such reverse junctions, so-called *tunnel junctions* are included. These tunnel junctions align the valence band at one side with the conduction band at the other side of the tunnel junction, as illustrated in Figure 13.8 (b). They have a high bandgap to prevent any parasitic absorption losses. Further, tunnel junctions are relatively thin and have an extremely narrow depletion zone. As a result, the slopes of the valence and conduction bands are so steep that the electrons from the *n*-layer can tunnel through the small barrier to the *p*-layer, where they recombine with the holes. It is important to have the tunnel junctions with a low resistance such that the voltage loss across them is low.

In the triple-junction cell, two tunnel junctions are present. First, a tunnel junction via which the holes in the *p*-layer of the top cell have to recombine with the electrons of the

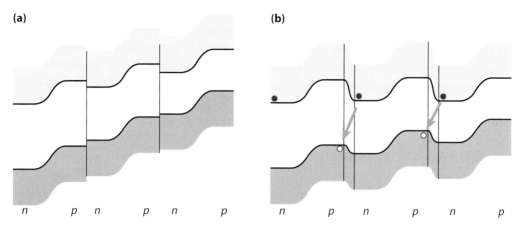

Figure 13.8: The band diagram of the III-V triple junction cell (a) without and (b) with tunnel junctions.

n-layer of the middle cell. Secondly, a tunnel junction where the holes in the p-layer of the middle cell recombine with the electrons of the n-layer of the bottom cell. The electrons in the top cell n-layer are collected at the front contact and the holes in the p-layer of the bottom cell are collected at the back contact. It is important to realize that the recombination current at the tunnel junctions represents the current density of the complete triple junction.

Let us take a look at the external parameters of a typical lattice-matched triple-junction solar cell from SpectroLab, a subsidiary of The Boeing Company. Because this cell was developed for space applications, it is not tested under AM1.5 conditions, but under AM0 conditions with an irradiance of 135 mW/cm². The total open circuit voltage is 2.6 V. The short circuit current density is 17.8 mA/cm², which corresponds to a spectral utilization up to the lowest bandgap (0.67 eV) of

$$J_{\text{tot}} = 3 \times 17.8 \text{ mA/cm}^2 = 53.4 \text{ mA/cm}^2. \tag{13.5}$$

The maximal theoretical current that could be generated when absorbing AM0 for all wavelengths shorter than 1,850 nm, which corresponds to 0.67 eV, is 62 mA/cm². The average EQE of the solar cell is thus 86%, which is a very impressive value. Further, this cell has a very high fill factor of 0.85, which means that it has a conversion efficiency of 29.5% under AM0 illumination.

Figure 13.9 shows the EQE of the subcells of a typical III-V triple-junction cell. We note that the shape of the spectral utilization of all the curves approaches the shape of block functions, which would be the most ideal shape. These block shapes are possible because the III-V materials have very sharp bandgap edges and high absorption coefficients. We see that the bottom cell generates much more current than the middle and top cells. Hence, a lot of current is lost in the bottom cell.

This ineffective use of the near-infrared part can be reduced using quadruple junctions instead of triple junctions. In these cells, an additional cell is placed in-between the middle and bottom cells of the triple junction. The spectral utilization can be even increased further by moving to multi-junction solar cells consisting of five or even six junctions. The

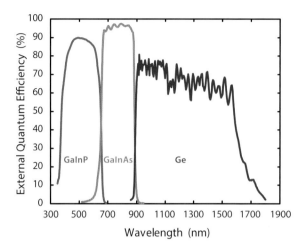

Figure 13.9: The external quantum efficiency of a three-junction III-V solar cell (data from SPECTROLAB).

major challenge for these cells is that lattice matching can no longer be guaranteed. Lattice-mismatched multi-junctions are called *metamorphic multi-junctions*. They require buffer layers that have a profiling in the lattice constant, going from the lattice constant of one *p-n* junction to the lattice constant of the next *p-n* junction. Using this technology, Spectrolab demonstrated 38.8% conversion efficiency of a 5-junction solar cell under 1-sun illumination [47].

The III-V PV technology is very expensive. Hence, such cells are mainly used for space applications and in concentrator technology, where high performance is more important than the cost. Further, using concentrator photovoltaics (CPV), which is discussed in Section 15.8, allows the solar cell area to be reduced drastically.

The current world record conversion efficiency for all solar cell technologies on lab scale is 46.0% for a 4-junction III-V solar cell under 508-fold concentrated sunlight conditions. This result was achieved by cooperation of Soitec and CEA-Leti, France, with the Fraunhofer Institute for Solar Energy Systems ISE [48].

13.2.2 Processing of III-V semiconductor materials

As already mentioned above, high quality III-V semiconductor materials can be deposited using *epitaxy* deposition methods. In this method, crystalline overlayers are deposited on a crystalline substrate, such that they adopt the crystal lattice structure of the substrate. The *precursor* atoms from which the layers are grown are provided by various elemental sources. For example, if GaAs is deposited with epitaxy, Ga and As atoms are directed to a growth surface under ultra high vacuum conditions. For growing III-V semiconductors, germanium substrates are usually are. On this substrate the GaAs crystalline lattice is grown layer-by-layer and adopts the structure of the crystalline substrate.

As the layer-by-layer growth process is very slow, it allows the deposition of compact materials without any vacancy defects. Furthermore, processing at high vacuum condi-

13. Thin-film solar cells

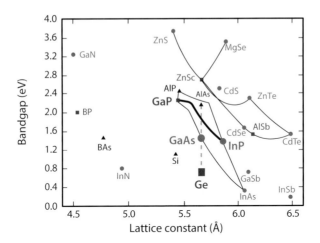

Figure 13.10: The bandgap and lattice constant of various III-V semiconductor materials.

tions prevents the incorporations of impurities. Hence, III-V semiconductors can be deposited up to a very high degree of purity. Dopants can be added to make it *n*- or *p*-type.

Typically, III-V semiconductor layers are deposited by using *metal organic chemical vapour deposition* (MOCVD). Typical precursor gases are trimethylgallium ($Ga(CH_3)_3$), trimethylindium ($In(CH_3)_3$), trimehtylaluminium ($Al_2(CH_3)_6$), arsine gas (AsH_3) and phosphine gas (PH_3). Surface reactions of the metal-organic compounds and hydrides, which contain the required metallic chemical elements, create the right conditions for the epitaxial crystalline growth. Epitaxial growth is a very expensive process; similar techniques are also used in the microchip production process.

A big challenge of depositing III-V semiconductor materials is that the lattice constants of the various materials are different, as seen in Figure 13.10. We see that every III-V semiconductor has a unique bandgap-lattice constant combination. Hence, interfaces between different III-V materials show a lattice mismatch, as illustrated in Fig. 13.11. Because of this mismatch, not every valence electron is able to make a bond with a neighbour. This problem can be solved by *lattice matching*, as was done in the triple-junction cell discussed in Section 13.2.1. To understand what this means, we take another look at the phase diagram shown in Fig. 13.11. The triple junction is processed on a *p*-type germanium substrate, the bottom cell is a Ge cell. On top it is best to place a junction that has a higher bandgap but the same lattice constant. As we can see in the plot, GaAs has exactly the same lattice constant as Ge. Therefore, between GaAs and Ge interfaces can be made without any coordination defects related to mismatched lattices. For reasonable current density matching the desired bandgap of the top cell should be around 1.8 eV. However, we see that no alloys, based on solely two elements, exist. But if we take a mixture of gallium, indium and phosphorus, we can make III-V alloys with a bandgap of 1.8 eV and with a matching lattice constant. Consequently, triple-junction cells based on GaInP, GaAs and Ge can be fully lattice-matched.

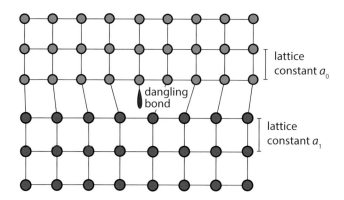

Figure 13.11: Illustrating the lattice mismatch at an interface between two different III-V materials.

13.3 Thin-film silicon technology

In this section we will discuss thin-film silicon solar cells, which can be deposited on glass substrates and even on flexible substrates.

13.3.1 Thin-film silicon alloys

Thin-film silicon materials are usually deposited with *chemical vapour deposition* (CVD) processes that we will discuss in more detail in Section 13.3.3. In chemical vapour deposition different *precursor gases* are brought into the reaction chamber. Due to chemical reactions, a layer is formed on the substrate. Depending on the precursors used and other deposition parameters such as the gas flow rate, pressure, and temperature, various different alloys with different electrical and optical parameters can be deposited. We will discuss the most important alloys in the following paragraphs.

We start with two alloys consisting of silicon and hydrogen: *hydrogenated amorphous silicon* (a-Si:H) and *hydrogenated nanocrystalline silicon* (nc-Si:H), which is also known as microcrystalline silicon. The term *hydrogenated* indicates that some of the valence electrons in the silicon lattice are *passivated* by hydrogen, which is indicated by the ':H' in the abbreviation. The typical atomic hydrogen content of these alloys is from 5% up to about 15%. The hydrogen passivates most defects in the material, resulting in a defect density around 10^{16} cm^{-3} [66], which is suitable for PV applications. Often, the term 'hydrogenated' is left out for simplicity. Pure amorphous silicon (a-Si) would have an extremely high defect density ($> 10^{19}$ cm^{-3}) [67], which would result in fast recombination of photoexcited excess carriers. Similarly, we can make alloys from *germanium* and hydrogen: *hydrogenated amorphous germanium* (a-Ge:H) and *hydrogenated nanocrystalline germanium* (nc-Ge:H).

Let us now take a look at alloys of silicon with four valence electrons with other elements with four valence electrons, *carbon and germanium*. In thin-film silicon solar cells, both hydrogenated amorphous and nanocrystalline *silicon-germanium alloys* (a-SiGe:H and nc-SiGe:H) are being used. Silicon is also mixed using the four valence electron material, carbon, leading to hydrogenated amorphous *silicon carbide* (a-SiC:H).

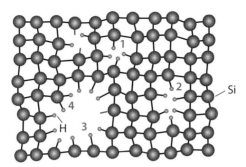

Figure 13.12: Illustrating the atomic structure of amorphous silicon with four typical defects: (1) monovacancies; (2) divacancies; (3) nanosized voids with monohydrides and (4) dihydrides. With kind permission of M. A. Wank [68].

Another interesting alloy is obtained when oxygen with six valence electrons is incorporated into the lattice: *hydrogenated nanocrystalline silicon oxide* is often used in thin-film silicon solar cells.

All these alloys can be doped, usually boron is used as a $p-$ dopant while phosphorus is the most common $n-$ dopant.

Many of the alloys mentioned above are present as *amorphous* materials. It is hence important to discuss the structure of amorphous lattices. In this discussion we will limit ourselves to amorphous silicon, since it is the best studied amorphous semiconductor and the general properties of the other amorphous alloys are similar. In Chapter 12 we thoroughly discussed crystalline silicon, which has an ordered lattice in which the orientation and structure is repeated in all directions. For amorphous materials this is not the case, but the lattice is disordered as illustrated in Figure 13.12. This figure shows a so-called *continuous random network* (CRN). On atomic length scales, also called *short-range order*, the atoms still have a tetrahedral coordination structure, just like crystalline silicon. But the silicon bond angles and silicon-silicon bond lengths are slightly distorted with respect to a crystalline silicon network. However, at larger length scales, also referred to as *long range*, the lattice does not look crystalline any more. Not all silicon atoms have four silicon neighbours, but some valence electrons form dangling bonds, similar to the unpassivated surface electrons already discussed in Chapter 12. Recent studies show that these dangling bonds are not distributed homogeneously throughout the amorphous lattice, but that they group in divacancies, multivacancies or even nanosized voids [69]. The surfaces of these volume deficiencies are passivated with hydrogen.

Another phase of hydrogenated silicon alloys is the *nanocrystalline* phase with a structure even more complex than in amorphous materials. Nanocrystalline hydrogenated silicon consists of small grains that have a crystalline lattice and are a few tens of nanometres big. These grains are embedded in a tissue of hydrogenated amorphous silicon. Figure 13.13 shows the various phases of thin-film silicon [70, 71]. On the left-hand side, a fully crystalline phase is shown, which is close to that of polycrystalline silicon, except that it contains many more cracks and pores. On the right-hand side, the phase represents the amorphous lattice. Please note again, that the terms microcrystalline and nanocrystalline refer to the same material. Going from right to left, the amorphous phase is changing

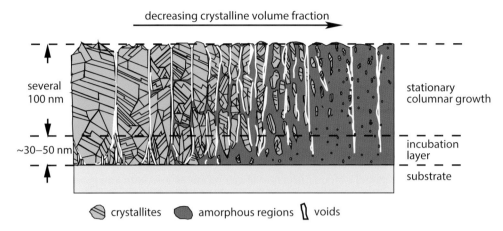

Figure 13.13: The various phases of thin-film silicon with a highly crystalline phase on the left and amorphous silicon on the right [70, 71].

into a mixed phase with a few small crystalline grains to a phase which is dominated by large crystalline grains and a small fraction of amorphous tissue. Research has shown that the best nanocrystalline bulk materials used in solar cells have a network close to the *amorphous-nanocrystalline silicon transition region* and its crystalline volume fraction is in the order of 60%.

The bandgap of nanocrystalline silicon is close to that of crystalline silicon (≈ 1.12 eV) due to the crystalline network in the grains. The bandgap of amorphous silicon is in the order of 1.6 up to 1.8 eV, which can be tuned by the amount of hydrogen incorporated into the silicon network. It is larger than that of crystalline silicon because of the distortions in bond angles and bond lengths. It is out of the scope of this book to discuss the reasons for this increase in bandgap in more detail. An important consequence of a disordered amorphous lattice is that the electron momentum is poorly defined in contrast to crystalline silicon. As we discussed in Chapter 12, both energy and momentum transfer are needed to excite an electron from the valence band to the conduction band. Hence, crystalline silicon is an indirect bandgap material. This is not true for amorphous silicon, which is a direct bandgap material. Therefore the absorptivity of a-Si:H is much higher than that of c-Si, as we can see in Figure 13.14. We see that the absorption coefficient for amorphous silicon in the visible spectrum is much larger than that of crystalline silicon. In some wavelength regions it is about two orders of magnitude larger, which means that much thinner silicon films can be used in reference to the typical wafers in crystalline silicon solar cells. In the figure, data for amorphous silicon-germanium are also shown. a-SiGe:H has lower bandgap and even higher absorption coefficient in the visible part of the spectrum. Its bandgaps are in the range of 1.4 up to 1.6 eV. Amorphous silicon carbide alloys have bandgaps of 1.9 eV and larger. Finally, nanocrystalline silicon oxides have bandgaps exceeding 2 eV.

13. Thin-film solar cells

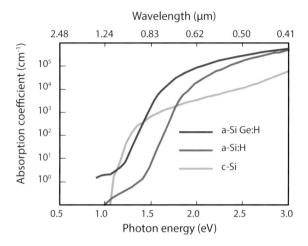

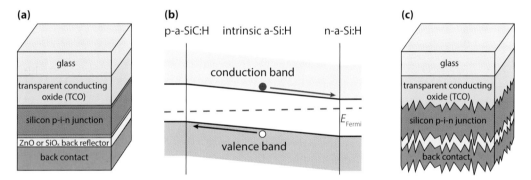

Figure 13.14: The absorption coefficients of different thin-film silicon materials.

Figure 13.15: Illustrating (a) the layer structure; and (b)the band diagram of an amorphous silicon solar cell. (c) Thin-film silicon solar cells have nanotextured interfaces.

13.3.2 The design of thin-film silicon solar cells

The first successful a-Si:H solar cell with an efficiency of 2.4% was reported by Carlson and Wronski in 1976 [72]. We will now discuss how such solar cells can be designed and what are the issues in designing them.

When compared to c-Si, hydrogenated amorphous and nanocrystalline silicon films have a relatively high defect density of around 10^{16} cm^{-3}. As mentioned before, because of the disordered structure not all valence electrons are able to make bonds with the neighbouring atoms. The dangling bonds can act as defects that limit the lifetime of the light-excited charge carriers. This is described with Shockley–Read–Hall recombination, which controls the diffusion length. Because of the high defect density the diffusion length of charge carriers in hydrogenated amorphous silicon is only 100 nm up to 300 nm. Hence, the transport of charge carriers in a thick absorber cannot rely on diffusion.

Therefore, amorphous silicon solar cells are *not* based on a *p-n* junction like wafer-based c-Si solar cells. Instead, they are based on a *p-i-n* junction, which means that an intrinsic (undoped) layer is sandwiched between thin *p*-doped and *n*-doped layers, as illustrated in Figure 13.15 (a). While the *i*-layer is several hundreds of nanometres thick, the doped layers are only about 10 nm thick. Between the *p*- and *n*-doped layers a built-in electric field across the intrinsic absorber layer is created. This is illustrated in the electronic band diagram shown in Figure 13.15 (b). If the layers are not connected to each other, the Fermi level in the *p*− and *n*− layers is closer to the valence and the conduction bands, respectively. For the intrinsic layer, it is in the middle of the bandgap. When the *p*, *i*, and *n* layers are connected to each other, the Fermi level has to be the same throughout the junction, if it is in the dark and under thermal equilibrium. This creates a slope over the electronic band in the intrinsic film as we can see in the illustration. This slope reflects the built-in electric field.

Another way to look at the field is to consider the *i* layer as a dielectric in-between two charged plates (the doped layers). Neglecting edge effects, the electric field in the *i* layer is given by

$$E = \frac{\sigma}{\epsilon\epsilon_0}, \qquad (13.6)$$

where σ is the charge density on the doped layers and ϵ is the electric permittivity of the *i*-layer material. Note that the electric field in the *i* layer is constant. Following Eq. (4.29), we find the voltage in the *i* layer to be

$$V(x) = -\int E(x)\,\mathrm{d}x = V_0 - \frac{\sigma}{\epsilon\epsilon_0}x. \qquad (13.7)$$

We see that the voltage and hence the energy in the *i* layer is linear, just as shown in the band diagram of Figure 13.15 (b).

Because of the electric field, the light-excited charge carrier will move through the intrinsic layer. As discussed in Section 6.5.1, the holes move up the slope in the valence band towards the *p* layer and the electrons move down the slope in the conduction band towards the *n* layer. Such a device, where electronic drift because of an electric field is the dominant transport mechanism is called a *drift device*. In contrast, a wafer-based crystalline silicon solar cell, as discussed in Chapter 12, can be considered as a *diffusion device*.

Note, that because of the intrinsic nature of the absorber layer the hole and electron densities are of the same order of magnitude. On the other hand, in the *p* layer the holes are the majority charge carriers and the dominant transport mechanism is diffusion. Similarly, in the *n* layer the electrons are the majority charge carriers and again diffusion is the dominant transport mechanism. Because of the low diffusion length, both *p* and *n* layers must be very thin.

Now we take another look at Figure 13.15 (a). The cell sketched there is deposited in *superstrate* configuration. This means that the layer that is passed first by the incident light in the solar cell is also deposited first in the production process. Also the term *p-i-n* layer refers to this superstrate configuration, as it indicates the order of the depositions: because in thin-film silicon holes have a significantly lower mobility than electrons, the *p* layer is in front of the *n* layer and hence the *p* layer is deposited first. As the generation rate is then higher close to the *p* layer, more holes can reach it.

In superstrate thin-film silicon solar cells, glass is usually used as a superstrate because it is highly transparent and can easily handle all the chemical and physical conditions in which all the depositions are carried out. Before the *p-i-n* layers can be deposited, a *transparent front contact* has to be deposited. Usually *transparent conducting oxides* are used, that we introduced in Section 13.1.

Another possible configuration is the *substrate configuration*. There, either the substrate acts like a back contact or the back contact is deposited on the substrate. Consequently, no light will pass through the substrate. Thin-film silicon solar cells deposited in the substrate configuration are also called *n-i-p* cells as the *n* layer is deposited before the *i* layer and the *p* layer.

Usually, thin-film silicon solar cells have no flat interfaces, as shown in Figure 13.15 (a), but nanotextured interfaces, as illustrated in Figure 13.15 (c). These textured interfaces scatter the incident light and hence prolong the average path length of the light through the absorber layer. Therefore, more light can be absorbed and the photocurrent can be increased. This is an example of *light management* that we discussed in Section 10.4.3.

Often, the *p* layer is not made from amorphous silicon. Instead, materials with a higher bandgap are used, such as silicon carbide or silicon oxides in order to minimize *parasitic absorption* mainly in the blue part of the spectrum. Usually, boron is used as a dopant. Also for the *n* layer, silicon oxides are often used, but sometimes still a-Si:H is utilized with phosphorus being the main dopant. The *n*-SiO$_x$:H is very transparent. Therefore it can also be used as a back reflector structure when its thickness is chosen such that destructive interference occurs at the *i-n* interface, minimizing the electric field strength and hence parasitic absorption in the *n* layer. Additionally, several tenths of nm thick TCO can be used for the same purpose. Further, a metallic back reflector that also acts as the electric back contact is used. Because of its attractive cost, aluminium is mainly used for this purpose. However, silver that has a higher reflectivity can also be used, but it is more expensive.

The bandgap of hydrogenated amorphous silicon is in the order of 1.75 eV, hence it only is absorptive for wavelengths shorter than 700 nm. The highest current densities achieved in single-junction amorphous silicon solar cells are 17 up to 18 mAcm^{-2}, whereas the maximum theoretic current that could be achieved up to 700 nm is in the order of 23 mAcm^{-2}. Thus, the EQE averaged over the spectrum is in the order of 74 to 77%.

The highest achieved open circuit voltages are in the order of 1.0 V. With respect to a bandgap of 1.75 eV, the bandgap utilization is quite low because of the high levels of Shockley–Read–Hall recombination and the relatively broad valence and conduction-band tails. The highest stabilized efficiency of single junction solar cells is 10.1% [47]. It was obtained by the research Oerlikon Solar Lab in Switzerland, which is currently a subsidiary of the Japanese Tokyo Electron Ltd.

Besides a-Si:H, nanocrystalline silicon films are also used for the intrinsic absorber layers in *p-i-n* solar cells as well. The spectral utilization of the nc-Si:H is better than that of amorphous silicon because of the lower bandgap of nc-Si:H. However, to utilize the spectral part from 700 up to 950 nm, thicker films are required, because of the indirect bandgap of the silicon crystallites. Typical intrinsic film thicknesses are between 1 μm and 3 μm. The current nc-Si:H record cell has a short circuit current density of 28.8 mAcm^{-2}, an open circuit voltage of 523 mV and an efficiency of 10.8% [47]. This result was achieved by the *Japanese Institute of Advanced Industrial Science and Technology* (AIST).

Neither a-Si:H nor nc-Si:H has an optimal spectral utilization. Therefore in thin-film technology the *multi-junction* approach is used, just like for III-V solar cells. Probably the

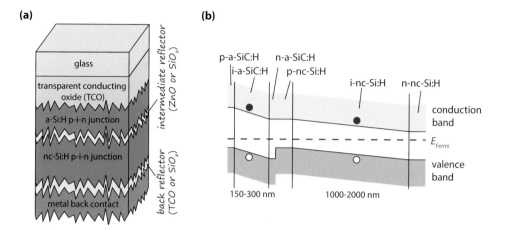

Figure 13.16: Illustrating (a) the layer structure; and (b) the band diagram of a micromorph silicon solar cell.

most studied concept is the *micromorph* concept illustrated in Figure 13.16 (a), which is a double-junction concept consisting of an a-Si:H and an nc-Si:H junction. As before, the solar cell with the highest bandgap is used as a top cell that converts the most energetic photons into electricity, while the lower bandgap material is used for the bottom cell and converts the lower energetic photons.

Figure 13.16 (b) shows a typical band diagram of such a micromorph solar cell, which is also called a *tandem solar cell*. On the left-hand side the electronic band diagram of the amorphous silicon top cell is shown; on the right-hand side the electronic band diagram of the nanocrystalline silicon bottom cell is shown. The blue and green short-wavelength light is absorbed in the top cell, where electron-hole pairs are generated. Similarly, the red and infrared long-wavelength light is absorbed in the bottom cell, where electron-hole pairs are also generated. Let us take a closer look at the two electron-hole pairs excited in the top and bottom cells, respectively. The hole generated in the amorphous top cell moves to the top p layer and the electron excited in the bottom cell drifts to the bottom n layer. Both can be collected at the front and back contacts. However, the electron excited in the top cell drifts to the top n layer and the hole generated in the bottom cell drifts to the p layer. Just as for the III-V multi-junction devices, the electrons and holes have to recombine at a *tunnel recombination junction* between the n layer of the top cell and the p layer of the bottom cell. Often a very thin and defect-rich layer is used for this purpose. Just as for the III-V multi-junction solar cells, the total current density is equal to that of the junction with the lowest current density. Therefore, for an optimized multi-junction cell all current densities in the various subcells have to be matched in order to achieve the best spectral utilization.

Figure 13.17 shows the J-V curves of a single junction a-Si:H solar cell and of a single junction nc-Si:H solar cell. Let us assume that the V_{oc} of the high bandgap a-Si:H top cell has an open circuit voltage of 0.9 V and a relatively low short circuit density of 15 mAcm^{-2}, whereas the low bandgap material of nc-Si:H has a lower open circuit voltage of 0.5 V and a higher short circuit current density of 25 mAcm^{-2}. If we make a tandem of both junctions,

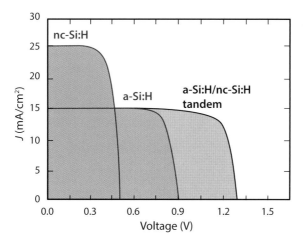

Figure 13.17: The *J-V* curve of a micromorph solar cell and its isolated subcells.

the resulting current density of the double junction is lower than the currents in both bottom cells. Because the open circuit voltage is approximately proportional to $\ln(J_{ph}/J_0)$, the open circuit voltages of the junctions in a tandem cell will be slightly lower than for similar single junction cells. The total current utilization of the tandem cell is determined by the bottom cell, i.e. 25 mAcm^{-2}. Given the examples of the single junctions here, the best current density matching of both cells would deliver 12.5 mAcm^{-2}. The current record tandem cell has an efficiency of 12.3% and was manufactured by the Japanese company *Kaneka* [47].

Just as for the III-V technology, also with thin-film technology, multi-junction cells with more than two junctions can be made. For example the former US company *United Solar Ovonic LLC* made a thin-film silicon-based triple-junction device with an a-Si:H top cell, an a-SiGe:H middle cell and an nc-Si:H bottom cell, illustrated in Figure 13.18 (a). It also shows that various other combinations for triple junctions can be made, for example a-Si:H/nc-Si:H/nc-Si:H. Figure 13.18 (b) shows the spectral utilization[3] of the three junctions and the total cell of the record device by United Solar. In contrast to the EQE of the multi-junction III-V cell (Figure 13.9), where the EQEs of the individual cells are block functions, here the individual EQEs show various overlaps: light with wavelengths below 450 nm is utilized by the top cell only, light at around 550 nm is utilized by the top and middle cells, light at around 650 nm is utilized by all three junctions, and light above 900 nm is utilized by the bottom cell only. Consequently, optimizing thin-film silicon multi-junction solar cells is a complex interplay between the various absorber thicknesses and light management concepts. The *initial efficiency* of the current record triple-junction cell by Uni Solar Ovonic is 16.3%.

The term *initial* refers to a big issue for amorphous thin-film silicon alloys: they suffer from *light-induced degradation* which leads to a reduction of the efficiency. For example, for the initial 16.3% record cell the efficiency drops below the 13.4% of the current *stabilized* record cell. This record cell consists of an a-Si:H/nc-Si:H/nc-Si:H stack and was produced

---

[3]The spectral utilization is given as EQE·AM1.5.

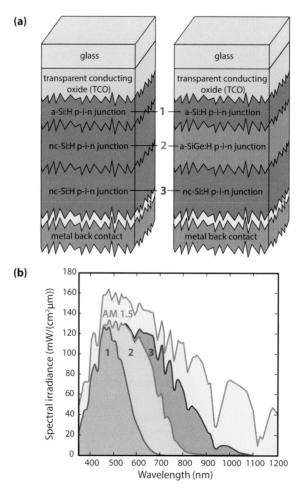

Figure 13.18: (a) Two possible structures; and (b) the typical spectral utilization (EQE·AM1.5) of a triple-junction thin-film silicon cell.

by the South Korean *LG Corp.* [47].

Light-induced degradation, which is also referred to as the *Staebler–Wronski effect* (SWE), is one of the biggest challenges for thin-film solar cells. It was discovered one year after the first a-Si:H solar cells were made in 1977 [73]. Because of the recombination of light-excited charge carriers, *metastable* defects are created in the absorber layers. The increased defect density leads to increased bulk charge carrier recombination, which mainly affects the performance of the amorphous solar cells. After about 1,000 hours of illumination, the efficiency of amorphous thin-film solar cells stabilizes at around 85–90% of the initial efficiency and stays stable for the rest of its lifetime. If the SWE could be tackled, thin-film silicon devices could easily achieve stable efficiencies well above 16%.

Just as for III-V cells, current density matching is very important for thin-film silicon multi-junction cells. First, nanotextured surfaces scatter the incident light in order to enhance the average photon path length and hence to increase the absorption in the various absorber layers of the multi-junction cell. Scattering becomes more important for the bottom cell because its nc-Si:H film is the thickest layer in the device and has to absorb most of the red and infrared, whereas nc-Si:H is not that absorptive. Secondly, intermediate reflector layers are used as a tool to redistribute the light between the junctions above and below them. More specifically, the top and bottom junctions are separated with a low reflective index material, such as nc-SiO$_x$:H. Because of the refractive index difference of the top absorber and the nc-SiO$_x$:H, more light is reflected back into the top cell. Thus, the a-Si:H top cell can be made thinner, which makes it less sensitive to light-induced degradation.

13.3.3 Making thin-film silicon solar cells

To get a better understanding of the fabrication of thin-film silicon solar cells, we take a look at the production process used at the Else Kooi Laboratory in Delft, the Netherlands. Before the deposition can start, the sample has to be cleaned in an *ultrasonic cleaning bath*. During cleaning, dirt and dust particles are removed, which could lead to a *shunt* between the front and back contacts in the final solar cell. Now, the TCO layer can be deposited. In Delft, we can deposit ZnO:Al or ITO with sputtering processes. After sputtering, ZnO:Al can be etched with acids in order to achieve a crater-like structure for light scattering. Alternatively, the sample may already be covered with a TCO layer, for example SnO$_2$:F from the Japanese *Asahi* company, which already has a pyramid-like structure for light scattering because of its deposition process. Also on SnO$_2$:F a 5–10 nm thick ZnO:Al layer is deposited that protects the SnO$_2$:F from being reduced by hydrogen-rich plasma present during the deposition of the thin-film silicon layers. During sputtering, the zinc oxide target is bombarded using an ionized noble gas like argon. The generated aluminium zinc oxide species are sputtered into the chamber and deposited onto the substrate.

Before the thin-film silicon layers are deposited, a thin Al strip is deposited on the side of the sample, that will act as an electric front contact when the cells are measured.

Processing of the different thin-film silicon layers often happens in multi-chamber setups that allow each layer to be processed in a separate chamber and therefore can prevent cross-contamination, e.g. from *p* and *n* dopants, which would reduce the quality of the layers. After the sample is mounted on a suitable *substrate holder*, it enters the setup via a *load lock*, in which the substrate is brought under low pressure before it's moved into the processing chambers. This avoids the processing chambers becoming contaminated with various unwelcome atoms and molecules present in ambient air. Then the sample is brought into a central chamber, from where all process chambers can be accessed.

The typical precursor gas for depositing the Si films is silane (SiH_4), often diluted with hydrogen (H_2). The precursor gases are brought into the process chamber at low pressure. In plasma-enhanced chemical vapour deposition (PECVD), a radio frequency (RF) or very high frequency (VHF) bias voltage between two electrodes is used to generate a plasma, which leads to dissociation of the SiH_4 atoms into radicals such as SiH_3, SiH_2 or SiH. These radicals react with the substrate, where a layer is growing. By increasing the H_2 content, the material becomes more nanocrystalline, as sketched in Figure 13.13.

As precursors for the doping, diborane (B_2H_6) is mainly used for the *p* layers and phosphine (PH_3) for the *n* layers. For silicon carbide layers, in addition to the silane, methane (CH_4) is used. Silicon oxide layers can be made by combining silane with carbon dioxide (CO_2).

After the *p-i-n* junction is deposited, the sample is covered with a *mask*, which defines the areas onto which the metallic back contacts are to be deposited. In Delft, silver is deposited with *evaporation*. Little pieces of silver are put in a boat that is heated by a very high current flowing through it. The silver melts and evaporated silver particles can move freely through a vacuum until they hit the sample, where a silver layer is deposited. This process is explained in more detail in Chapter 14.

Now the solar cells are ready – every metallic square defines a little solar cell. The small metal strip is the front contact and the back of the metallic square is the electric back contact. Naturally, such a configuration is not suitable for large-scale thin-film PV modules. The production of thin-film modules is discussed in Chapter 15.

13.3.4 Crystalline silicon thin-film solar cells

We conclude the section on thin-film silicon technology with a brief discussion on *crystalline silicon thin-film solar cells*. The aim is to combine the advantages of crystalline silicon technology and thin-film silicon technology [74] by using high-quality crystalline films of only several tenths of micrometres thickness as absorbers. These films are positioned on a substrate or a superstrate.

Different approaches are investigated for reaching this goal: Kerf-less wafering techniques are investigated where thin wafers can be prepared without any kerf loss. The wafers are then transferred onto a glass substrate. With this technique, efficiencies of 10.0% were obtained with a 50 μm thick absorber [74].

Alternatively, films of large grain nanocrystalline silicon or amorphous silicon can be deposited. The amorphous films can be crystallized after deposition. This crystallization process can be categorized between *solid-phase* (SPC) and *liquid-phase* crystallization (LPC). SPC can be done for example in a thermal annealing step. Because of the low electrical quality of the defect-rich SPC silicon films, the maximally achieved efficiency is 10.5% which was achieved by the company CSG Solar [74]. In LPC, the amorphous Si film is molten using, for example, an electron beam or a laser. Then, the molten silicon recrystallizes with grain sizes up to several centimetres in the growth direction and up to several millimetres orthogonal to the growth direction. These large grains lead to open-circuit voltages comparable to the wafer-based Si solar cells with reported maximum values of

13. Thin-film solar cells

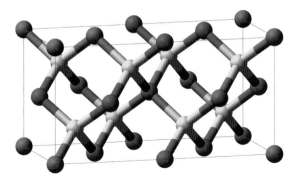

Figure 13.19: The crystal structure of *copper indium diselenide*, a typical chalcopyrite. The colours indicate copper (red), selenium (yellow) and indium (blue). For *copper gallium diselenide*, the In atoms are replaced by Ga atoms [78].

656 mV [75] and efficiencies of 11.8% have been demonstrated [76].

The highest efficiencies are reached with silicon layers grown in an epitaxy process, just as for the high performance III-V solar cells. The epitaxy films then are transferred onto glass. The current record cell made with this method has an efficiency of 20.1% on a 43 µm thick substrate and was fabricated by the American company SOLEXEL [77].

13.4 Chalcogenide solar cells

The third class of thin-film solar cells that we discuss are the large class of *chalcogenide solar cells*, where our focus will mainly be on copper indium gallium selenide (CIGS) and cadmium telluride (CdTe) solar cells. The term *chalcogenides* refers to all chemical compounds consisting of at least one *chalcogen* anion from the group 16 (also known as *group VI*) with at least one or more electropositive elements. Five elements belong to group 16: oxygen (O), sulphur (S), selenium (Se), tellurium (Te), and the radioactive polonium (Po). Typically, oxides are not included in discussions of chalcogenides. Because of its radioactivity, compounds with Po are of very limited relevance for semiconductor physics.

13.4.1 Chalcopyrite solar cells

The first group of chalcogenide solar cells that we discuss are *chalcopyrite solar cells*. The name of this class of materials is based on *chalcopyrite* (copper iron disulphide, CuFeS$_2$). Like all the chalcopyrites, it forms *tetragonal* crystals, as illustrated in Figure 13.19.

Many chalcopyrites are semiconductors. As they consist of elements from groups I, III, and VI, they are also called I-III-VI semiconductors or *ternary* semiconductors. In principle, all these combinations can be used:

$$\begin{pmatrix} Cu \\ Ag \\ Au \end{pmatrix} \begin{pmatrix} Al \\ Ga \\ In \end{pmatrix} \begin{pmatrix} S \\ Se \\ Te \end{pmatrix}_2.$$

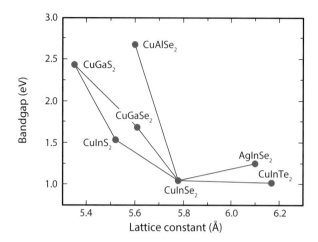

Figure 13.20: The bandgap vs the lattice constant for several chalcopyrite materials. Data taken from [79].

Figure 13.20 shows the bandgap vs the lattice constant for several chalcopyrite materials.

The most common chalcopyrite used for solar cells is a mixture of *copper indium diselenide* ($CuInSe_2$, CIS) and *copper gallium diselenide* ($CuGaSe_2$, CGS). This mixture is called *copper indium gallium diselenide* [$Cu(In_xGa_{1-x})Se_2$, CIGS], where the x can vary between 0 and 1. Several research groups and companies use a compound that also contains sulphur; it is called *copper indium gallium diselenide/disulphide* [$Cu(In_xGa_{1-x})(Se_yS_{1-y})_2$, CIGSS], where y is a number between 0 and 1.

The physical properties of CIGS(S) are rather complex and many different views exist on these properties among scientists. $CuInSe_2$ has a bandgap of 1.0 eV, the bandgap of $CuInS_2$ is 1.5 eV and $CuGaSe_2$ has a bandgap of 1.7 eV. By tuning the In:Ga ratio x and the Se:S ratio y, the bandgap of CIGS can be tuned from 1.0 eV to 1.7 eV. As CIGS(S) is a direct bandgap semiconductor material, it has a large absorption coefficient, hence an absorber thickness of 1–2 μm is sufficient to absorb a large fraction of the light with energies above the bandgap energy. Also the typical electron diffusion length is in the order of a few micrometers. CIGS(S) is a *p*-type semiconductor, the *p*-type character resulting from intrinsic defects in the material that among others are related to Cu deficiencies. The many different types of defects in CIGS(S) and their properties are a topic of ongoing research.

Figure 13.21 (a) illustrates a typical CIGS solar cell structure deposited on glass, which acts as a substrate. On top of the glass a *molybdenum layer* (Mo) of typically 500 nm thick is deposited, which acts as the electric back contact. Then, the *p*-type CIGS absorber layer is deposited with a thickness up to 2 μm. Onto the *p*-CIGS layer, a thin *n*-CIGS layer is deposited, for example an indium/gallium rich $Cu(In_xGa_{1-x})_3Se_5$ alloy. The $p-n$-junction is formed by stacking a thin *cadmium sulphide* (CdS) *buffer layer* of around 50 nm thickness onto the CIGS layers. The *n*-type region is extended with the TCO layer, that is also of *n*-type. First an intrinsic *zinc oxide* (ZnO) layer is deposited, followed by a layer of Al-doped ZnO. The Al is used as an *n* dopant for the ZnO. Similar to thin film silicon technology, the *n*-type TCO acts as the transparent front contact for the solar cell.

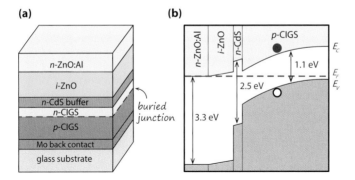

Figure 13.21: The (a) layer structure; and (b) band diagram of a typical CIGS solar cell.

Figure 13.21 (b) shows the electronic band diagram of a CIGS solar cell. The light enters the cell from the left, via the ZnO. The p-type CIGS absorber layers used in industrial modules typically have a bandgap of 1.1–1.2 eV, which is achieved using $Cu(In_xGa_{1-x})Se_2$ with $x \approx 0.3$ [80]. The n-type CdS buffer layer has a bandgap of 2.5 eV. The bandgaps of the n- and p-type materials are different, which means that such CIGS cells can be considered as heterojunctions. The bandgap of ZnO is very large with values of 3.2 eV or even higher, which minimizes the parasitic absorption losses in this device [81].

The defect density at the surface is higher than in the bulk, which could be a loss mechanism for the minority charge carriers. This recombination can be reduced by placing an n-type CIGS layer between the p-type CIGS and the n-type CdS layers, as already mentioned above. The $p - n$ junction within the CIGS is called a *buried junction*; there the electron-hole pairs are separated. In p-CIGS, which is Cu-deficient, the dominant recombination mechanism is Shockley–Read–Hall recombination in the bulk. In contrast, in Cu-rich CIGS films the SRH recombination at the CIGS/CdS interface becomes dominant.

A very important issue in the development of CIGS solar cells is the role of *sodium* (Na). A low contamination with sodium appears to reduce the recombination in p-type CIGS materials because of better recombination of the grain boundaries. The reduced recombination rate results in a higher bandgap utilization and thus a higher open circuit voltage. Typically, the optimal concentration of sodium in the CIGS layers is about 0.1%. Often, as Na source soda-lime glass is used, which is also the substrate for the solar cell. If no soda-lime glass is used, the Na has to be intentionally added during the deposition process. The scientific question of why Na significantly improves several properties of the CIGS films is still not completely solved. Currently, the influence of *potassium* (K) is also being heavily investigated.

Fabrication of CIGS solar cells

CIGS films can be deposited with various different technologies. Because many of these activities are developed within companies, not much detailed information is available on these processing techniques. One approach is co-evaporation under vacuum conditions. Using various crucibles of copper, gallium, indium and selenium, the precursors are co-evaporated onto a heated substrate. The second approach is sputtering onto a substrate at room temperature. After that, the substrate is *thermally annealed* under the presence of selenium vapour, such that the CIGS structure can be formed. Alternatively, a selenium-rich layer can be deposited on top of the initially deposited alloy, followed by an annealing step. Because of the variety and complexity of the reactions taking place during the *'selenization'* process, the properties of CIGS are difficult to control. Companies that use or have used co-evaporation processes are Würth Solar, Global Solar and Ascent Solar Technologies. Among CIGS companies using sputter approaches are Solar Frontier, Avancis, MiaSolé and Honda Soltec.

An alternative approach to produce CIGS layers is based on a *wafer bonding* technique. Two different films are deposited onto a substrate and a superstrate, respectively. Then, the films are pressed together under high pressure. During annealing, the film is released from the superstrate and a CIGS film remains on the substrate.

Non-vacuum techniques can be based on depositing nanoparticles of the precursor materials on a substrate after which the film is sintered. During the *sintering process* films are made out of powder. To achieve this, the powder is heated up to a temperature below the melting point. At such high temperatures, atoms in the particles can diffuse across the particle boundaries. The particles thus fuse together forming one solid. This process has to be followed by a *selenization step*.

An important advantage of the CIGS PV technology is that it has achieved the highest conversion efficiencies among the different thin-film techologies except the III-V technology. The current record for lab-scale CIGS solar cells processed on glass is 20.8% and was achieved by the German research institute ZSW [47]. This record cell has a V_{oc} of 0.757 V, a J_{sc} of 34.77 mA/cm^2, and a FF of 79.2%. The world record on flexible substrates is held by the Swiss Federal Laboratories for Materials Science and Technology (EMPA). On flexible polymer foil they achieved a conversion efficiency of 20.4% [82].

For making CIGS modules, interconnections are made as we discussed already in detail in Chapter 15. As is generally true for the different PV technologies, the record efficiencies of modules are significantly lower than that of lab-scale cells. The record efficiencies of 1 m^2 modules are in the order of 13%, whereas the aperture-area efficiencies are just above 14% as confirmed by NREL. The German manufacturer Manz AG has presented a 15.9% aperture-area efficiency and a total area efficiency of 14.6%. The Japanese company Solar Frontier claims a 17.8% aperture-area efficiency on a small module of 900 cm^2 size.

Despite the very high conversion efficiencies, the CIGS technology faces several challenges. As CIGS is deposited in a complex deposition process, it is challenging to perform large area deposition with a high *production yield*, i.e. with a high percentage of modules coming off the production line that fulfil the product specifications.

Kesterites

Figure 13.1 shows the abundance in the Earth's crust for several elements. As we can see, indium is a very rare element. However, it is a crucial element of CIGS solar cells. Because of its scarcity, In might be *the* limiting step in the upscaling of the CIGS PV technology to future terawatt scales. In addition, the current thin-film display industry depends on In as well, as ITO is integrated in many display screens.

As a consequence, other chalcogenic semiconductors are investigated that do not contain rare elements. An interesting class of materials are the *kesterites* which are *quaternary* or *pentary* semiconductors consisting of four or five elements, respectively. When mineral kesterite ($Cu_2(ZnFe)SnS_4$), where zinc and iron atoms are interchangeable, is not used as a semiconductor, kesterite without iron (Cu_2ZnSnS_4) is used. It also is known as *copper zinc tin sulphide* (CZTS) and is a I_2-II-IV-VI_4 semiconductor. Other kesterites are for example *copper zinc tin selenide* ($Cu_2ZnSnSe_4$, CZTSe), or ones using a mixture of sulphur and selenium, $Cu_2ZnSn(SSe)_4$ (CZTSS).

In contrast to CIGS, CZTS is based on non-toxic and abundantly available elements. The current record efficiency is 12%. It is achieved with CZTSS solar cells on lab scale by IBM [47].

13.4.2 Cadmium telluride solar cells

In this section we will discuss the *cadmium telluride* (CdTe) technology, which currently is the thin-film technology with the lowest demonstrated cost per W_p. We start with discussing the physical properties of CdTe, which is a II-VI semiconductor because it consists of the II valence electron element cadmium (Cd) and the VI valence electron element tellurium (Te). Like the III-V semiconductors discussed in Section 13.2, CdTe forms a *zincblende* lattice structure where every Cd atom is bonded to four Te atoms and vice versa.

The bandgap of CdTe is 1.44 eV, a value which is close to the optimal bandgap for single junction solar cells. CdTe is a direct bandgap material, consequently only a few micrometres of CdTe are required to absorb all the photons with an energy higher than the bandgap energy. In order for the light-excited charge carriers to be collected efficiently at the contacts, their diffusion coefficient has to be in the order of the thickness.

CdTe can be *n*-doped by replacing the II-valence electron atom Cd with a III-valence electron atom like aluminium, gallium or indium. *n* doping can be achieved as well by replacing a VI-valence tellurium atom with a VII-valence electron element like fluorine, chlorine, bromine and iodine atoms. The III- and VII-valence atoms act as shallow donors. A tellurium vacancy also acts like a donor.

p-doping of CdTe can be achieved by replacing Cd with a I-valence electron atom like copper, silver or gold. It can also be achieved by replacing Te atoms with V-valence electron elements such as nitrogen, phosphorus or arsenic. These elements act as shallow acceptors. But a Cd vacancy also acts as an acceptor. In solar cells, *p*-doped CdTe is used. However, it is difficult to obtain CdTe with a high doping level.

Figure 13.22 (a) shows the structure of a typical CdTe solar cell. First, transparent front contact is deposited onto the glass superstrate. This can be *tin oxide* or *cadmium stannate*, which is a Cd-Sn-oxide alloy. On top of that the *n* layer is deposited, which is a *cadmium sulphide* layer, similar to the *n*-buffer layer in CIGS solar cells (Section 13.4.1). Then, the *p*-type CdTe absorber layer is deposited with a typical thickness of a few micrometres. Making a good back contact on CdTe is rather challenging because the material properties of CdTe restrict the choice of acceptable metals. Heavily doping the contact area with a semiconductor material improves the contact qualities; however, achieving high doping levels in CdTe is problematic. Copper-containing contacts have been used as back contacts, however, in the long term they may face instability problems due to diffusion of Cu through the CdTe layer up to the CdS buffer layer. Nowadays, stable *antimony telluride* layers in combination with molybdenum are used.

The band diagram of a CdTe solar cell is shown on Figure 13.22 (b). The *p*-type semi-

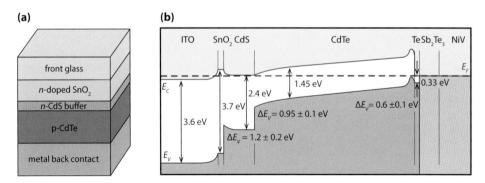

Figure 13.22: The (a) layer structure; and (b) band diagram of a typical CIGS solar cell. Band diagram data taken from [83].

conductor CdTe has a bandgap of 1.44 eV whereas the *n*-type CdS has a bandgap of 2.4 eV. Consequently, the junction is a heterojunction, similar to what we have seen for CIGS PV devices. The light-excited minority electrons in the *p*-layer are separated at the heterojunction and collected at the TCO-based front contact. The holes are collected at the back contact.

Manufacturing CdTe solar cells

Usually, the CdS/CdTe layers are processed using the *closed-space sublimation method*. In this method, the source and the substrate are placed at a short distance from each other, ranging from a few millimetres up to several centimetres, under vacuum conditions. Either the source can consist of granulates or powders of CdTe. Both the source and the substrate are heated, where the source is kept at a higher temperature than the substrate. This temperature difference induces a driven force on the precursors, which are deposited on the substrate. As such, the bulk *p*-type CdTe is formed. In the process chamber, an inert carrier gas like argon or nitrogen can be used.

The US company First Solar Inc. is the leading company in the CdTe technology [84]. Like the German company Antec Solar, they use the closed space sublimation method. Another company producing CdTe solar cells is the German Calyxo. First Solar is by far the largest CdTe manufacturer. In 2008, First Solar had an annual production rate of 500 MW. By 2006 and 2007 it was already one of the biggest solar module manufacturers in the world.

The record conversion efficiency of lab-scale solar CdTe solar cells is 19.6% [47] and was obtained by GE Global Research in 2013. The open circuit voltage of this cell is 857.3 mV, the short circuit current density is 28.59 mA/cm^2, and the FF is 80.0%. The current record module based on CdTe technology has a conversion efficiency of 16.1% at an area of 7,200 cm^2.

The current cost of modules from First Solar is in the order of 0.68 to 0.70 USD/W$_p$ and is expected to drop to 0.40 USD/W$_p$, hence keeping the cost per W$_p$ lower than that of modules based on crystalline silicon wafers.

An important aspect that needs to be addressed is the toxicity of cadmium. However,

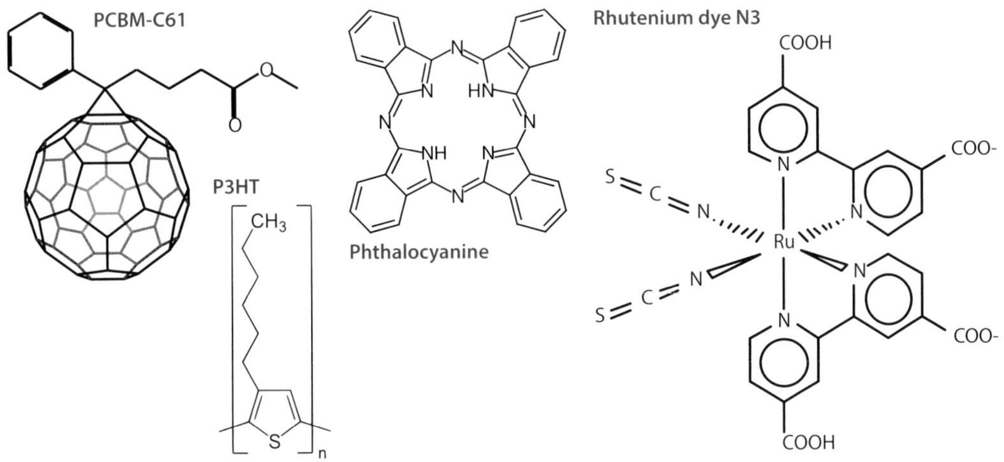

Figure 13.23: Some examples of organic molecules used for organic photovoltaics.

insoluble Cd compounds like CdTe and CdS are much less toxic than the elementary Cd. Nonetheless, it is very important to prevent cadmium entering into the ecosystem. It is an important question whether CdTe modules could become a major source of Cd pollution. As of 2014, First Solar has an installed *nameplate capacity* of 1.8 GW_p/year[85]. With this capacity, about 2% of the total industrial Cd consumption would be taken by First Solar. Nevertheless, recycling schemes have been set up for installed CdTe solar modules. For example, First Solar has a recycling scheme in which a deposit of 5 dollar cents per W_p is included to cover the cost of the recycling at the end of the module lifetime.

Maybe an even bigger challenge for the CdTe technology is the supply of tellurium. Let us take another look at Figure 13.1 that shows the abundance of the various elements in the Earth's crust. Tellurium is one of the rarest stable solid elements with an abundance of about 1 µg/kg, which is comparable to that of platinum [86]. Because of its scarcity, the supply of Te might be the limiting step to upscaling the CdTe PV technology to future terawatt scales. On the other hand, Te as source material has only had a few users so far. Thus, for Te no dedicated mining has been explored so far. In addition, new supplies of tellurium-rich ores were found in Xinju, China. In conclusion, at the moment it remains unclear to what extent the CdTe PV technology might be limited by the supply of tellurium.

13.5 Organic photovoltaics

So far we discussed inorganic thin-film semiconductor materials such as III-V semiconductors, amorphous and nanocrystalline silicon, CIGS and CdTe solar cells. In this section we will take a look at *organic photovoltaics*. The used absorber materials are either *conductive organic polymers* or *organic molecules* that are based on carbon, which may form a cyclic, acyclic, linear or mixed compound structure.

Figure 13.23 shows some examples of organic materials that can be used for PV ap-

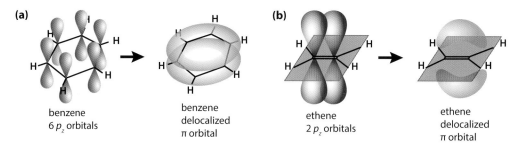

Figure 13.24: The chemical structure and π orbitals of (a) benzene; and (b) ethene.

plications: P3HT, phthalocyanine, PCBM and ruthenium dye N3. All these materials can be considered as large conjugated systems, which means that carbon atoms in the chain have an alternating single or a double bond and every atom in the chain has a p orbital available. In such conjugated compounds, the p orbitals are delocalized, which means that they can form one big mixed orbital. Hence, the valence electron of the original p orbital is shared over all the orbitals. A classical example would be the benzene molecule, which is a cyclic conjugated compound. As we can see in Figure 13.24 (a), this molecule has six carbon atoms and six p orbitals, which mix and form two circular orbitals that contain three electrons each. These electrons do not belong to one single atom but to a group of atoms.

In contrast, a methane molecule (CH_4) is tetrahedrally coordinated, which means that it has four equivalent sp^3 hybrid bonds with a bond angle of 109.5°. An ethene (C_2H_4) molecule has three equivalent sp^2 hybrid bonds with a bond angle of 120° plus an electron in a p^z orbital. Two neighbouring p^z orbitals form a so-called π orbital, as illustrated in Figure 13.24 (b).

In Chapter 6 we have discussed that two individual sp^3 hybrid orbitals can make an anti-bonding and a bonding state. The same is valid for the two p^z orbitals forming a molecular π orbital. They can make bonding and anti-bonding π states. Therefore, conjugated molecules can have similar properties as semiconductor materials.

At room temperature, most electrons are in the bonding state, which is also called the *highest occupied molecular orbital* (HOMO). The anti-bonding state can be considered as the *lowest unoccupied molecular orbital* (LUMO). As the conjugated molecules are getting longer, the HOMO and LUMO will broaden and act similarly to valence and conduction band in conventional semiconductors. The energy difference between the HOMO and LUMO levels can be considered as the bandgap of the polymer material.

To discuss whether an organic material is *p* type or *n* type, we first have to introduce the concept of the *vacuum level*, shown in Figure 13.25 (a). The vacuum level is defined as the energy of a free stationary electron that is outside of any material or, in other words, *in vacuo*. This level often is used as the level of alignment for the energy levels of two different materials. The *ionization energy* is the energy required to excite an electron from the valence band or HOMO to the vacuum state. The *electron affinity* is the energy obtained by moving an electron from the vacuum just outside the semiconductor or conjugated polymer to the bottom of the conduction band or LUMO. When a material has a low ionization potential, it can release an electron out of the material relatively easy, i.e. it can act as an *electron donor*. On the other hand, when a material has a high electron affinity, it can easily accept

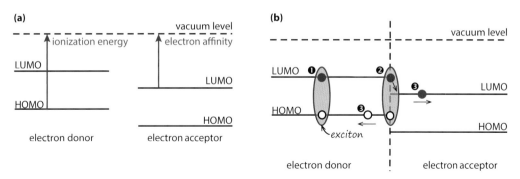

Figure 13.25: Illustrating (a) the energy levels in organic PV materials; and (b) the separation of electrons and holes in an exciton at the acceptor-donor interface..

an additional electron in the LUMO or conduction band, it thus acts as an *electron acceptor*.

As we have discussed earlier, in inorganic semiconductors an electron can be excited from the valence band to the conduction band leaving a hole in the valence band. Such an electron-hole pair is only weakly bound and both entities are easily separated and can diffuse away from each other. In organic materials this is not the case. Absorption of a photon with sufficient energy results in the creation of an *exciton*, illustrated in Figure 13.25 (b) ❶. An exciton is an excited electron-hole pair that is still in a bound state because of the mutual Coulomb forces between the particles. Such excitons can diffuse through the material, but they have a low lifetime in organic materials, recombining back to the ground state within a few nanoseconds. Hence, the diffusion length of such excitons is in the order of only 10 nm.

If an electron donor and an electron acceptor material are brought together, an interface is formed between those two. The HOMO and LUMO of both polymers can be aligned considering their energy levels with reference to the vacuum level, as illustrated in Fig. 13.25 (b). At the interface we see a difference in the HOMO and LUMO levels. Because of this difference, an electrostatic force exists between the two materials. If the materials are chosen such that the difference is large enough, these local be injected into the electron acceptor and a hole remains in the electron donor material ❸.

Figure 13.26 (a) shows the structure of a typical organic solar cell. Here, we consider a solar cell consisting of both organic acceptor- and donor-type materials. Similarly to semiconductor materials, a heterojunction based on two different materials or conjugated compounds can be constructed. As mentioned before, the typical diffusion length in the organic materials is only about 10 nm. Hence, the thickness of the solar cell in principle is strongly limited by the diffusion length, while it has to be at least 100 nm to absorb a sufficient fraction of the light. Therefore, organic solar cells are based on *bulk heterojunction* photovoltaic devices, illustrated in Figure 13.26 (b), where the electron-donor and electron-acceptor materials are mixed together and form a *blend*. In this way, typical length scales in the order of the exciton diffusion lengths can be achieved. Hence, a large fraction of the excitons can reach an interface, where they are separated into an electron and a hole. The electrons move through the acceptor material to the electrode and the holes move through the donor material to be collected at the other electrode. The holes are usually collected at

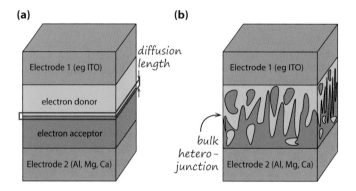

Figure 13.26: Illustrating (a) the layer structure of organic solar cells; and (b) an organic solar cell with a *bulk heterojunction*.

a TCO electrode, for example *indium tin oxide* (ITO). The electrons are collected at a metal back electrode.

Chemical engineering allows tuning of the bandgap. Advantages of organic solar cells are that they have low production costs and can be integrated into flexible substrates. Important disadvantages are a low efficiency, low stability and low performance compared to inorganic PV cells.

Although organic solar cells themselves can be cheap, they require expensive encapsulation materials to protect the organic materials from humidity, moisture and air, which limits their industrial application. To the authors' knowledge, no company is currently producing organic PV modules. In the past, the US based company Konarka Technologies produced small PV modules with efficiencies in the range of 3 up to 5%, which worked for a couple of years only.

The reported efficiency record for organic solar cells is 11.1% and was achieved by Mitsubishi chemicals [77].

13.6 Hybrid organic-inorganic solar cells

In this section we discuss two hybrid concepts, where the junction consists of both inorganic and organic compounds. These concepts are *dye-sensitized solar cells* and *perovskite solar cells*.

13.6.1 Dye-sensitized solar cells

An alternative solar cell concept based on organic materials is the dye-sensitized solar cell (DSSC), which is a photoelectrochemical system. It contains *titanium dioxide* (TiO_2) nanoparticles, organic *dye* particles, an electrolyte and platinum contacts. Figure 13.27 (a) shows an illustration of a DSSC. The dye sensitizer, which acts like an electron donor, is the photo-active component of the solar cell; the TiO_2 nanoparticles act as electron acceptors. Similar to organic bulk heterojunction cells, the dye is mixed with TiO_2.

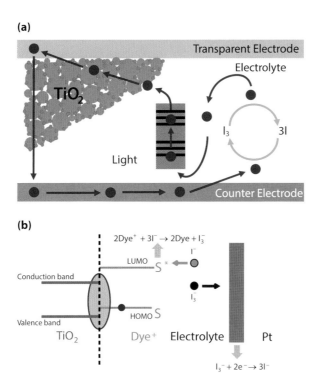

Figure 13.27: (a) A sketch of a dye-sensitized solar cell; and (b) the relevant energy levels in its operation.

As a photoactive dye sensitizer, *ruthenium polypyridine* is used. The operation is illustrated in Figure 13.27 (b). If a photon is absorbed by the ruthenium polypyridine, it can excite an electron from its ground state, the S state, to an excited state, referred to as S^*. In this case the S can be considered as a HOMO and the S^* state can be considered as a LUMO. The S^* energy level is above the energy level of the conduction band of the TiO_2. As a result, the light-excited electrons are injected into the TiO_2 nanoparticles, while the dye-sensitizer molecule remains positively charged. The electrons move through the TiO_2 to the TCO based back contact in a diffusion-based transport mechanism. Via the electric circuit, the electrons move to the front contact. The front contact is electrically connected to the dye via an *electrolyte*. Electrolytes are solutions or compounds that contain ionized entities that can conduct electricity. The typical electrolyte used for DSSC contains *iodine*. The positively charged oxidized dye molecule is neutralized by a negatively-charged iodide. Three negatively-charged iodides neutralize two dye molecules and create one negatively-charged tri-iodide. This negatively-charged tri-iodide moves to the counter electrode where it is reduced using two electrons into three negatively charged iodine. To facilitate the chemical reactions, these *photoelectrochemical* cells require a *platinum* back contact.

The performance of a DSSC depends on the HOMO and LUMO levels of the dye material, the Fermi level of TiO_2 nanoparticles and the *redox potential* of the iodide and triodide reactions. The current record efficiency of dye-sensitized PV devices on lab scale is 11.9% and was achieved by Sharp [77].

The major advantage of DSSC are the low production costs. A disadvantage is the stability of the electrolyte under various weather conditions: At low temperatures the electrolyte can freeze, which stops the device from generating power and might even result in physical damage. High temperatures result in thermal expansion of the electrolyte, which make encapsulating modules more complicated. Another challenge is the high cost of the platinum electrodes, hence replacing the platinum with cheaper materials is a topic of ongoing research. In addition, more stable and resistive electrolyte materials must be developed. Finally, research is needed on improved dyes that enhance the spectral and bandgap utilization of the solar cells.

As yet, no dye-sensitized PV products are available commercially.

13.6.2 Perovskite solar cells

A rapidly emerging PV technology is that of *perovskite solar cells* that has seen a tremendous increase in initial efficiency in recent years. Perovskite cells with 2.2% were reported in 2006. Cells with 6.5% were reported in 2011 and since then the reported efficiencies have rapidly increased with an actual (October 2014) certified record efficiency of 17.9% [77, 87].

The mineral perovskite is named after the Russian mineralogist Lev A. Perovski (1792–1856) and has the chemical formula $CaTiO_3$. Minerals with the general formula ABX_3, where X is an anion, and both A and B are cations, are called perovskites. A is larger than B [87]. Figure 13.28 (a) shows the general cubic crystal structure of perovskites.

For photovoltaics, organic-inorganic perovskites are used, where the large cation A is organic; often methylammonium ($CH_3NH_3^+$) is used. X is a halogen, such as iodine chlorine, or bromine, often in a mixed halide material. The cation B usually contains lead (Pb). While tin (Sn) can also be used, which theoretically gives even more ideal bandgaps,

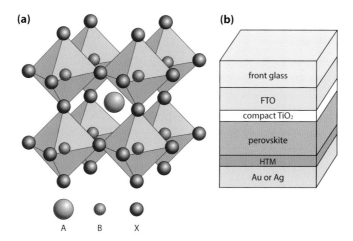

Figure 13.28: (a) A sketch of the perovskite crystal structure with the anion X and the cations A and B indicated (adapted by permission from Macmillan Publishers Ltd: M. A. Green, A. Ho-Baillie, and H. J. Snaith, Nature Photonics, vol. 8, pp. 506–514, copyright (2008)) [87]. (b) The layer structure of a thin-film-based perovskite solar cell.

the stability is usually lower. When using these compounds (A: $CH_3NH_3^+$, B: Pb, X: I), the total compound is called *methylammonium lead triiodide* and has the chemical formula ($CH_3NH_3PbI_3$).

Figure 13.28 (b) shows the structure of a typical thin-film perovskite-based solar cell [87]. A compact titanium dioxide (TiO_2) is deposited onto fluorine-doped tin oxide (FTO), thus allowing electrons to be transported to the FTO layer. The perovskite layer is then deposited onto this TiO_2 layer and covered by a hole-transporting material (HTM). On top of the HTM, silver or gold is placed as a back metal. Besides thin-film architectures, meso-porous architectures are also used, where the perovskite is present in a porous TiO_2 layer.

Perovskite materials used for solar cells have several properties that enable the high efficiencies: they have a very strong absorption of the incident light combined with low non-radiative carrier recombination. Further, their development can build on more than 20 years of experience from the development of organic and dye-sensitized solar cells [87].

All current high efficiency perovskite devices contain lead. This might be a problem because Pb is toxic. However, as we have seen above, CdTe technology is currently very successful, despite the toxic cadmium that is required to make these PV devices. Another issue is degradation of the cells due to ultraviolet radiation and/or moisture. This degradation can be quite fast [87] and is an issue that clearly must be solved if this technology is to be applied industrially.

Measuring the *J-V* characteristics of perovskite cells needs to be done carefully, because of *hysteresis*: depending on the voltage scan direction (from high to low or from low to high voltages) and the scan speed, the shape of the measured *J-V* curve can change significantly. This can lead to an overestimation or an underestimation of the device performance [88].

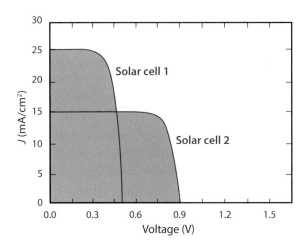

Figure 13.29

13.7 Exercises

13.1 Figure 13.29 shows the *I-V* curves of two single-junction solar cells. If you were to make a multi-junction solar cell with these two cells, which one would you use as a top cell?

13.2 In Figure 13.10 we have the bandgap vs the lattice constant of various III-V materials. Which of the following statements is true?

(a) The junctions of a lattice-matched triple-junction solar cell can be based on the semiconductor materials GaInP, GaAs and Si.

(b) If the bottom cell of a 4-junction cell is based on Ge, the first junction above the bottom cell is based on GaInAs.

(c) A combination of three junctions based on the semiconductor materials GaInP, GaAs and Ge can only result in a metamorphic triple-junction solar cell.

(d) The semiconductor material InAs is a logic choice to be used as a top cell in a triple junction.

13.3 Why can't an a-Si:H solar cell rely on diffusion to separate the photogenerated carriers as much as a c-Si solar cell does?

(a) Because the higher bandgap (around 1.7 eV) prevents the photogenerated carriers from diffusing through the intrinsic layer of the a-Si:H cell.

(b) Because the absorption coefficient of a-Si:H is 70 times higher than that of c-Si in the visible region of the solar spectrum.

(c) Because the diffusion length in a-Si:h is around 300 nm, while in c-Si it is around 300 μm.

13.4 Consider a *p-i-n* a-Si:H solar cell. The various films are, from top to bottom: ZnO, a-Si:H (p-i-n), ZnO and Ag. Which deposition technologies and in which chronological order (from the first processing step to the last) are used to process the solar cell? Use Table 13.1.

Table 13.1

	(a)	(b)	(c)	(d)
front ZnO	sputtering	evaporation	sputtering	CBD
a-Si:H (p-i-n)	PE-CVD	CBD	MBE	MBE
back ZnO	sputtering	MBE	PE-CVD	PE-CVD
back Ag	evaporation	evaporation	evaporation	evaporation

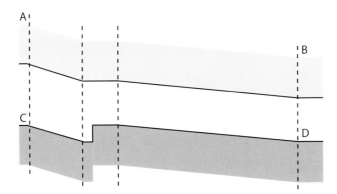

Figure 13.30

13.5 Figure 13.30 shows the band diagram of a tandem cell based on an amorphous and a microcrystalline silicon junction. In which position (A, B, C or D) are the electrons and holes, generated in the microcrystalline silicon junction, collected?

13.6 In this exercise we study thin-film silicon tandem solar cells and a simple solar simulator. The spectral irradiance $I_{e\lambda}$ of the solar simulator is shown in Fig. 13.31 (a). It is given as

$$I_{e\lambda} = 7.5 \times 10^{15} \text{ Wm}^{-2}\text{m}^{-2} \times \lambda - 2.25 \times 10^9 \text{ Wm}^{-2}\text{m}^{-1}$$
for 300 nm $< \lambda <$ 500 nm,

$$I_{e\lambda} = -1.5 \times 10^{15} \text{ Wm}^{-2}\text{m}^{-2} \times \lambda + 2.25 \times 10^9 \text{ Wm}^{-2}\text{m}^{-1}$$
for 500 nm $< \lambda <$ 1500 nm,

where the wavelength λ is expressed in metres. The EQE of the tandem cell with junction A and junction B under short circuited ($V = 0$ V) conditions is shown in Figure 13.31 (b).

(a) Calculate the total irradiance of the solar simulator.
(b) What is the photon flux of the solar simulator?
(c) Which junction acts like a top cell in the tandem cell? A or B?
(d) What is the bandgap of the absorber layer of junction A?
(e) Calculate the short circuit current density J_{sc} of junction A if the solar cell is measured under the spectrum provided by the solar simulator.
(f) Junction B has a different absorber layer than junction A. Above its bandgap, the solar cell B has an EQE of 0.60 that remains constant. Calculate the short-circuit current density J_{sc} of junction B if the solar cell is measured under the solar simulator.

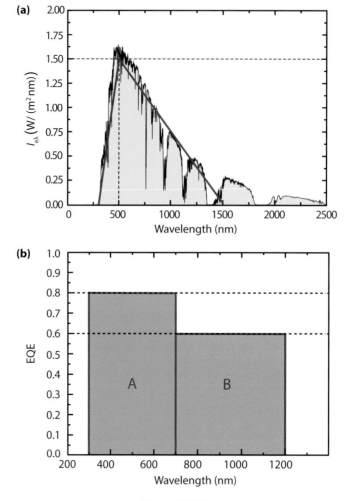

Figure 13.31

13. Thin-film solar cells

13.7 During the manufacture of a CIGS device, the metal back contact is deposited first, while in the case of the CdTe solar cells, the back contact is deposited in the last processing step. What are the configurations of both cells?

   (a) Both the CI(G)S and CdTe solar cells are made in a superstrate structure.
   (b) The CI(G)S solar cell is made in a superstrate structure, whereas the CdTe solar cell is made in a substrate structure.
   (c) The CI(G)S solar cell is made in a substrate structure, whereas the CdTe solar cell is made in a superstrate structure.
   (d) Both CI(G)S and CdTe solar cells are made in a substrate structure.

13.8 In a CdTe solar cell ...

   (a) ...light is mainly absorbed in the n-type CdS layer; this layer makes a junction with the p-type CdTe layer.
   (b) ...light is mainly absorbed in the n-type CdTe layer; this layer makes a junction with the p-type CdS layer.
   (c) ...light is mainly absorbed in the p-type CdS layer; this layer makes a junction with the n-type CdTe layer.
   (d) ...light is mainly absorbed in the p-type CdTe layer; this layer makes a junction with the n-type CdS layer.

13.9 What are the demonstrated advantages of CdTe technologies for solar cells?

   A. Low cost potential.
   B. Recycling of materials for further use.
   C. High module efficiencies in reference to crystalline silicon wafer-based modules.

   (a) Only A.
   (b) Only C.
   (c) Both A and B.
   (d) Both A and C.

13.10 What are the known disadvantages of CdTe materials used in solar cells?

   A. Cd is a toxic element.
   B. Not possible to recycle the materials.
   C. Te is not abundantly available which limits the possibility of scaling up.

   (a) Only A.
   (b) Only C.
   (c) Both A and B.
   (d) Both A and C.

13.11 A First Solar module FS-380 has the following specifications: Area $A = 0.72$ m^2, nominal power $P_{\text{nom}} = 80.0$ W, $V_{\text{oc}} = 61.7$ V, $I_{\text{sc}} = 1.76$ A, $V_{\text{MPP}} = 50.7$ V, $I_{\text{MPP}} = 1.58$ A. What is the efficiency of this module under STC?

13.12 In dye-sensitized solar cells (DSSC) the absorption of photons occurs in ...

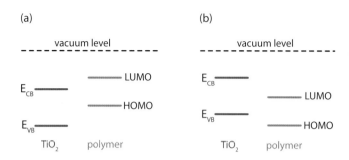

Figure 13.32

(a) ... the wide bandgap of the semiconductor TiO_2.
(b) ... the dye.
(c) ... the electrolyte.
(d) ... the platinum/TCO contact.

13.13 Which statement about DSSCs is true?

(a) The electrons that pass through the external load oxidize the iodide to yield triodide.
(b) The electrons that pass through the external load reduce the iodide to yield triodide.
(c) The electrons that pass through the external load oxidize the triodide to yield iodide.
(d) The electrons that pass through the external load reduce the triodide to yield iodide.

13.14 Consider an organic solar cell consisting of a TiO_2 semiconductor with an ionization energy of 7.8 eV and a polymer with an electron affinity of 3.4 eV. The bandgap of TiO_2 is 3.5 eV.

(a) Which of the band diagrams in Figure 13.32 corresponds to the organic solar cell?
(b) What is the driving force for electron injection from the polymer to the semiconductor?
(c) The electron diffusion coefficient is $D_n = 4 \times 10^{-5}$ cm^2/s and the thickness of the TiO_2 layer is 4 μm. What is the lifetime of the injected electrons?

14

A closer look to some processes

No matter whether solar cells are to be made from crystalline wafers as discussed in Chapter 12 or with thin-film technology as treated in Chapter 13, many processes will be involved. In this chapter, we discuss some important processes that are used for the fabrication of solar cells. Note that the list of processes treated in this chapter is by no means conclusive. The content of this chapter was provided by G. Papakonstantinou and is used with his kind permission [89].

14.1 Plasma-enhanced chemical vapour deposition

Plasma-enhanced chemical vapour deposition (PECVD) can be employed for the growth of thin-film silicon materials, such as amorphous silicon, nanocrystalline silicon, and alloys such as silicon carbide or silicon oxide. The layers can be made intrinsic, n- or p-doped. PECVD allows the deposition of thin films at low operating temperatures, below 200 °C. The principal properties of these films are high quality, appropriate adhesion, desirable coverage of textured structures and uniformity.

Figure 14.1 (a) illustrates the PECVD process. Depending on the desired material, a mixture of different gases such as silane (SiH_4), methane (CH_4), hydrogen (H_2), and carbon dioxide (CO_2) is introduced in an ultrahigh vacuum (UHV) reaction chamber to be converted into glow discharge (plasma). The conversion is regulated by an oscillating electric field between two conductive electrodes through an RF signal of 13.56 MHz or a VHF signal of 50–80 MHz.

Because of the plasma, reactive radicals, ions, neutral atoms, molecules and electrons are generated. The composition of these reactive and high energy entities is the result of collisions in the gas phase, which ensure the maintenance of a low temperature of the substrate. To be more specific, these atomic and molecular particles form an interaction

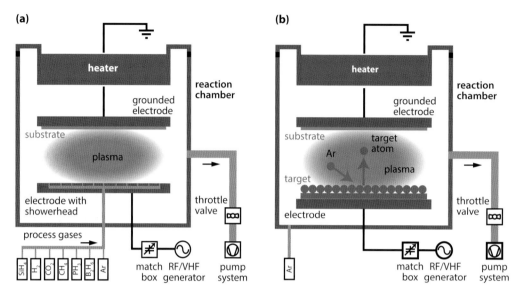

Figure 14.1: Illustrating (a) plasma-enhanced chemical vapour deposition (PECVD); and (b) physical vapour deposition (PVD) with sputtering. Adapted with kind permission from P. Babal [90].

with the substrate, which is of limited potential compared to the plasma, allowing the formation of the thin-film layer since diffusion towards the substrate can occur only by neutral or positively-charged particles.

The quality of the deposited layer can be controlled by the pressure and temperature in the reaction chamber, the gas flow rates of the different process gases, and the power coupled into the plasma by the RF or VHF generator.

14.2 Physical vapour deposition

The term *physical vapour deposition* (PVD) refers to vacuum deposition technologies that produce the source gas by a non-chemical method [91]. Primarily, PVD is applied to metals and compounds such as Ti, Al, Cu, TiN, and TaN for the fabrication of contacts [31]. Thermal evaporation, electron beam evaporation, plasma spray deposition, and sputtering are the most common production methods [91]. We discuss two PVD technologies: sputtering and evaporation.

14.2.1 Sputtering

Sputtering is a common physical vapour deposition (PVD) method, where energy and momentum from an accelerated atom or ion are transmitted to atoms on a target via sequential collisions. Consequently, these atoms can escape from the target and shift to the gas phase, hence evaporate. Sputtering can be accomplished with two different power supply configurations that employ magnetrons, reactive direct current (DC) and radio frequency (RF).

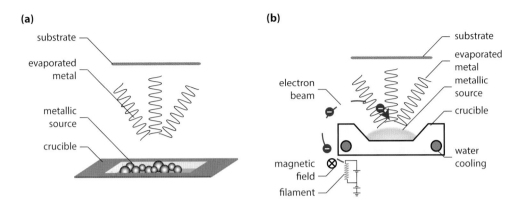

Figure 14.2: (a) Physical vapour deposition (PVD) with resistive evaporation;(b) PVD with electron beam evaporation. With kind permission from G. Papakonstantinou [89].

DC sputtering is restricted to conducting materials, whereas RF sputtering may be also used for dielectrics.

Figure 14.1 (b) illustrates the process of sputtering. The sputtering material is deposited onto the substrate (anode) by bombarding the target (cathode) with ions. The vacuum chamber is kept under a well-defined pressure with inert argon gas (Ar), such that an argon plasma can be inflamed by an oscillating electromagnetic field between the anode and the cathode. This process generates and accelerates positively-charged argon ions (Ar^+) that bombard the target surface. Due to this bombardment, atoms are ejected from the target and can form a layer on the substrate. In addition to the Ar, other gases like nitrogen (N) can be used in order to produce nitridic coatings.

Sputtering can be used to deposit transparent conducting oxide layers such as indium tin oxide or aluminium-doped zinc oxide. While for the former a target consisting of ZnO and 2% Al is used, for the latter the target consists of 10% SnO_2 and 90% In_2O_3. However, sputtering can also be used for depositing metallic layers that are used as electrical front and/or back contacts.

14.2.2 Evaporation

Evaporation is a PVD technique that can be used for depositing a great variety of layers, such as metallic layers of Al, Cu, or Ag. We can distinguish between resistive evaporation and electron beam evaporation, as illustrated in Figure 14.2. In general, deposition of a metallic layer with evaporation is realized by heating the metallic source material above its melting point, which subsequently condensates on the cooler surface of the substrate.

For *resistive evaporation*, the metallic source is loaded in an open boat (crucible), which is heated resistively by applying a high current. For *electron beam evaporation*, the metallic source is loaded in a water-cooled crucible and is irradiated by an intense electron beam, which heats the source. The electron beam is emitted by a tungsten filament and directed by strong magnetic fields. In both methods, the metallic source evaporates and its atoms follow straight trajectories targeting the substrate.

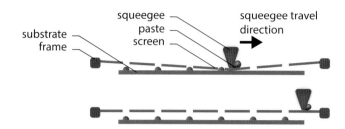

Figure 14.3: Schematic representation of the screen printing process. With kind permission from G. Papakonstantinou [89].

Evaporation is performed under a high vacuum, in which the pressure ranges from 10^{-3} to 10^{-6} Pa. This pressure range is sufficient to control the thickness and the oxidation level of the deposited layer [91]. Usually, the metallic sources are located at the bottom of the chamber, such that the metallic vapour flows upwards. If the substrate is mounted on a rotating holder, the homogeneity of the deposited layer can be improved.

14.3 Screen printing technology

Screen printing (SP) is another method to fabricate metal front and back contacts. Unlike evaporation, SP is a method in which the metallic source is liquid in the form of a viscous paste. SP is the most extensively used technique in industry [92]. It is a fast, well-established, reliable and cost-efficient method that has dominated the mass production of solar cells. In principle, all shapes and patterns can be realized with the SP technology [93].

As illustrated in Figure 14.3, screen printers consist of three basic structures: the frame, the screen and the squeegee. These structures are responsible for transferring the metallic paste onto the surface of the substrate in the shape of the grid pattern. The rigid frame serves as a supporting structure for the screen. The screen consists of an interwoven thin-wire mesh, which is fixed to the frame under high tension. It has openings in the shape of the grid pattern. Prior to the printing process, the substrate and the screen are not in intimate contact. During the printing process, the squeegee moves across the surface of the screen and sweeps the metallic paste. While the squeegee moves, it applies a force onto the screen and coerces the metallic paste to pass through its openings. As a result, the squeegee causes the screen to deflect and come locally into contact with the substrate at the position of the squeegee. At the same time, the screen peels off from the substrate behind the travelling squeegee, whereas the paste remains on the substrate. In this manner, the desired metallization pattern is reproduced on the surface of the substrate.

The printing step is followed by the *contact formation*. For conventional c-Si solar cells, contact formation with heavily doped silicon regions requires a high temperature firing process in a furnace at about 800 °C [94, 95]. For silicon heterojunction (SHJ) solar cells, which we discussed in Section 12.5.3, there are two distinct differences. First, SHJ solar cells are processed with PECVD at moderate temperatures to prevent degradation of thin a-Si:H layers. Secondly, the metallic paste is deposited onto a TCO layer [95]. Therefore, it is mandatory to use specific pastes that ensure good adhesion, high conductivity and good

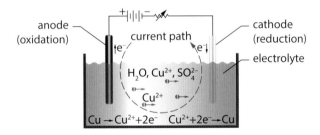

Figure 14.4: Schematic of the electroplating unit with a copper sacrificial anode immersed in CuSO$_4$ solution. (Adapted from [98].)

contact formation with the underlying TCO layer at temperatures below 200 °C.

14.4 Electroplating technology

14.4.1 Working principle

Electroplating is an electrodeposition process that allows dense, uniform, and adherent coatings, usually of metal or alloys, to be deposited upon a surface using an electric current [96]. The working principle of electroplating is based on the reduction-oxidation reaction, which is driven by an external DC source. As illustrated in Figure 14.4, the fundamental building unit of the electrodeposition process is an *electrolytic cell*, which consists of two electrodes, the cathode and the anode, immersed in a solution – the electrolyte – and connected to a DC power source. The substrate to be plated is the cathode. The anode completes the electrical circuit and may also fulfil a second function, which is to replenish the solution with metal that has been removed by being deposited on the cathode. Such an anode is called sacrificial (soluble); it is made of the material that is being deposited. Further, the anode can be permanent (insoluble); then it is typically made of platinum-coated titanium [97].

Upon the application of a voltage difference across the two electrodes in an electrolytic cell, a current flow is established, which causes a reaction of metal ions (M^{z+}) with electrons (e^-) to form a metal (M) at the cathode:

$$M^{z+} + ze^- \rightarrow M. \quad (14.1)$$

Electrons will flow from the anode through the external circuit back to the cathode. Thus, the anode becomes positively charged and the positive metal ions (cations) are repelled from the anode, migrating to the electrolyte. Simultaneously, the cathode becomes negatively charged and attracts the M^{z+} ions that are contained in the electrolyte. The M^{z+} ions are supplied either from the metal salt added to the solution or by the sacrificial anode. In this manner, the negatively-charged cathode reduces the cations by providing e^- and thus the metal atoms will be deposited onto the surface of the negatively charged substrate. Consequently, the anode will dissolve and a current will flow through the electrolyte by virtue of the movement of these cations. The metal mass m deposited on the surface of the substrate is proportional to the current I passing through the electrolytic cell and to the duration of the electroplating process t. In the ideal case, the metal mass is equal to

$$m = \frac{It A_w}{zF} \quad (14.2)$$

where A_w is the atomic weight, $F = 96485.3365$ As/mol is the Faraday constant and z is the number of electrons transferred per deposited atom.

14.4.2 Copper electroplating

While the metal contacts in silicon solar cells are made from aluminium or silver most of the time, there is a great interest in using copper. This interest is primarily driven by the need for improved performance, which is possible due to the high conductivity of Cu (5.8×10^7 Sm^{-1}), which in turn is only slightly lower than that of Ag (6.1×10^7 Sm^{-1}) and higher than that of Al (3.5×10^7 Sm^{-1}). Additionally, the replacement of the relatively expensive silver by the cheaper copper for the metallization of Si devices may lower their manufacturing costs [99, 100]. It is not possible to effectively deposit thick Cu layers on complex geometries with PVD or CVD techniques [101].

However, Cu electroplating (CE) is an inexpensive and straight forward process. Besides performance and cost-related reasons, CE can be carried out at ambient conditions of temperature and pressure, which, in combination with its ability to produce high quality layers [101], makes CE a very promising candidate for the fabrication of metal contacts in wafer-based solar cells.

Even though CE itself is a simple process, the implementation of Cu as a metallization material is challenging. This is primarily because Cu diffuses extremely fast into Si which leads to the creation of impurities and hence defects [102]. These defects act as traps for Shockley–Read–Hall recombination (see Section 7.3) and lead to a decrease of the minority-carrier lifetime and ultimately degrade the electrical characteristics of the Si devices [103]. Furthermore, studies on the impact of Cu contamination in Si have shown a substantial increase in the leakage current of *p-n* junctions [102]. The introduction of Cu as metallization material is possible by introducing thin-film diffusion barriers that prevent Cu from diffusing into Si. These diffusion barriers can also act as seed layers of high conductivity that facilitate CE. Several investigations have shown that titanium (Ti), nickel (Ni), tantalum (Ta) or alloys of these metals with nitrogen such as TaN or TiN are efficient Cu seed layers and Cu diffusion barriers [99, 103, 104]. They can be deposited with PVD and CVD methods. Their utilization for the fabrication of the front metal contacts with subsequent CE in c-Si and SHJ solar cells has improved their performance [100].

15
PV modules

In this chapter we will discuss the most important issues concerning *PV modules*. Before starting with the actual discussion, we have to introduce some important terms. Figure 15.1 (a) shows a crystalline silicon (c-Si) *solar cell*; these were discussed in Chapter 12. In Sections 15.6 we will only consider modules made from c-Si solar cells. In a *PV module* many solar cells are connected, as illustrated in Figure 15.1 (b). The names PV module and solar module are often used interchangeably. As illustrated in Figure 15.1 (c), a *solar panel* consists of several PV modules that are electrically connected and mounted on *one* supporting structure. Finally, a *PV array* consists of several solar panels. An example of such an array is shown in Fig. 15.1 (d). This array consists of two strings of two solar panels each, where *string* means that these panels are connected *in series*.

15.1 Series and parallel connections in PV modules

If we make a solar module out of an ensemble of solar cells, we can connect the solar cells in different ways. First, we can connect them in a *series connection* as shown in Figure 15.2 (a). In a series connection the voltages add up. For example, if the open circuit voltage of one cell is equal to 0.6 V, a string of three cells will produce an open circuit voltage of 1.8 V. For solar cells with a classical front metal grid, a series connection can be established by connecting the busbars at the front side with the back contact of the neighbouring cell, as illustrated in Figure 15.2 (b). For series-connected cells, the currents do not add up. In contrast, the current of the whole string is determined by the cell that delivers the smallest current. Hence, the total current in a string of solar cells is equal to the smallest current generated by one single solar cell.

Figure 15.2 (d) shows the *I-V* curve of solar cells connected in series. If we connect two solar cells in series, the voltages add up while the current stays the same. The resulting

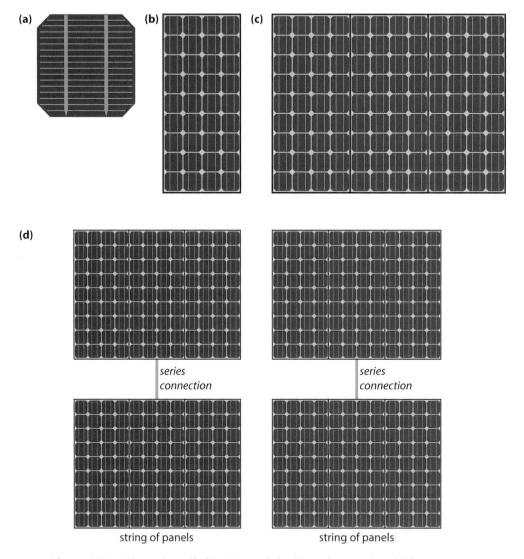

Figure 15.1: (a) A solar cell; (b) a PV module, (c) a solar panel; and (d) a PV array.

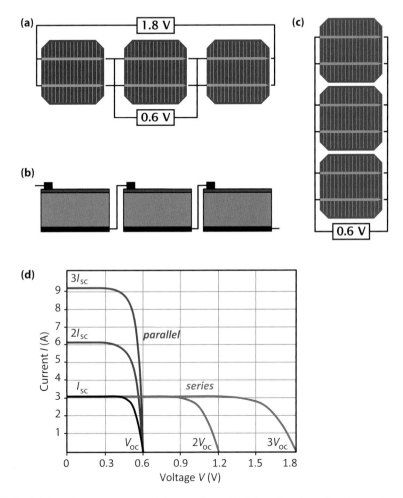

Figure 15.2: (a) A series connection of three solar cells; (b) realisation of such a series connection for cells with a classical front metal grid. (c) A parallel connection of three solar cells. (d) *I-V* curves of solar cells connected in series and parallel.

open circuit voltage is twice that of the single cell. If we connect three solar cells in series, the open circuit voltage becomes three times as large, whereas the current is still that of one solar cell.

Secondly, we can connect solar cells in *parallel* as illustrated in Figure 15.2 (c), which shows three solar cells connected in parallel. If cells are connected in parallel, the voltage is the same over all solar cells, while the currents of the solar cells add up. If we connect, for example, three cells with the same *I-V* characteristics in parallel, the current becomes three times as large, while the voltage is the same as for a single cell, as illustrated in Figure 15.2 (d).

Several strings of series-connected solar cells can be connected in parallel, which is sometimes done in PV modules for rural applications. In principle, several groups of parallel-connected cells can also be connected in series. However, this is not usually done in reality because when cells are connected in parallel, the currents become higher which increases resistivity losses in the cables.

The reader may have noticed that we used *I-V* curves, i.e. the *current-voltage* characteristics, in the previous paragraphs. This is different to Chapters 4–14, where we used *J-V* curves instead, i.e. the *current density–voltage* characteristics. The reason for this switch from *J* to *I* is that PV modules usually are characterized by short circuit and maximum power point currents instead of current densities. As the area of a module is a constant, the shapes of the *I-V* and *J-V* curves of a module are similar.

For a total module, therefore, the voltage and current output can be partially tuned via the arrangements of the solar cell connections. Figure 15.3 (a) shows a typical PV module that contains 36 solar cells connected in series. If a single junction solar cell has a short circuit current of 5 A, and an open circuit voltage of 0.6 V, the total module would have an output of $V_{oc} = 36 \times 0.6\,\text{V} = 21.6\,\text{V}$ and $I_{sc} = 5\,\text{A}$.

However, if two strings of 18 series-connected cells are connected in parallel, as illustrated in Figure 15.3 (b), the output of the module will be $V_{oc} = 18 \times 0.6\,\text{V} = 10.8\,\text{V}$ and $I_{sc} = 2 \times 5\,\text{A} = 10\,\text{A}$. In general, for the *I-V* characteristics of a module consisting of m identical cells in series and n identical strings in parallel, the voltage multiplies by a factor m while the current multiplies by a factor n. Modern PV modules often contain 60, 72 or even 96 solar cells that are usually all connected in series in order to minimize resistive losses and to enable high voltages that are required for an efficient operation of the inverter, which we will discuss in Section 19.2.

15.2 PV module parameters

As for solar cells, a set of parameters can be defined to characterize a PV module. The most common parameters are the *open circuit voltage* V_{oc}, the *short circuit current* I_{sc} and the *module fill factor* FF_M. On a module level, we have to distinguish between the *aperture area efficiency* and the *module efficiency*. The *aperture area*, also known as the *active area*, is defined as the area of the PV-active parts only. The total module area is given as the aperture area plus the dead area consisting of the interconnections and the edges of the module. Clearly, the aperture area efficiency is larger than the module efficiency.

Characterizing the efficiency and the fill factor of a PV module is less straightforward than measuring voltage and current. In an ideal world with perfectly matched solar cells

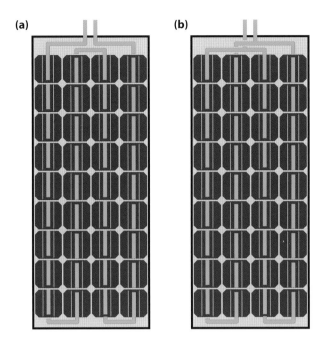

Figure 15.3: A PV module consisting of (a) a string of 36 solar cells connected in series; and (b) two strings each of 18 solar cells that are connected in parallel.

and no losses, one would expect that the efficiency and fill factor at both module and cell levels to be the same. This is not the case in real life. As mentioned above, the cells are connected to each other using interconnects that induce resistive losses. Further, there might be small mismatches between the interconnected cells. When $m \times n$ cells are interconnected, the cell with the lowest current in a string of m cells in series determines the module current.

The reason for mismatch between individual cells are inhomogeneities that occur during the production process. Hence, in practice PV modules perform a little worse than one would expect from ideally matched and interconnected solar cells. This loss in performance translates to a lower efficiency at module level. If the illumination across the module is not constant or if the module heats up non-uniformly, the module performance reduces even further. Often, differences between cell and module performance are mentioned in datasheets that are provided by the module manufacturers. For example, the datasheet of a Sanyo HIT-N240SE10 module gives a cell level efficiency of 21.6%, but a module level efficiency of only 19%.

15.3 Bypass diodes

PV modules have so-called *bypass diodes* integrated. These diodes are necessary, because in real-life conditions, PV modules can be *partially shaded*, as illustrated in Figure 15.4 (a). The shade can be from an object nearby, like a tree, a chimney or a neighbouring

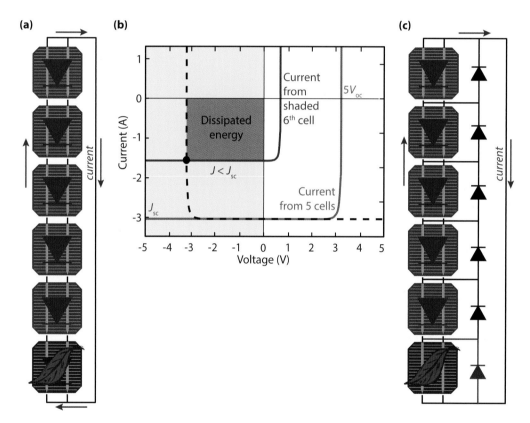

Figure 15.4: (a) A string of six (short circuited) solar cells of which one is partially shaded. (b) This has dramatic effects on the I-V curve of this string. (c) Bypass diodes can solve the problem of partial shading.

building. It also can be caused by a leaf that has fallen onto the module. Partial shading can have significant consequences for the output of the solar module. To understand this, we consider the situation in which one solar cell in the module is shaded. For simplicity, we assume that all six cells are connected in series. This means that the current generated in the shaded cell is significantly reduced. In a series connection the current is limited by the cell that generates the lowest current; this cell thus dictates the maximum current flowing through the module.

In Figure 15.4 (b) the theoretical *I-V* curve of the five unshaded solar cells and the shaded solar cell is shown. The five unshaded solar cells act like a reverse bias source on the shaded solar cell, which can be graphically represented by reflecting their I-V curve through the $V=0$ axis (see dashed line in Figure 15.4 (b)). Hence, the shaded solar cell is operated at the intersection of its *I-V* curve and the reflected curve. As this operating point is in its reverse-bias area, it does not generate energy, but starts to dissipate energy and heats up. The temperature can increase to such a critical level that the encapsulation material cracks, or other materials wear out. Further, high temperatures generally lead to a decrease of the PV output. In addition, a large reverse bias applied to the cell may induce *junction breakdown*, which can potentially damage the cell.

These problems occurring from partial shading can be prevented by including bypass diodes in the module, as illustrated in Figure 15.4 (c). As discussed in Chapter 8, a diode blocks the current when it is under negative voltage, but conducts a current when it is under positive voltage. If no cell is shaded, no current is flowing through the bypass diodes. However, if one cell is (partially) shaded, the bypass diode starts to pass current through because of the biasing from the other cells. As a result, current can flow around the shaded cell and the module can still produce the current equal to that of an unshaded single solar cell. In real PV modules, not every solar cell is equipped with a bypass diode, but groups of cells share one diode. For example, a module of 60 cells, connected in series forming one string, can contain three bypass diodes, where each diode is shared by a group of 20 cells.

15.4 Fabrication of PV modules

As we discussed in Section 15.5, a PV module must withstand various influences in order to survive a lifetime of 25 years or even longer. To ensure such a long lifetime, the PV module must be built of well chosen components. Figure 15.5 shows the typical components of a crystalline silicon PV module. For real PV modules, the layer stack may consist of different materials depending on the manufacturer. The major components are [105]:

- *Soda-lime* glass with a thickness of several millimetres, which provides mechanical stability while being transparent for the incident light. It is important that the glass has a low iron content as iron leads to absorption of light, which can lead to losses. Further, the glass must be *tempered* in order to increase its resistance to impacts.
- The solar cells are sandwiched in-between two layers of *encapsulants*. The most common material is *ethylene-vinyl-acetate* (EVA), which is a thermoplastic polymer (plastic). This means that it goes into shape when it is heated but that these changes are reversible.
- The *back layer* acts as a barrier against humidity and other stresses. Depending on the manufacturer, it can be another glass plate or a composite polymer sheet. A material

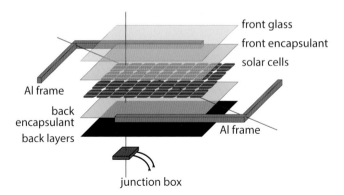

Figure 15.5: The components of a typical c-Si PV module.

combination that is often used is PVF-polyester-PVF, where PVF stands for *polyvinyl fluoride*, which is often known by its brand name *Tedlar®*. PVF has a low permeability for vapours and is very resistive against weathering. A typical polyester is *polyethylene terephthalate* (PET). Polyester isolates the module electrically and Tedlar protects polyester from wind and weather conditions.

- A *frame* usually made from aluminium is put around the whole module in order to enhance the mechanical stability.

- A *junction box* is usually placed at the back of the module. In it, the electrical connections to the solar cell are connected with the wires that are used to connect the module to the other components of the PV system.

One of the most important steps during module production is *laminating*, which we will briefly explain for the case where EVA is used as an encapsulant [105]. For lamination, the whole stack consisting of front glass, the encapsulants, the interconnected solar cells, and the back layer are brought together in a laminator, which is heated above the melting point of EVA, which is around 120 °C. This process is performed *in vacuo* in order to ensure that air, moisture and other gases are removed from within the module stack. After some minutes, when the EVA is molten, pressure is applied and the temperature is increased to about 150 °C. During the *curing* process, a curing agent, which is present in the EVA layer, starts to cross-link the EVA chains, leading to transverse bonds between the EVA molecules being formed. As a result, EVA has elastomeric, rubberlike properties.

The choice of the layers that light traverses before entering the solar cell is also very important from an optical point of view. If these layers have an increasing refractive index, they act as anti-reflective coatings and thus can enhance the amount of light that is incoupled in the solar cell and finally absorbed, which increases the current produced by the solar cell.

15.5 PV module lifetime testing

The typical lifetime of PV systems is about 25 years. During these years little maintenance should be required on the system components, especially the PV modules which are required to be maintenance-free. Manufacturers typically guarantee a power between 80% and 90% of the initial power after 25 years. During the lifetime of 25 years or more, PV modules are exposed to various external stresses from various sources [106]:

- *temperature* changes between night and day as well as between winter and summer;
- *mechanical stress* for example from wind, snow and hail;
- stress by agents transported via the *atmosphere* such as dust, sand, salty mist and other agents;
- *moisture* originating from rain, dew and frost;
- *humidity* originating from the atmosphere;
- *irradiance* consisting of direct and indirect irradiance from the Sun; mainly the highly-energetic UV radiation is challenging for many materials.

Modules are qualified and type-approved according to IEC 61215 [107] for modules based on crystalline silicon solar cells and IEC 61646 [108] for thin-film modules. The aim of both IEC 61215 and IEC 61646 is to judge, within reasonable time and costs, whether modules are suitable for long term operation, experiencing the stresses mentioned above. Since the modules cannot be tested during a period of many years, accelerated stress testing must be used. The required tests are [109]:

- *Thermal cycles* for studying whether thermal stress leads to broken interconnects, broken cells, electrical bond failure, or adhesion of the junction box.
- *Damp heat* testing to see whether the modules suffer from corrosion, delamination, loss of adhesion and elasticity of the encapsulant, or adhesion of the junction box.
- *Humidity freeze* testing in order to test delamination, or adhesion of the junction box.
- *UV testing*, because UV light can lead to delamination, loss of adhesion and elasticity of the encapsulant, or ground fault due to backsheet degradation. Mainly, UV light can lead to a discolouration of the encapsulant and back sheet, which means that they turn yellow. This can lead to losses in the amount of light that reaches the solar cells.
- *Static mechanical loads* in order to test whether strong winds or heavy snow loads lead to structural failures, broken glass, broken interconnect ribbons or broken cells.
- *Dynamic mechanical loads*, which can lead to broken glass, broken interconnect ribbons or broken cells.
- *Hot spot* testing in order to see whether hot spots due to shunts in cells or inadequate bypass diode protection are present.
- *Hail testing* to see whether the module can handle the mechanical stress induced by hail.
- *Bypass diode thermal testing* to study whether overheating of these diodes causes degradation of the encapsulant, backsheet or the junction box.

Meeting the requirements of IEC 61215/IEC 61646, however, is not a 25 year guarantee. The actual lifetime is not only determined by the module design and the module used, but also the climate (e.g. hot-dry or warm high-humidity) and the specific application (e.g. integrated in a roof or free-standing). Research is ongoing to determine which accelerated tests are required in order to guarantee a specific lifetime (e.g. 25 years) depending on the climate and the specific application.

In addition to IEC 61215/IEC 61646 qualification and type approvals safety qualifications according to IEC 61730 (Part 1 [110] and Part 2 [111]) are also required for PV modules used in grid-connected PV systems and in most other applications as well.

For special applications dedicated standards are available e.g. IEC 61701 (salt-mist corrosion testing [112]) for applications in a salty environments and IEC 62716 (ammonia testing [113]) for applications on or near stables.

Qualification tests are carried out by independent organizations like TÜV Rheinland in Germany.

15.6 Thin-film modules

Making thin-film modules is very different from making modules from c-Si solar cells. While for c-Si technology producing solar cells and PV modules are two distinct steps, in thin-film technology producing cells and modules cannot be separated from each other. To illustrate this we look at a PV module where the thin-films are deposited in *superstrate configuration* on glass, as illustrated in Figure 15.6. For making such a module, a transparent front contact, a stack of (photo)active layers that also contain one or more semiconductor junctions, and a metallic back contact are deposited onto each other. In industrial production, the glass plates onto which these layers are deposited can be very large, with sizes significantly exceeding 1×1 m^2.

Such a stack of layers deposited onto a large glass plate in principle forms one very large solar cell that will produce a very high current. Since all the current would have to be transported across the front and back contacts, which are very thin, resistive losses in the module are an even bigger problem than for c-Si modules. Therefore, the module is produced such that it consists of many very narrow cells of about 1 cm width and the length being equal to the module length. These cells then are connected in series across the width of the module. On the very left and right of the module, metallic busbars collect the current and conduct it to the bottom of the module where they are connected with external cables.

The series connection is established with *laser scribing*. In total, *three* laser scribes are required for separating two cells from each other and establishing a series connection between them. The first laser scribe, called P1, is performed after the transparent front contact is deposited, as shown in Figure 15.6 (a). The wavelength of the laser is such that the laser light is absorbed in the front contact and the material is evaporated, leaving a 'gap' in the front contact, as shown in Figure 15.6 (b). Then the photoactive layers are deposited onto the front contact and also fill the gaps. Then, the second laser scribe, called P2, is performed, as illustrated in Figure 15.6 (c). The laser wavelength has to be chosen such that it is not absorbed in the transparent front contact but in the absorber layer. For example, if the absorber consists of amorphous silicon, green laser light can be used. The

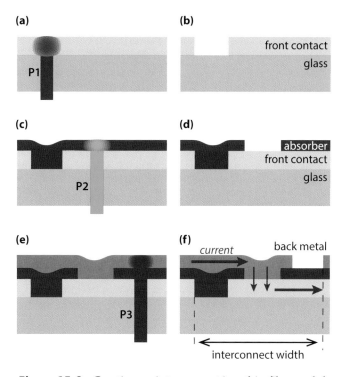

Figure 15.6: Creating an interconnect in a thin-film module.

P2 scribe leaves a gap in the absorber layer, as illustrated in Figure 15.6 (d). The next step is the deposition of the metallic back contact that also fills the P2 gap. Finally, the third laser scribe (P3) is performed as illustrated in Figure 15.6 (e). The wavelength for this scribe has to be chosen such that it is neither absorbed in the front contact nor in the absorber stack, so it maybe, for example, infrared. The P3 scribe shoots a gap into the back contact, as shown in Figure 15.6 (f).

To understand the action of the laser scribes, we take a look at Figure 15.6 (f): the P1 scribe filled with absorber material forms a barrier, since the absorber is orders of magnitudes less conductive than the transparent front contact. Similarly, the P3 scribes forms an insulating gap in the metallic back contact. However, the P2 scribe that is also filled with metal forms a highly conducting connection between the front and back contacts – this is where the actual series connection is performed.

For example, for making CIGS-based PV modules, first the molybdenum back contact is deposited on top of the glass substrate and the cell areas are defined by P1 laser scribes. Then the CIGS p layer and the CdS n layer are deposited including a P2 laser scribe step. Finally the intrinsic and n-doped zinc oxide is deposited, followed by a final P3 laser scribe step. Now the front TCO electrode is connected with the molybdenum back contact of the next solar cell.

The performance of such an interconnect established via laser scribes and hence the total module performance is determined by several things. First, the P2 scribe has to be highly conductive. This means that it has to be wide enough and that there must be no barrier at the interface between the front contact and the metal of the P2 scribe. Further, the P1 and P3 scribes must form good barriers to separate the cells from each other effectively. Thirdly, the region between the P1 and P3 scribes does not contribute to the current generated by the module. Therefore, the ratio between this width and the total cell width (including the scribes) has to be as small as possible in order to keep the active area of the module as large as possible. Another issue is the fact that the three laser scribes are performed in different steps of production and thus often with different machines. Further, the distance between the scribes might be different at the different processes when they are performed at different temperatures. Thus, aligning the glass plates in all the production steps is extremely important for manufacturing high quality thin-film modules.

The production steps and also the exact processing of the laser scribes is of course dependent on which thin-film technology is used and even on the manufacturer itself. However, the basic principles and the action behind these processes is valid in general.

One advantage of thin-film PV technology is that flexible modules can be made. For example, at the Dutch company HyET Solar, thin-film silicon layers are deposited onto a temporary aluminium substrate [114] that is etched away during the production process. This results in a low weight flexible substrate, which can be integrated for example into curved rooftop elements. A big advantage is that such very light modules can be installed on weak roofs that would be unable to support heavy PV panels. Further, if such flexible modules are directly integrated into roofing elements, installation costs could be reduced significantly. Often, installation costs are the largest contributor to the non-modular costs of a PV system. Currently, only thin-film silicon technologies have demonstrated flexible modules with reasonable efficiencies.

15.7 Some examples

Table 15.1 shows some parameters of PV modules using different PV technologies. While the first module is a typical c-Si module as discussed in Chapter 12, the other three are examples of different thin-film technologies discussed in Chapter 13.

Table 15.1: Specifications of different PV modules.

			SunPower X21-345	Avancis POWERMAX 140	Kaneka U-EA120	First Solar FS-392
Technology			c-Si	CIS	a-Si/nc-Si	CdTe
Rated power	P_{MPP}	(W_p)	345	140	114	92.5
Rated current	I_{MPP}	(A)	6.02	2.98	2.18	1.94
Rated voltage	V_{MPP}	(V)	57.3	47	55	47.7
Short circuit current	I_{SC}	(A)	6.39	3.31	2.6	2.11
Open circuit voltage	V_{OC}	(V)	68.2	61.5	71	60.5
Dimensions		(m)	1.56	1.6	1.2	1.2
		(m)	1.05	0.67	1.00	0.60
Max warranty on P_{MPP}		(yrs)	25	25	25	25

15.8 Concentrator photovoltaics (CPV)

In this section we discuss *concentrator photovoltaics*, which stand out a bit with respect to the other topics discussed in this chapter. However, as CPV systems finally have to be realized as CPV modules, this topic fits well into the chapter on PV modules.

As we have discussed in Section 10.4, the voltage of solar cells increases (logarithmically) with increasing irradiance. Hence, illuminating a solar cell with concentrated sunlight helps to increase the efficiency, as long as Auger recombination does not become dominant. This concept is especially interesting for high end solar cell technologies such as the III-V technology discussed in Section 13.2. Indeed, the current world record conversion efficiency for all solar cell technologies on lab scale is 46.0% for a four-junction III-V solar cell under 508-fold concentrated sunlight conditions [48]. At the end of 2014, about 330 MW_p of grid-connected CPV systems were installed [115].

15.8.1 Theoretical aspects of solar concentration

Before we look at a few examples of different solar concentrators, we briefly consider the theoretical limits of solar concentration. As we discussed in Chapter 5, the *radiance* is one of the most important radiometric properties. According to Eq. (5.9), it is given as

$$L_e = \frac{1}{\cos\theta} \frac{\partial^4 P}{\partial A \, \partial \Omega}, \tag{15.1}$$

hence it is the radiant flux (or power) per unit solid angle and per unit projected area. Note that L_e is a directional quantity: for a surface illuminated by the Sun, it has very high values for the solid angle range from which the Sun illuminates this surface, but very low for the other directions. Dividing by the refractive index n squared, we determine the so-called *basic radiance*,

$$L_e^* = \frac{L_e}{n^2}. \tag{15.2}$$

A very important property of the basic radiance is that it is conserved in ideal (non-absorbing) optical systems, it can only decrease but never increase in real optical systems. This can be explained with the fact that L_e^* is linked to the *étendue* of the optical system. The étendue of an optical system can never decrease. This interesting property of the étendue can be understood because of its link to the *second law of thermodynamics* [116].

From this theoretical concept it follows that an (optical) concentrator keeps the basic radiance unchanged or slightly reduces it due to attenuation of the light as it traverses the concentrator system. However, the solid angle, from which a solar cell under a concentrator receives sunlight, is increased. Therefore, we can estimate the maximal concentration of a solar concentrator system.

For this estimation we have to determine the irradiance that is received by the solar cell under no concentration. Using Eq.(5.10) we find

$$I_e(\text{1-Sun}) = \int_{\Omega_{\text{Sun}}} L_e^{\text{Sun}} \cos\theta \, d\Omega, \tag{15.3}$$

where $\Omega_{\text{Sun}} \approx 68.5\,\mu\text{sr}$ is the solid angle of the Sun as seen on Earth, as we derived in Section 5.5. For normal incidence onto the solar cell we have $\cos\theta = 1$ and hence

$$I_e(\text{1-Sun}) \approx L_e^{\text{Sun}} \Omega_{\text{Sun}}. \tag{15.4}$$

Under maximal concentration, the solar cell receives light from the whole hemisphere. Hence,

$$I_e(\text{max. conc.}) = \int_{2\pi} L_e^{\text{Sun}} \cos\theta \, d\Omega = L_e^{\text{Sun}} \int_{2\pi} \cos\theta \, d\Omega = L_e^{\text{Sun}} \times \pi. \quad (15.5)$$

To calculate the theoretically maximal concentration factor we divide Eq. (15.5) by Eq. (15.4) and find

$$C_{\text{solar}}^{\text{max}} = \frac{\pi}{\Omega_{\text{Sun}}} \approx \Omega_{\text{Sun}} \frac{\pi}{6.85 \times 10^{-5}} = 45,862. \quad (15.6)$$

Hence, it is theoretically not possible to concentrate sunlight by more than a factor of about 45,000.

15.8.2 Types of concentrator PV

One can distinguish between *high concentration PV* (HCPV) with typical concentration ratios between 300 and 1,000, and *low concentration PV* with concentration ratios below 100 [115] (LCPV). HCPV is applied for high end III-V solar cells and requires two-axis tracking of the Sun, while for LCPV, which is applied to c-Si and other solar cells, single-axis tracking is sufficient [115].

Different optical elements can be used to concentrate the sunlight [117]. First, there are *reflective* optical elements such as parabolic concentrators, compound parabolic concentrators (CPCs) comprising two parabolic surfaces, and V-through concentrators. *Refractive lenses* can also be used. As normal lenses would become very thick, most of the time Fresnel lenses are used, where the lens is collapsed back to zero thickness at a number of points [117]. A disadvantage of Fresnel lenses is that they show more optical losses than conventional lenses. To reach high concentration ratios exceeding 200, secondary optics is required, which also may increase the acceptance angle of the incident sunlight [117]. Most modern HCPV systems use Fresnel lenses as primary optics [115]. Currently, microconcentrators with diameters ranging from μm to a few mm are also being investigated [118]. More information on the different concepts and building blocks for concentrator systems can be found in [117].

15.8.3 Other aspects of concentrator PV

As we will discuss in Chapter 18, sunlight arrives at a PV module as direct and diffuse radiation. As for *concentrated solar power* (which we discuss in Chapter 22), only direct radiation can be concentrated and hence converted into electricity in a CPV system. Hence, CPV systems must be mounted at arid places with a lot of direct sunlight and a low percentage of diffuse radiation.

As mentioned above, CPV requires single- or even dual- tracking of the sunlight in order to concentrate it onto the solar cell independently of the position of the Sun. The tracking systems guarantee optimal light concentration on the small PV device during the entire day. Mounting the modules on a tracking system adds costs to the system which must be recovered by the increased efficiency of the CPV system.

Especially for HCPV, the irradiance received by the module is very high. Without cooling, this would lead to very high temperatures of several hundreds of °C and hence would

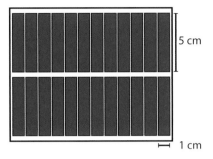

Figure 15.7

destroy the cell. Either *passive cooling* or *active cooling* can be applied. Passive cooling for example is done with heat sinks that are thermally connected to the solar cells and transport the heat to the surrounding air via convection. If the system is actively cooled via a cooling liquid that transports the heat away from the solar cell, this waste heat can, in principle, be reused [115]. The higher the temperature of the cooling liquid, the more valuable the waste heat. However, a high temperature of this liquid would also mean that the solar cell is operating at a high temperature which reduces its efficiency, as we will discuss in Section 20.3.

15.9 Exercises

15.1 We have identical crystalline silicon solar cells available, with a short circuit current of 4 A and an open circuit voltage of 0.6 V at STC. These cells are used to make a PV module with 54 cells connected in series.

  (a) What is the open circuit voltage of the module at STC? Assume that the interconnection losses are negligible.

  (b) What is the short circuit current of the module at STC?

  (c) Assume that the module does not have any bypass diodes. If the module is partially shaded and one of the cells is only able to produce 2 A, what will be the new short circuit current under these conditions?

15.2 Consider a small module based on interconnected tandem cells of hydrogenated amorphous and microcrystalline silicon as depicted in Figure 15.7. The V_{oc} of a single tandem cell is 1.35 V and the short circuit density J_{sc} is 12 mA/cm². Which combination of the external parameters can reflect the external parameters of the module depicted in the figure above?

  (a) $V_{oc}^{mod} = 16.2$ V, $I_{oc}^{mod} = 60$ mA.
  (b) $V_{oc}^{mod} = 16.2$ V, $I_{oc}^{mod} = 120$ mA.
  (c) $V_{oc}^{mod} = 32.4$ V, $I_{oc}^{mod} = 720$ mA.
  (d) $V_{oc}^{mod} = 32.4$ V, $I_{oc}^{mod} = 120$ mA.

15.3 Consider the following statements concerning interconnecting solar cells to make a solar module. Which of the statements is false?

(a) The open circuit voltage of a solar module with series-connected solar cells is the sum of the open circuit voltages of all the individual solar cells.

(b) The short circuit current of a solar module with series connected solar cells is the sum of the short circuit currents of all the individual solar cells.

(c) The use of bypass diodes in a solar module can help in reducing the undesired effects of a dysfunctional solar cell. The bypass diode is connected in parallel to a solar cell but with a polarity opposite to that of the solar cell.

15.4 Let us assume a monocrystalline silicon solar cell with the following specifications: $I_{sc} = 5$ A, $V_{oc} = 0.6$ V. 72 identical cells with these specifications are to be interconnected to create a PV module, with all the solar cells connected in series.

(a) What is the open circuit voltage of the PV module?

(b) What is the short circuit current of the PV module?

(c) Now assume that two solar cells are damaged, meaning that they completely stop generating power. Fortunately, bypass diodes are connected across these damaged solar cells. Assume that the bypass diodes are ideal (they have a zero voltage drop when conducting). What is the measured open circuit voltage (in V) of the above PV module with the two damaged solar cells?

15.5 Consider a PV module made up of 40 cells in series, of which 20 solar cells are shaded 30% of their area and 10 solar cells are shaded 60% of their area. Each individual cell has a V_{oc} of 0.68 V and I_{sc} of 6 A. Assume that each solar cell in the module has its own bypass diode with a constant forward voltage of 10 mV. Also assume that there is no leakage current from the bypass diodes and the module temperature is 25 °C.

(a) Draw I-V curves of this module. Indicate specific (x,y) intercepts and necessary numbers in the curves.

(b) Calculate the total power loss from the bypass diodes.

(c) For the same module and conditions as above, if the breakdown voltage of a shaded solar cell is 12 V, what will be the maximum number of solar cells that can be connected in series with one bypass diode?

16
Third generation concepts

The term *third generation photovoltaics* refers to all novel approaches that aim to overcome the Shockley–Queisser (SQ) single bandgap limit, preferably at a low cost. The SQ limit – discussed extensively in Chapter 10 – is a thermodynamic approach to estimate the maximum efficiency of a single-junction solar cell in dependence of the bandgap of its semiconductor material. For its derivation we assume that the AM1.5 spectrum is incident on a solar cell. Further, we do not allow the solar cell to increase in temperature but we force it to keep the ambient temperature of 300 K. This means that all the energy absorbed by the solar cell can escape the solar cell either by the generated current or by radiative recombination of charge carriers. Under these assumptions the efficiency limit is around 33% in the bandgap range from 1.0 eV up to 1.8 eV.

Now we are going to look at some very fundamental limitations of classical single-junction solar cells.

Firstly, in single-junction solar cells only one bandgap material is used. Hence, a large fraction of the energy of the most energetic photons is lost as heat as illustrated in Figure 10.7. The energy of highly energetic photons could be utilized better if absorbers with a high bandgap were used or if they could create more than one excited electron in the conduction band.

Secondly, most solar cell concepts are based on an incident irradiance level of 1 sun. However, higher irradiance means more current generation and also higher voltage levels, resulting in a higher overall efficiency.

Thirdly, the photons with energies below the bandgap are not absorbed. Hence, they do not result in charge-carrier excitation, as illustrated in Figure 10.7.

If any of these fundamental limitations could be solved, PV concepts with conversion efficiencies exceeding the Shockley–Queisser limit could be developed. In this chapter, we will discuss several third generation concepts: multi-junction solar cells, concentrator photovoltaics, spectral up and down conversion, multi-exciton generation, intermediate band solar cells and hot carrier solar cells. Note, that besides multi-junction and the concentrator approach, none of these concepts have yet resulted in high efficiency solar cells. The other concepts are still in the fundamental research phase and it will be some time before they are implemented in PV modules.

16.1 Multi-junction solar cells

The first limitation discussed in the list above can be tackled using *multi-junction* solar cells. Depending on the author, these solar cells are seen as part of the second or third generation, and indeed we already briefly discussed them in Chapter 13 on thin-film solar cells when looking at the III-V technology (Section 13.2) and thin-film silicon technology (Section 13.3). For completeness, we will summarize this technology here.

In multi-junction cells, several cell materials with different bandgaps are combined in order to maximize the amount of the sunlight that can be converted into electricity, as we already illustrated in Figure 10.9. To realize this, two or more cells are stacked onto each other. The top cell has the highest bandgap and will absorb and convert the short wavelength (blue) light. Light with wavelengths longer than the bandgap wavelength can traverse the top cell and be absorbed in the cells with lower bandgaps below. The bottom cell has the lowest bandgap and absorbs the long wavelength (red and near-infrared) light. In order to optimize the performance of multi-junction solar cells with two electrical terminals, matching the currents of all the subcells (current matching) is crucial. Multi-junction cells with more terminals do not have this restriction, but their production is more complicated.

In thin-film silicon tandem cells, an a-Si:H top cell is stacked onto an nc-Si:H bottom cell. In order to achieve current matching, the top cell is much thinner than the bottom cell. The cell can be further optimized by using an intermediate reflector between the top and the bottom cell in order to reflect the blue light back into the top cell while letting the red light pass to the bottom cell. The reported record efficiency of a-Si:H/nc-Si:H tandem cells is 12.3% and for a-Si:H/nc-Si:H/nc-Si:H it is 13.4% [77].

Multi-junction cells containing III-V semiconductors are at present the most efficient solar cells. The current world record efficiency is 46.0% for a four-junction GaInP/GaAs; GaInAsP/GaInAs cell that is used in a concentrator PV system [48]. Because of the high concentration factor of 508 suns the overall efficiency increases and hence we also tackle the second limitation mentioned in the list above. As a result the SQ limit can be exceeded by more than 10%.

16.2 Spectral conversion

Single-junction solar cells have the limitation that every photon can only generate one collected electron-hole pair at most. This limitation can be tackled by *spectral conversion*. The main idea of spectral conversion is to add an additional layer to the solar cell consisting of a material that can alter the incident spectrum. Materials that are considered for spectral conversion are organic dyes, quantum dots, lanthanide ions, and transition metal ion systems [119].

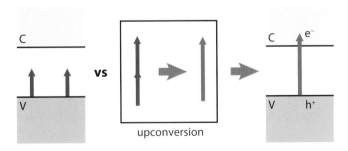

Figure 16.1: Illustrating the principle of spectral upconversion.

16.2.1 Spectral upconversion

Spectral upconversion works by combining the energies of two absorbed low-energy photons into one higher energy state, which can then radiate an upconverted photon into the solar cell active layer, as illustrated in Figure 16.1.

Two distinct material systems have been employed to produce photon upconversion for solar cells: firstly, *triplet-triplet annihilation upconverters* (TTA-UC), which use pairs of organic chromophores with rationally ordered energy levels [120], and secondly, *rare earth* upconverters, which make use of atomic transitions within lanthanide ions, from elements such as erbium and ytterbium [121].

TTA-UC systems, which are reliant on molecular absorbers, work better at shorter wavelengths between about 500 to 800 nm, while rare earths work in the region around 1,500 nm, where the ions are absorbed. Hence, the choice of the upconverter technology also depends on the bandgap of the solar cell to which it is connected. For both approaches, enhancement in solar cell efficiency has been demonstrated [122–124].

The upconverting layer should be placed at the back of the solar cell as low energy photons can pass through the solar cell absorber to this layer and can be converted there to high energy photons that are absorbed by the solar cell in a second pass [119]. In this way, upconverting layers with low efficiencies might also help to increase the photocurrent of the solar cell. Further, placing the upconverting layers at the back prevents them from being irradiated by high energy photons which might reduce their lifetime. Due to parasitic absorption it is not advised to put these layers at the front of the solar cell.

16.2.2 Spectral downconversion

The idea of *spectral downconversion* is to split one high energy photon into multiple lower energy photons [125], as illustrated in Figure 16.2. A high energy photon is absorbed at the front of the solar cell and converted into at least two photons with lower energies. If the energy of the initial photon is $E_{ph} > 2E_g$ and the energies of the resulting two photons are still larger than that of the bandgap of the absorber material, both photons can be absorbed and used for exciting charge carriers. As a result, a high energy photon, for example in the blue, can result in two excited electrons. In other words, the maximum theoretical EQE of 100% at the wavelength of the blue photon can be increased to 200%. If the photon had sufficient energy to be split into three photons with sufficient energy, a theoretical EQE of

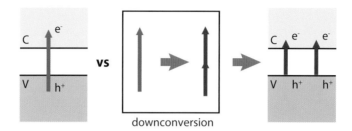

Figure 16.2: Illustrating the principle of spectral downconversion.

300% could be obtained. In contrast to upconversion, a downconverting layer has to be at the front of the solar cell, as highly energetic photons are always absorbed in the absorber layer. Hence, parasitic absorption might be a problem in this technology.

One possibility that is under investigation for realizing spectral downconversion is to use so-called *quantum dots* (QDs). These are small spherical nanoparticles made of semiconductor materials with typical diameters of a few nanometres, as illustrated in Figure 16.3 (a). These semiconductor particles still behave like a semiconductor material; however, due to so-called *quantum confinement* the bandgap of the semiconductor quantum dots can be larger than that of the same semiconductor in a bulk configuration. The bandgap of the QDs can be tuned by varying their size. The smaller the particles, the larger the bandgap. This enables interesting opportunities for bandgap engineering, such as multi-junction solar cells based on junctions with different QDs of different sizes in every junction.

If QDs were to be used for downconversion, an ensemble of nanoparticles would be embedded in a host material, where the particles are in very close proximity to one another. Figure 16.3 (b) shows the electronic bandgap diagram of two nanoparticles. Now, a high energy photon is absorbed by one QD and hence one electron is excited into the conduction band of the particle. In contrast to a bulk semiconductor, the excess energy of the photon is not necessarily lost as heat; but it can be transferred as a *quantized energy package* to a neighbouring quantum dot. Here a second electron is excited into the conduction band of the second quantum dot. As a result, two electron-hole pairs have been created out of one photon. If non-radiative recombination mechanisms like Auger recombination and SRH recombination are sufficiently suppressed, the electron-hole pairs in both quantum dots can radiatively recombine such that each of the two QDs emits one reddish photon. In summary, one incident bluish photon is converted into two reddish photons, which can be absorbed by a PV material.

Figure 16.4 shows some experimental results by Jurbergs et al. on downconversion based on silicon quantum dots in a narrow spectral range [126]. The horizontal axis represents the photon emission wavelength. At around 790 nm a downconversion efficiency of 60% is achieved. The EQE exceeds 100% in the blue region from 3.1 up to 3.4 eV. The major challenge is have QD layers with a spectral response exceeding 100% at lower photon energies, because the solar spectrum contains far more photons in that spectral range. In practice, only small enhancements in efficiency due to up/downconverters have been reported [127].

16. Third generation concepts

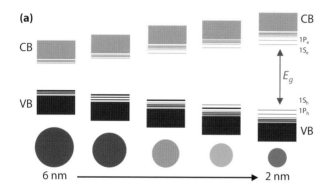

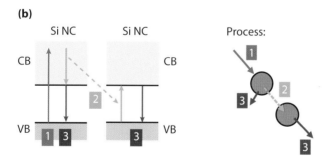

Figure 16.3: Illustrating (a) the effect of the quantum dot (QD) size on the bandgap; and (b) spectral downconversion with QDs.

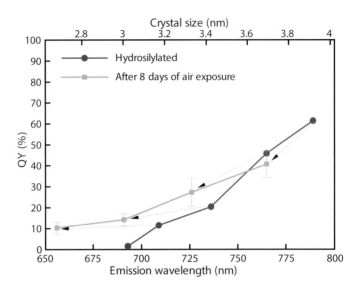

Figure 16.4: Downconversion at Si quantum dots (reprinted with permission from D. Jurbergs, E. Rogojina, L. Mangolini, and U. Kortshagen, Applied Physics Letters, vol. 88, 233116. Copyright (2006), AIP Publishing LLC) [126].

16.3 Multi-exciton generation

Another approach to enhance the charge-carrier excitation by a single energetic photon is called *multiple exciton generation* (MEG), where more than one electron-hole pair is generated from high energy photons. In contrast to spectral downconversion, here, two or more excitons are generated in the MEG layer, which then are transported to the PV-active layers. It is important to note that they are not converted back to lower energy photons.

Like downconversion, MEG can be realized with quantum dots, as illustrated in Figure 16.5. Again, in one particle an electron is excited into the conduction band and the excess energy is transferred to a neighbouring QD, where a second electron is excited into the conduction band of the second. However, here the charge carriers in the two electron-hole pairs are separated before they can recombine such that one incident photons results in more than one generated electron [129]. Just as for spectral downconversion, quantum efficiencies exceeding 100% are theoretically possible when one incident photon creates statistically more than one charge carrier. Figure 16.6 shows that EQEs exceeding 100% can be achieved as demonstrated by Semonin et al. [128]. Here, the absorber layer consists of PbSe quantum dots.

Another way of performing downconversion is to utilize *singlet fission*, which for example can be done using tetracene, a polycyclic aromatic hydrocarbon. In fact, singlet fission is the equivalent to MEG in organic materials, where a high energy singlet excited state fissions into two low energy triplet states on neighbouring molecules [130, 131].

16. Third generation concepts

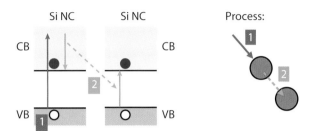

Figure 16.5: Illustrating multi-exciton generation with quantum dots (adapted by permission from Macmillan Publishers Ltd: D. Timmerman I. Izeddin, P. Stallinga, I. N. Yassievich, and T. Gregorkiewicz, Nature Photonics, vol. 2, pp. 105–109, copyright (2008)).

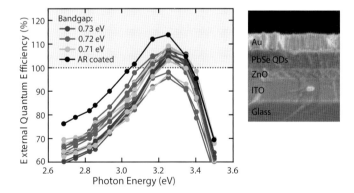

Figure 16.6: Multi-exciton generation using PbSe quantum dots with EQEs exceeding 100% [128]. From O. E. Semonin, J. M. Luther, S. Choi, H.-Y. Chen, J. Gao, A. J. Nozik, and M. C. Beard, Science 334, 1530 (2011). Reprinted with permission from AAAS.

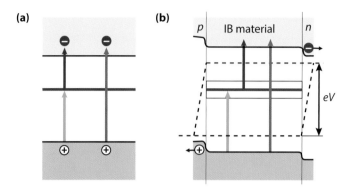

Figure 16.7: Band diagram with (a) the intermediate band depicted; and (b) the quasi-Fermi levels exceeding the energy of the low energy photons.

16.4 Intermediate band solar cells

The concept of *intermediate band solar cells* (IB) tries to tackle the problem photons with energies below the bandgap that cannot be utilized for current generation. As shown in Figure 16.7 (a), in intermediate band cells energy levels are created artificially in the bandgap of the absorber material [132]. As in conventional single-junction solar cells, photons with sufficient energy can excite an electron from the valence band into the conduction band. However, in difference to conventional semiconductors photons with energies below the bandgap can excite an electron from the valence band in to the intermediate band. A second low energy photon is required to excite the electron from the intermediate band into the conduction band. As illustrated in Figure 16.7 (b), the absorption of two photons with energies smaller than the bandgap energy can therefore result in quasi-Fermi level splitting exceeding the energy of each of these photons.

Various studies have been performed on how intermediate band cells can be realized, as for example summarized in [132]. In such solar cells, a layer with the intermediate states is placed in-between p and n layers. For example quantum dots can be used to realize the intermediate states. Further, various bulk materials have been studied for realizing the intermediate states. One major problem of experimental IB cells is that the voltages are lower than those of reference cells without IB structures.

16.5 Hot carrier solar cells

The idea of *hot carrier solar cells* is to reduce the energy losses due to relaxation and hence thermalization. As illustrated in Figure 16.8, this should be achieved by collecting electron-hole pairs of high energy photons just after light excitation and before they have a chance to relax back to the edges of the electronic bands. In the figure, the population of the charge carrier levels reflects the situation just after the excitation of a photon by the absorption. This distribution is not in thermal equilibrium as many electrons are excited into a position further up the conduction band and the holes are excited down to lower levels in the valence band. These charge carriers are called *hot* electrons and holes [26, 133].

16. Third generation concepts

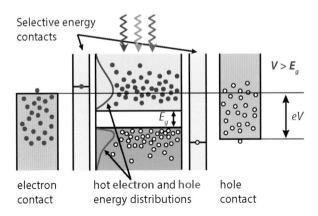

Figure 16.8: The working principle of a hot carrier solar cell.

It only takes a few picoseconds (10^{-12} s) for the hot charge carriers to relax back to the edges of the electronic bands. The idea of hot carrier cells is to collect the charge carriers as long as they are still *hot*. Hence, an energy larger than the bandgap energy can be utilized per excited charge carrier and the average bandgap utilization would exceed the bandgap.

The fundamental challenge is to collect the hot carriers before they relax back to the edges of the electronic bands. Such a concept would require *selective contacts*, which only select electrons above a particular energy level in the conduction band and holes below a certain energy level in the valence band. At the moment the main challenge is to increase the lifetime of the hot charge carriers, such that they have time to move from the absorber material to the selective contacts [134].

16.6 Exercises

16.1 Figure 16.9 shows the EQE of a triple-junction cell with junctions A, B and C under short circuited ($V = 0$ V) condition.

(a) What is the bandgap of the absorber layer of junction A?

(b) What is the bandgap of the absorber layer of junction B?

(c) What is the bandgap of the absorber layer of junction C?

(d) Which of the following statements is true?

  i. Junction C acts as the top cell, junction B as the middle cell, and junction A as the bottom cell.

  ii. Junction B acts as the top cell, junction C as the middle cell, and junction A as the bottom cell.

  iii. Junction A acts as the top cell, junction B as the middle cell, and junction C as the bottom cell.

(e) Each junction is illuminated under standard test conditions. Given the photon fluxes

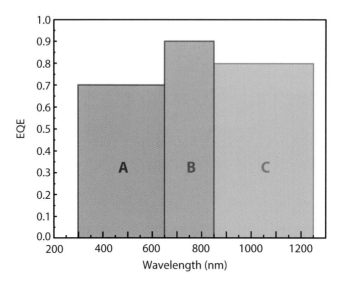

Figure 16.9

below, calculate the short circuit current density of each (separate) junction.

$$\Phi_{ph} = 9.3 \times 10^{20} \text{ m}^{-2}\text{s}^{-1} \text{ for 300 nm} < \lambda < 650 \text{ nm},$$
$$\Phi_{ph} = 8.4 \times 10^{20} \text{ m}^{-2}\text{s}^{-1} \text{ for 650 nm} < \lambda < 850 \text{ nm},$$
$$\Phi_{ph} = 1.4 \times 10^{20} \text{ m}^{-2}\text{s}^{-1} \text{ for 850 nm} < \lambda < 1250 \text{ nm}.$$

(f) The V_{oc} of each junction can be roughly estimated by the equation

$$V_{oc} = \frac{E_g(\text{eV})}{2q},$$

where the bandgap energy E_g is given in eV. Assuming a fill factor of 75%, calculate the efficiency of the triple-junction solar cell.

16.2 Which of the following mechanisms can be used to split a high energy photon into two (or more) lower energy photons?

(a) Multi-junction.

(b) Upconversion.

(c) Downconversion.

(d) Intermediate band.

16.3 Imagine you want to fabricate a triple-junction solar cell that consists of three quantum dot-based solar cells (see Figure). Each cell consists of quantum dots of a certain diameter: 4 nm, 6 nm and 10 nm. In which order (from top to bottom) would you put these cells in order to absorb the largest part of the solar spectrum?

16.4 Figure 16.10 shows a simplified AM1.5 solar spectrum with a total irradiance of 1000 W/m^2. The spectrum is divided into three spectral ranges,

16. Third generation concepts

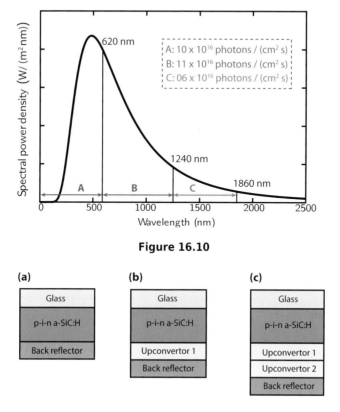

Figure 16.10

Figure 16.11

A 0 nm < λ < 620 nm,
B 620 nm < λ < 1240 nm,
C 1240 nm < λ < 1860 nm.

In the figure, also the photon flux in each spectral range is shown.
Hydrogenated silicon carbide (a-SiC:H) is a type of amorphous semiconductor material that has been recently studied for PV applications. This material has a relatively large bandgap of 2.0 eV. Imagine we integrate this material in a single junction *p-i-n* solar cell as shown in Fig. 16.11 (a).

(a) In which spectral range does this solar cell convert light into charge carriers?

(b) What is the J_{sc} of the solar cell if only 65% of the absorbed photons result in a current?

(c) The V_{oc} of each junction can be roughly estimated by the equation

$$V_{oc} = \frac{E_g(\text{eV})}{2q},$$

where the bandgap energy E_g is given in eV. Assuming a fill factor of 80%, calculate the efficiency of the solar cell.

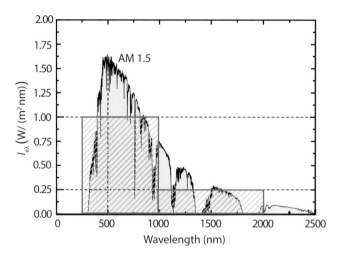

Figure 16.12: AM1.5 spectrum and its simplified representation in blue.

(d) An upconverter is a material that can convert two low energy photons into a higher energy photon. Placing an upconverter in our solar cell can help to reduce the spectral mismatch, since it can convert some photons with energy lower than 2 eV, which are not absorbed by the a-SiC:H cell, into photons with energy higher than 2 eV. Figure 16.11 (b) depicts this possibility.

In the upconverter 1, two photons are converted into one photon with 100% conversion efficiency. If all photons with energy above that of the bandgap of a-SiC:H are absorbed in the a-SiC:H layer, in which spectral range from Figure 16.10 can the photons be upconverted so that they contribute to the current in the cell as well? A, B, or C?

(e) In that case what would be the short-circuit current density and the efficiency of the solar cell illustrated in Fig. 16.11 (b)? Assume again that 65% of the absorbed photons result in a current.

(f) Now imagine that in upconverter 2, three photons are converted into one photon with 100% conversion efficiency, as illustrated in Figure 16.11 (c). If all photons with energy above that of the bandgap of a-SiC:H are absorbed in the p-i-n cell, and converter 1 absorbs only the photons in the spectral range determined above, in which spectral part can the photons be upconverted by converter 2 so that they contribute to the current in the cell as well? A, B, or C?

(g) In that case what would be the short circuit current density and the efficiency of the solar cell illustrated in Figure 16.11 (c)? Assume that 65% of the absorbed photons result in a current.

16.5 Figure 16.12 shows the AM1.5 solar spectrum illustrated by the yellow region. A rough approximation of the AM1.5 solar spectrum is represented by the blue region. The spectral irradiance of this region is divided in two spectral ranges,

$$I_{e\lambda} = 1.00 \times 10^9 \text{ Wm}^{-2}\text{m}^{-1} \text{ for } 250 \text{ nm} < \lambda < 1,000 \text{ nm},$$
$$I_{e\lambda} = 0.25 \times 10^9 \text{ Wm}^{-2}\text{m}^{-1} \text{ for } 1,000 \text{ nm} < \lambda < 2,000 \text{ nm}.$$

(a) Demonstrate that the irradiance of the above simplified spectrum is equal to 1,000 W/m².

(b) Not all energy of an absorbed photon (E) is used to excite the electron from the valence to the conduction band (E_g). $E - E_g$ is lost as heat. Consider an ideal solar cell, i.e. $EQE = 1$ for all energies above the bandgap. For simplicity, we neglect possible reflection of light from the window layer(s). Show that the irradiance lost as heat can be described by:

$$I_{e,\,\text{heat-loss}} = I_e - \int I_{e\lambda}\left(\frac{\lambda}{\lambda_g}\right) d\lambda.$$

(c) Consider a single-junction solar cell based on a semiconductor material with a band gap of 0.67 eV. Again, consider EQE = 1 for energies above the bandgap. How much of the irradiance is lost as heat (expressed in W/m²)?

(d) The single-junction solar cell in the above question is now integrated as bottom cell in a double-junction solar cell. The bandgap of the absorber material of the top cell is 1.24 eV. Again, consider EQE = 1 for energies above the bandgap. How much of the irradiance is lost as heat (expressed in W/m²) in this double junction solar cell?

(e) A triple junction has been processed using various III-V alloys. The EQE of the individual cells can be considered as a block function with EQE = 0.8. The bandgap of the bottom cell is 0.67 eV. The short circuit current density of the triple junction is J_{sc} = 15 mA.cm⁻². What are the smallest values for both bandgaps expressed in eV of the middle and top cells?

16.6 The spectral irradiance of the AM1.5 solar spectrum is shown in Figure 16.13 (a). A rough approximation of the AM1.5 spectrum is represented by the blue region, which will be used as the input solar spectrum in this particular exercise. The spectral irradiance of the simplified spectrum is given by:

$$I_{e\lambda} = 0.50 \times 10^9 \text{ Wm}^{-2}\text{m}^{-1} \text{ for } 250 \text{ nm} < \lambda < 2,250 \text{ nm}.$$

The band diagram of an intermediate band solar cell is depicted in Figure 16.13 (b). The bandgap is 1.5 eV, the intermediate band is 1.0 eV above the valence band and 0.5 eV below the conduction band. Assume that every photon above the bandgap is absorbed and creates charge carriers, i.e. EQE = 100 %. All photons between 1.0 eV and 1.5 eV excite an electron from the valence to the intermediate band, whereas all photons between 0.5 eV and 1.0 eV excite an electron from the intermediate band to the conduction band. Reflection losses can be neglected. For questions (a) to (c), *ignore* the intermediate band depicted in Figure 16.13 (b).

(a) Plot the photon utilization efficiency (PUE) of the solar cell.

(b) What amount of irradiance expressed in Wm⁻² is used to excite an electron from the valence to the conduction band under the condition as described in (a)?

(c) How much photon flux is absorbed and what is the short circuit current density, J_{sc}?

For questions (d) to (h), also *consider* the intermediate band depicted in Figure 16.13 (b).

(d) What is the photon flux absorbed in the spectral range of 1.0 eV up to 1.5 eV?

(e) What is the photon flux absorbed in the spectral range of 0.5 eV up to 1.0 eV?

(f) Plot the PUE of the intermediate band solar cell.

(g) What is the total short circuit current density, J_{sc} of the intermediate band solar cell?

(h) What amount of irradiance expressed in Wm⁻² is used to excite an electron from the valence to the conduction band in the intermediate band solar cell?

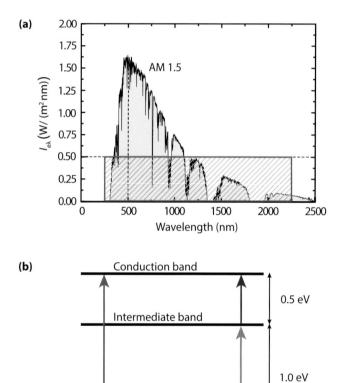

Figure 16.13: (a) AM1.5 spectrum and its simplified representation in blue; and (b) band diagram of an intermediate band solar cell.

Part IV

PV systems

17
Introduction to PV systems

17.1 Introduction

After discussing the fundamental scientific theories required for solar cells in Part II and taking a look at modern PV technology in Part III, we will now use the knowledge gained to discuss complete PV systems. Besides PV modules that were covered in Chapter 15, a PV system contains many different components. For successfully planning a PV system it is crucial to understand the function of the different components and to know their major specifications. Further, it is important to know the effect of the location on the (expected) performance of a PV system.

17.2 Types of PV systems

PV systems can be small and very simple, consisting of just a PV module and load, as in the direct powering of a water pump motor which only needs to operate when the Sun shines. On the other hand, PV systems can also be built as large power plants with a peak power of several MW; these are connected to the electricity grid. Many systems are placed on residential homes. When a whole house needs to be powered and is not connected to the electricity grid, the PV system must be operational day and night. It may also have to feed both AC and DC loads, have reserve power, and may even include a backup generator. Depending on the system configuration, we can distinguish three main types of PV systems: stand-alone, grid-connected, and hybrid. The basic PV system principles and elements remain the same. Systems are adapted to meet particular requirements by varying the type and quantity of the basic elements. A modular system design allows easy expansion when power demands change.

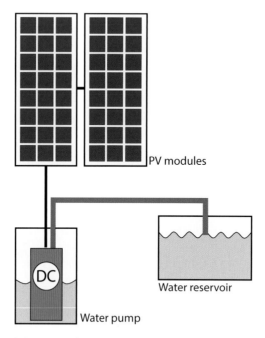

Figure 17.1: Schematic of a simple DC PV system to power a water pump.

17.2.1 Stand-alone systems

Stand-alone systems, which are also called *off-grid* PV systems, rely on solar power only. These systems can consist of the PV modules and a load only or they can include batteries for energy storage. When using batteries charge controllers are included, which disconnect the batteries from the PV modules when they are fully charged, and may disconnect the load to prevent the batteries from being discharged below a certain limit. The batteries must have enough capacity to store the energy produced during the day to be used at night and during periods of poor weather. Figure 17.1 shows a schematic representation of a simple DC PV system without a battery. In Figure 17.2 a complex PV system with both DC and AC loads is shown.

17.2.2 Grid-connected systems

Grid-connected PV systems have become increasingly popular for applications in the built environment. As illustrated in Figure 17.3, they are connected to the grid via inverters, which convert the DC power into AC electricity. In small systems such as those installed in residential homes, the inverter is connected to the distribution board, from where the PV-generated power is transferred into the electricity grid or to AC appliances in the house. In principle, these systems do not require batteries, since they are connected to the grid, which acts as a buffer into which an oversupply of PV electricity is transported. The grid also supplies the house with electricity in times of insufficient PV power generation. However, more and more grid-connected systems also contain batteries in order to increase *self*

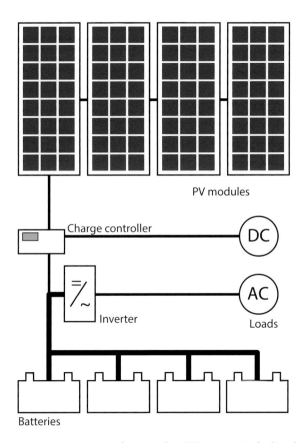

Figure 17.2: Schematic representation of a complex PV system including batteries, power conditioners, and both DC and AC loads.

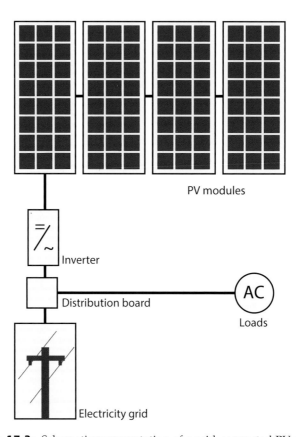

Figure 17.3: Schematic representation of a grid-connected PV system.

Figure 17.4: The 25.7 MW$_p$ Lauingen Energy Park in Bavarian Swabia, Germany [136].

consumption, i.e. the amount of PV-generated electricity that is consumed by the household [135].

Large PV fields act as power stations from which all the PV-generated electricity is directly transported to the electricity grid. They can reach peak powers of up to several hundreds of MW$_p$. Figure 17.4 shows a 25.7 MW$_p$ system installed in Germany.

17.2.3 Hybrid systems

Hybrid systems combine PV modules with a complementary method of electricity generation such as a diesel, gas or wind generator. A schematic representation of a hybrid system is shown in Figure 17.5. In order to optimize the different methods of electricity generation, hybrid systems typically require more sophisticated controls than stand-alone or grid-connected PV systems. For example, in the case of a PV/diesel system, the diesel engine must be started when the battery reaches a given discharge level, and stopped when the battery reaches an adequate charging state. The backup generator can be used to recharge batteries only or to supply the load as well.

17.3 Components of a PV system

Due to the limited size of the solar cell it only delivers a limited amount of power under fixed current voltage conditions that are not practical for most applications. In order to use solar electricity for practical devices which require a particular voltage and/or current for their operation, a number of solar cells have to be connected together to form a *solar module*, also called a *PV module*. For large-scale generation of solar electricity, solar panels are connected together into a *PV array*.

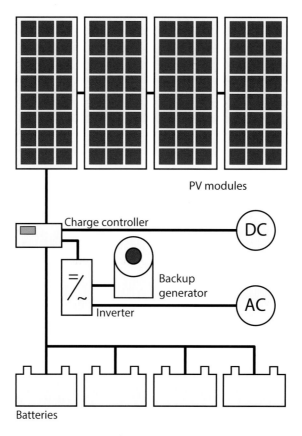

Figure 17.5: Schematic representation of a hybrid PV system that has a diesel generator as an alternative electricity source.

Although the solar panels are the heart of a *PV system*, many other components are required for a working system, as we discussed very briefly above. Together, these components are called the *balance of system* (BOS). Which components are required depends on whether the system is connected to the electricity grid or whether it is designed as a stand-alone system. The most important components belonging to the BOS are:

- A *mounting structure* in order to fix the modules and direct them towards the Sun.
- *Energy storage* as a vital part of stand-alone systems, because it assures that the system can deliver electricity during the night and in periods of bad weather. Usually, *batteries* are used as energy storage units.
- *DC-DC converters* in order to convert the module output, which will have a variable voltage depending on the time of the day and weather conditions, to a compatible output voltage that can be used as input for an inverter in a grid-connected system.
- *Inverters* that are used in grid-connected systems to convert the DC electricity originating from the PV modules into AC electricity that can be fed into the electricity grid. Many inverters have a DC-DC converter included to convert the variable voltage of the PV array to a constant voltage that is the input for the actual DC-AC converter. Also stand-alone systems may have an inverter that is connected to the batteries. The design of such an inverter differs considerably from that for a grid-connected system.
- *Charge controllers* that are used in stand-alone systems to control charging and often also discharging of the battery. They prevent the batteries from being overcharged and also from being discharged via the PV array during night. High end charge controllers also contain DC-DC converters together with a maximum power point tracker in order to make the PV voltage and current independent from the battery voltage and current.
- *Cables* that are used to connect the different components of the PV system with each other and to the electrical load. It is important to choose cables of sufficient thickness in order to minimize resistive losses.

Though not part of the PV system itself, the *electric load*, i.e. all the connected electric appliances, has to be taken into account during the planning phase. Further, it has to be considered whether the loads are AC or DC.

The different components of a PV system are schematically presented in Figure 17.6 and will be discussed in detail in Chapter 19.

17.4 Exercise

17.1 Which component is *not* part of a purely grid-connected PV system: PV panel, inverter, battery, distribution panel?

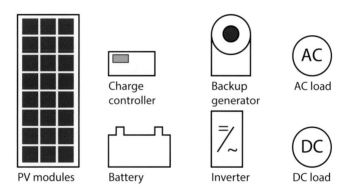

Figure 17.6: Different components of a PV system.

18
Location issues

In Chapter 5 we discussed solar radiation on Earth. There, we introduced the AM1.5 spectrum, which is used to evaluate the performance of solar cells and modules in laboratories and industry. The AM1.5 spectrum represents the solar irradiance if the centre of the solar disc is at an angle of 41.8° above the horizon.

Of course the Sun is not always at this position; the position is dependent on the time of the day and the year, and also on the location on Earth. In this chapter we will discuss how to calculate the position of the Sun at every location on Earth at an arbitrary time and date. Furthermore we will discuss scattering of sunlight when it traverses the atmosphere and how this influences the direct and diffuse spectra. We will also discuss the influence of the mounting angle and position of a PV module on the irradiance at the module.

18.1 The position of the Sun

When planning a PV system it is crucial to know the position of the Sun in the sky at the location of the solar system at a given time. In this section we explain how this position can be calculated.

Since celestial objects like the Sun, the Moon and the stars are very far away from the Earth it is convenient to describe their motion projected on a sphere with arbitrary radius and concentric to the Earth. This sphere is called the *celestial sphere*. The position of every celestial object thus can be parameterized by two angles. For photovoltaic applications it is most convenient to use the *horizontal coordinate system*, where the horizon of the observer constitutes the *fundamental plane*. In this coordinate system, the position of the Sun is expressed by two angles that are illustrated in Figure 18.1. The *altitude a* is the angular elevation of the centre of the solar disc above the horizontal plane. Its angular range is $a \in [-90°, 90°]$, where negative angles correspond to the object being below the hori-

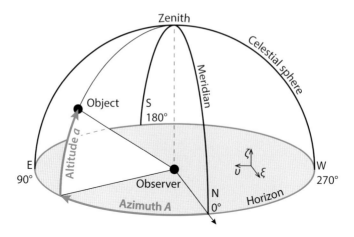

Figure 18.1: Illustrating the definition of the *altitude a* and the *azimuth A* in the horizontal coordinate system. Note that North is at the bottom of the figure.

zon and thus not visible. The *azimuth A* is the angle between the line of sight projected on the horizontal plane and due North. It is counted eastward, such that $A = 0°, 90°, 180°, 270°$ correspond to due *North, East, South* and *West*, respectively. Its angular range is $A \in [0°, 360°]$. In a different convention also used by the PV community, due South corresponds to $0°$ and is counted westward, the angles then are between $-180°$ and $180°$. In this book we will use the convention where $A = 0°$ corresponds to due North. Figure 18.1 also shows the *meridian*, which is a great circle on the celestial sphere passing through the celestial North and South poles as well as the zenith.

Instead of using the spherical coordinates a and A, we could also use Cartesian coordinates, that we here call ξ (xi), υ (upsilon) and ζ (zeta) and that are also depicted in Figure 18.1. The principal direction is parallel to ξ. The Cartesian coordinates are connected to the spherical coordinates via

$$\begin{pmatrix} \xi \\ \upsilon \\ \zeta \end{pmatrix} = \begin{pmatrix} \cos a \cos A \\ \cos a \sin A \\ \sin a \end{pmatrix}. \tag{18.1}$$

Note that on the celestial sphere $\xi^2 + \upsilon^2 + \zeta^2 = 1$ for all points.

In the horizontal coordinate system, the position of the Sun is given by the *solar altitude* a_S and the *solar azimuth* A_S. Their derivation is rather complex; it is shown in Appendix E, which also contains an example and a discussion on the *equation of time*.

Now we take a look at the solar paths throughout the year at four locations, shown in Figures 18.2–18.5: Delft, the Netherlands ($\phi_0 = 52.01°$ N), the North Cape, Norway ($\phi_0 = 71.17°$ N), Cali, Colombia ($\phi_0 = 3.42°$ N), and Sydney, Australia ($\phi_0 = 33.86°$ S). Note that all the times are given in the apparent solar time (AST, see Section E.2). While in Delft and on the North Cape, the Sun at noon is always south of the zenith, in Sydney it is always north. In Cali, close to the Equator, the Sun is either south or nNorth, depending on the time of the year. Since the North Cape is north of the Arctic Circle, the Sun does not set around 21 June. This phenomenon is called the *midnight Sun*. On the other hand, the Sun always stays below the horizon around 21 December–this is called the *polar night*.

18. Location issues

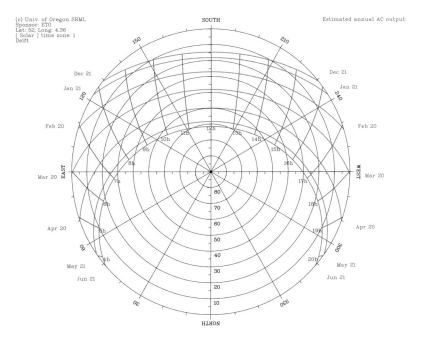

Figure 18.2: The Sun path in apparent solar time in Delft, the Netherlands ($\phi_0 = 52.01°$ N). The Sun path was calculated with the Sun path chart program by the *Solar Radiation Monitoring Lab.* of the *University of Oregon* (with kind permission from F. Vignola, University of Oregon) [137].

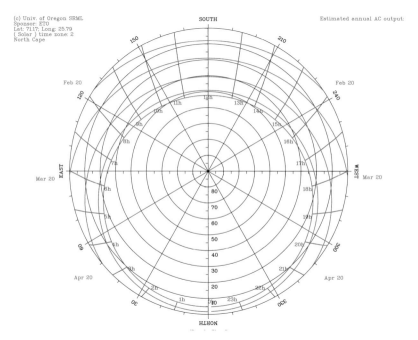

Figure 18.3: The Sun path in apparent solar time on the North Cape, Norway ($\phi_0 = 71.17°$ N). The Sun path was calculated with the Sun path chart program by the *Solar Radiation Monitoring Lab.* of the *University of Oregon* (with kind permission from F. Vignola, University of Oregon) [137].

18. Location issues

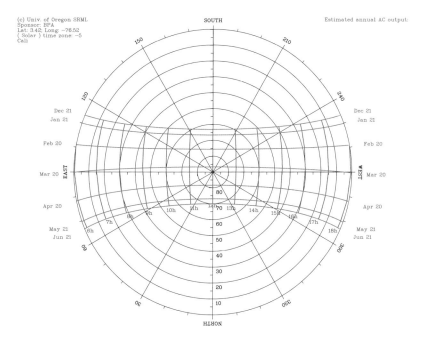

Figure 18.4: The Sun path in apparent solar time in Cali, Colombia ($\phi_0 = 3.42°$ N). The Sun path was calculated with the Sun path chart program by the *Solar Radiation Monitoring Lab.* of the *University of Oregon* (with kind permission from F. Vignola, University of Oregon) [137].

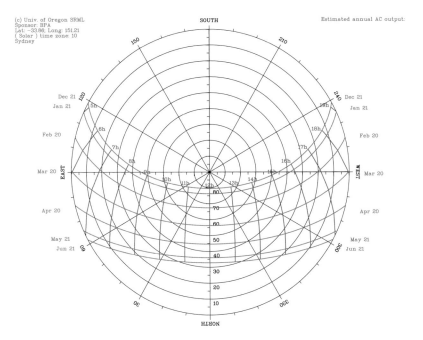

Figure 18.5: The Sun path in apparent solar time in Sydney, Australia ($\phi_0 = 33.86°$ S). The Sun path was calculated with the Sun path chart program by the *Solar Radiation Monitoring Lab.* of the *University of Oregon* (with kind permission from F. Vignola, University of Oregon) [137].

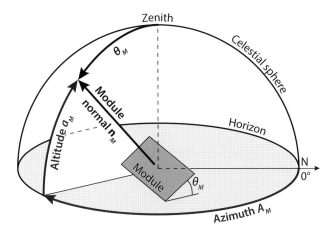

Figure 18.6: Illustrating the angles used to describe the orientation of a PV module installed on a horizontal plane.

18.2 Irradiance on a PV module

18.2.1 The angle of incidence (AOI)

In this section we discuss the implications of the changing position of the Sun on the irradiance present on solar modules. For this discussion we assume that the solar module is mounted on a horizontal plane and that it is tilted at an angle θ_M, as illustrated in Figure 18.6. The angle between the projection of the normal of the module onto the horizontal plane and due North is A_M. We then can describe the position of the module by the direction of the module normal in horizontal coordinates (A_M, a_M), where the altitude is given by $a_M = 90° - \theta_M$. Let the Sun now be at the position (A_S, a_S). Then the direct irradiance on the module G_M^{dir} is given by the equation

$$G_M^{\text{dir}} = I_e^{\text{dir}} \cos \gamma, \tag{18.2}$$

where I_e^{dir} is the *direct normal irradiance* (DNI); $\gamma = \sphericalangle(A_M, a_M)(A_S, a_S)$ is the angle between the surface normal and the incident direction of the sunlight or–in other words–the *angle of incidence* (AOI).

As derived in Appendix E.3, $\cos \gamma$ is given by

$$\cos \gamma = \cos a_M \cos a_S \cos(A_M - A_S) + \sin a_M \sin a_S. \tag{18.3}$$

Thus we obtain for the irradiance

$$\begin{aligned} G_M^{\text{dir}} &= I_e^{\text{dir}} \left[\cos a_M \cos a_S \cos(A_M - A_S) + \sin a_M \sin a_S \right] \\ &= I_e^{\text{dir}} \left[\sin \theta \cos a_S \cos(A_M - A_S) + \cos \theta \sin a_S \right]. \end{aligned} \tag{18.4}$$

Note that this equation only holds when the Sun is above the horizon ($a_S > 0$) and the azimuth of the Sun is within $\pm 90°$ of A_M, $A_S \in [A_S - 90°, A_S + 90°]$. Otherwise, $G_M^{\text{dir}} = 0$.

If the azimuth of the solar position is the same as the azimuth of the module normal $A_M = A_S$, Eq. (18.4) becomes

$$G_M^{\text{dir}} = I_e^{\text{dir}} [\cos a_M \cos a_S + \sin a_M \sin a_S]$$
$$= I_e^{\text{dir}} \cos(a_M - a_S). \qquad (18.5)$$

When using the tilt angle $\theta = 90° - a_M$ we find

$$G_M^{\text{dir}} = I_e^{\text{dir}} \sin(\theta + a_S). \qquad (18.6)$$

Very often, PV modules are not installed on a horizontal plane, but on a tilted roof, which makes the calculation of $\cos\gamma$ more complex. The derivations for modules on a tilted roof as well as some examples are shown in Appendix E.4.

18.2.2 Shading

As we have seen in Section 15.3 shading can have a detrimental effect on the power output of a PV module. Therefore, mutual *shading* between different rows of PV modules must be kept in mind when designing a PV system.

As shown in Appendix E.5 the length of the shadow of a module of length l is given by

$$d = l [\cos\theta_M + \sin\theta_M \cot a_S \cos(A_M - A_S)], \qquad (18.7)$$

where θ_M and A_M are the tilt and azimuth of the PV module, respectively, as illustrated in Figure 18.6. a_S and A_S are the altitude and azimuth of the Sun, respectively.

For a first estimation, one can use the rule of thumb that the distance between two rows of modules should be at least three times the length l of the module.

Example

A PV system is to be installed on a flat roof in Naples (Italy). The area of the roof that can be utilized for installing the PV system is 10×10 m^2. The roof is oriented such that the sides are parallel to the East-West and North-South directions, respectively.

The owner of the roof decides to use **Yingli PANDA 60** *modules with dimensions of $1650 \times 990 \times 40$ mm^3. The modules are installed facing south with a tilt of $30°$.*

He wants to install **as many modules as possible** *under the condition that on the shortest day of the year* **no mutual shading must occur for the duration of six hours**.

Should the modules be mounted with the long or short side touching the ground? How many modules can be mounted in this case?

Answer: *The shortest day of course is 21 December. The solar positions on this day at 9:00 h and 15:00 h are*

Time	Altitude (°)	Azimuth (°)
9:00	13.59	138.55
15:00	13.13	222.17

Because of the equation of time, the Sun is not at its highest point at exactly 12:00 noon. We see that the solar altitude at 15:00 is just slightly lower than at 9:00. Thus, when using 9:00 for calculating the length of the shadow, the duration without mutual shading will be slightly shorter than six hours. Thus, we use the position at 15:00 for the calculation.

The length of the shadow can be calculated with Eq. (18.7)

$$d = l\left[\cos\theta_M + \sin\theta_M \cot a_S \cos(A_M - A_S)\right].$$

We have $\theta_M = 30°$, $A_M = 180°$, $a_S = 13.13°$, and $A_S = 222.17°$.

If the module is mounted with the ground on the short side, we have $l=1650$ mm. Hence, we find $d=4050$ mm. The area directly beneath the module at the last row is

$$d' = l\cos\theta_M = 1{,}429 \text{ mm.}$$

Thus, we can mount **three rows** behind each other, because

$$2d + d' = 2 \times 4{,}050 + 1{,}429 = 9{,}529 \text{ mm,}$$

which is less than 10 m. In one row fit 10 modules because $10 \times 990 = 9{,}900$ mm. Thus, we can place 30 modules.

If the module is mounted with the ground on the long side, we have $l=990$ mm. Hence, we find $d=2{,}430$ mm. The area directly beneath the module at the last row is

$$d' = l\cos\theta_M = 857 \text{ mm.}$$

Thus, we can mount **four rows** behind each other, because

$$3d + d' = 3 \times 2430 + 857 = 8147 \text{ mm,}$$

which is less than 10 m. In one row fit 6 modules because $6 \times 1{,}650 = 9{,}900$ mm. Thus, we can place 24 modules.

18.3 Direct and diffuse irradiance

As sunlight traverses the atmosphere, it is partially scattered, leading to attenuation of the *direct beam* component. On the other hand, the scattered light will also partially arrive at on the Earth's surface as *diffuse light*. For PV applications it is important to be able to estimate the strength of the direct and diffuse components.

First, we discuss a simple model that allows us to estimate the irradiance on a *cloudless sky* independent of the *air mass* and hence the altitude of the Sun. As we have seen in Section 5.5, the air mass is defined as

$$\text{AM} := \frac{1}{\cos\theta} = \frac{1}{\sin a_S}, \qquad (18.8)$$

where the angle between the Sun and the zenith θ is connected to the solar altitude via $\theta = 90° - a_S$. This equation, however, does not take the curvature of the Earth into account. If the curvature is taken into account, we find [138]

$$\text{AM}(a_S) = \frac{1}{\sin a_S + 0.50572(6.07995 + a_S)^{-1.6364}}. \qquad (18.9)$$

To estimate the direct normal irradiance at a certain solar altitude a_S and altitude of the observer h, we can use the following empirical equation [139]

$$I_e^{\text{dir}} = I_e^0 \left[(1 - ch) \times 0.7^{(\text{AM}^{0.678})} + ch \right], \tag{18.10}$$

with the constant $c = 0.14$. The solar constant is given as $I_e^0 = 1361 \text{ Wm}^{-2}$. Even during clear sky conditions the diffuse irradiance is about 10% of the direct irradiance. Thus the global irradiance on a module perpendicular to the Sun can be approximated as [140]

$$I_e^{\text{global}} \approx 1.1 \times I_e^{\text{dir}}. \tag{18.11}$$

For a high diffusion percentage this approach no longer works very well. A more accurate model was developed in the framework of the *European Solar Radiation Atlas* [141]. In that model the direct irradiance for clear sky is given by

$$I_e^{\text{dir}} = I_e^0 \varepsilon \exp\left[-0.8662 \, T_L(\text{AM2}) \, m \, \delta_R(m) \right]. \tag{18.12}$$

I_0 is the *solar constant* that takes a value of 1,361 Wm^{-2}. The factor ε corrects for deviations of the Sun-Earth distance from its mean value. a_S is the solar altitude angle. $T_L(\text{AM2})$ is the *Linke turbidity factor* with which the haziness of the atmosphere is taken into account. In this equation its value at an Air Mass 2 is used. m is the relative optical air mass, and finally $\delta_R(m)$ is the integral Rayleigh optical thickness. The different components can be evaluated as follows.

The *correction factor* ε is given by

$$\varepsilon = \frac{I_e(R)}{I_e^0} = \frac{\text{AU}^2}{R^2}. \tag{18.13}$$

The distance between the Earth and Sun as a multiple of astronomic units (AU) is given in Eq. (E.6). Hence,

$$\varepsilon = (1.00014 - 0.01671 \cos g - 0.00014 \cos 2g)^{-2}, \tag{18.14}$$

where g is the *mean anomaly* of the Sun, which is explained in Appendix E. This leads to annual variations of about $\pm 3.3\%$.

The *Linke turbidity factor* approximates absorption and scattering in the atmosphere and takes both absorption by water vapour and scattering by aerosol particles into account. It is a unitless number and typically takes on values between two for very clear skies and seven for heavily polluted skies.

The relative optical air mass m expresses the ratio of the optical path length of the solar beam through the atmosphere to the optical path through a standard atmosphere at sea level with the Sun at zenith. In can be approximated as a function of the solar altitude a_S by

$$m(a_S) = \frac{\exp(-z/z_h)}{\sin a_S + 0.50572(a_S + 6.07995)^{-1.6364}}. \tag{18.15}$$

Here, z is the site elevation and z_h is the scale height of the Rayleigh atmosphere near the Earth surface, given by 8,434.5 m.

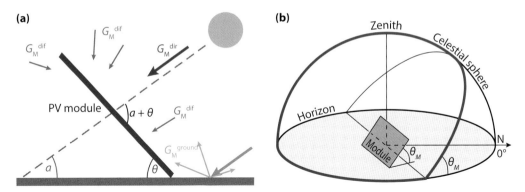

Figure 18.7: (a) The three contributions to the irradiance on a PV module G_M. (b) Illustrating the definition of the *sky view factor* (SVF), which is the fraction of the celestial hemisphere enclosed by the thick red line.

Finally, the Rayleigh optical thickness $\delta_R(m)$ is given by

$$\frac{1}{\delta_R(m)} = 6.62960 + 1.75130\,m - 0.12020\,m^2 \\ + 0.00650\,m^3 - 0.00013\,m^4. \quad (18.16)$$

In their paper, Rigollier et al. also take the effect of refraction at very low altitudes into account [141]. This however, is not relevant for our application.

They also present an expression for the *diffuse horizontal irradiance* (DHI) of the light, which is given by

$$I_e^{\text{dif}} = I_e^0 \varepsilon T_{rd}[T_L(\text{AM2})] F_d[a_S, T_L(\text{AM2})], \quad (18.17)$$

where T_{rd} is the diffuse transmission function at zenith, which is a second-order polynomial of $T_L(\text{AM2})$. T_{rd} has typical values between 0.05 for very clear skies and 0.22 for a very turbid atmosphere. F_d is a diffuse angular function, given as a second-order polynomial of $\sin a_S$. For more details we refer to the paper [141].

18.3.1 Computation of the irradiance on the module

In Section 18.2 we found that the direct irradiance on a module G_M is given by

$$G_M^{\text{dir}} = I_e^{\text{dir}} \cos \gamma, \quad (18.18)$$

i.e. it is given as the *direct normal irradiance* I_e^{dir} (DNI) times the cosine of the angle of incidence γ (AOI).

For PV modules installed on Earth two other factors also contribute to G_M, as illustrated in Figure 18.7 (a): first, the diffuse component from the sky, G_M^{dif}. It is proportional to the *sky view factor* (SVF), i.e. the portion of the sky from which the module can receive diffuse radiation, as illustrated in Figure 18.7 (b). It is given by

$$\text{SVF} = \frac{1 + \cos \theta_M}{2}. \quad (18.19)$$

For the detailed calculation of G_M^{dif}, different *sky models* can be used [142–146].

Secondly, the module receives radiation that is reflected from the ground, which can be approximated by

$$G_M^{\text{ground}} = \text{GHI} \times \alpha \times (1 - \text{SVF}). \tag{18.20}$$

GHI is the *global horizontal irradiance*, which is given by

$$\text{GHI} = \text{DNI} \times \cos(a_S) + \text{DHI}, \tag{18.21}$$

where a_S is the altitude of the Sun. DNI and DHI can be measured in meteorological stations with a pyrheliometer and a pyranometer, respectively. The factor α is the *albedo* of the ground, i.e. the reflection coefficient of the ground. The lower the albedo, the more light is absorbed by the ground. For example, the albedo of forest is between 0.05 and 0.10, that of snow is 0.6 and the albedo of urban areas is between 0.05 and 0.20 [147]. More example can be found in [148].

In summary, the irradiance on a PV module is given by

$$G_M = G_M^{\text{dir}} + G_M^{\text{dif}} + G_M^{\text{ground}}. \tag{18.22}$$

If only the measured GHI is available, sky models in combination with Eq. (18.21) allows the retrieval of the irradiance components that are necessary for the evaluation of G_M.

18.4 Exercise

18.1 Let us consider three locations: Vancouver, New Delhi and Buenos Aires. We want to install solar systems in these three locations at the adequate tilt and orientation. Please relate each location with the optimum tilt angle and orientation (North or South) of the given choices: North, South, South, around 20°, around 40°, around 30°.

19
Components of PV systems

In this chapter we discuss all the components of PV systems, except PV modules that were already treated in Chapter 15. We start with discussing *maximum power point tracking* in Section 19.1, which is followed by a treatment of *photovoltaic converters* in Section 19.2. Further, we we look at *batteries* in Section 19.3, charge controllers in Section 19.4, and *cables* in Section 19.5.

19.1 Maximum power point tracking

In this section we discuss the concept of *maximum power point tracking* (MPPT). An MPP tracker is never an actual component itself but always connected to a DC-DC converter in a PV inverter or a charge controller. However, as it is a very important concept we will discuss it here. MPPT is unique to the field of PV Systems, and hence brings a very special application of power electronics to the field of photovoltaics. The concepts discussed in this section are equally valid for cells, modules, and arrays, although MPPT usually is employed at PV module/array level.

As discussed earlier, the behaviour of an illuminated solar cell can be characterized by an *I-V* curve. Interconnecting several solar cells in series or in parallel merely increases the overall voltage and/or current, but does not change the shape of the *I-V* curve. Therefore, for understanding the concept of MPPT, it is sufficient to consider the *I-V* curve of one solar cell. The *I-V* curve is dependent on the module temperature and on the irradiance, as we will discuss in detail in Section 20.3. For example, an increasing irradiance leads to an increased current and slightly increased voltage, as illustrated in Figure 19.1. The same figure shows that an increasing temperature has a detrimental effect on the voltage.

Now we take a look at the concept of the *operating point*, which is the defined as the particular voltage and current at which the PV module operates at any given point in time.

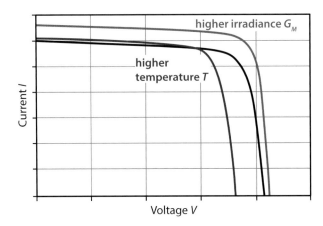

Figure 19.1: Effect of increased temperature T or irradiance G_M on the I-V curve.

For a given irradiance and temperature, the operating point corresponds to a unique (I, V) pair which lies on the I-V curve. The power output at this operating point is given by

$$P = I \times V. \tag{19.1}$$

The operating point (I, V) corresponds to a point on the power-voltage $(P$-$V)$ curve, shown in Figure 19.2. To generate the highest power output at a given irradiance and temperature, the operating point should such correspond to the maximum of the $(P$-$V)$ curve, which is called the *maximum power point* (MPP).

If a PV module (or array) is directly connected to an electrical load, the operating point is dictated by that load, as we have already seen in Section 9.3. To achieve the maximal power out of the module, it thus is imperative to force the module to operate at the maximum power point. The simplest way of forcing the module to operate at the MPP is to force the voltage of the PV module to be that at the MPP (called V_{MPP}).

However, the MPP is dependent on the ambient conditions. If the irradiance or temperature change, the I-V and the P-V characteristics will change as well and hence the position of the MPP may shift. Therefore, changes in the I-V curve have to be tracked continuously such that the operating point can be adjusted to be at the MPP after changes of the ambient conditions.

This process is called *maximum power point tracking* or MPPT. The devices that perform this process are called *MPP trackers*. We can distinguish between two categories of MPPT:

- *Indirect* MPPT, where the position of the MPP is estimated via a hard-coded algorithm.

- *Direct* MPPT, where the actual I-V data is used to determine the position of the MPP.

All the MPPT algorithms that we discuss in this section are based on finding and tuning the voltage until V_{MPP} is found. Other algorithms, which are not discussed in this section, work with the power instead and aim to find I_{MPP}.

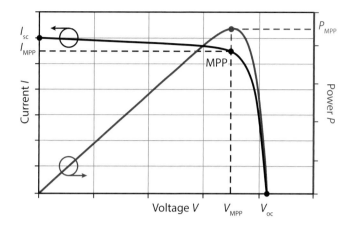

Figure 19.2: A generic *I-V* curve and the associated *P-V* curve. The maximum power point (MPP) is indicated.

19.1.1 Indirect MPPT

First, we discuss *indirect* MPP tracking, where simple assumptions are made for estimating the MPP based on a few measurements.

Fixed voltage method

For example, in the *fixed voltage* method (also called *constant voltage method*), the operating voltage of the solar module is adjusted only on a seasonal basis. This model is based on the assumption that for the same level of irradiance, higher MPP voltages are expected during winter than during summer. This method is not very accurate; it works best at locations with minimal irradiance fluctuations between different days.

Fractional open circuit voltage method

One of the most common *indirect* MPPT techniques is the *fractional open circuit voltage* method. This method exploits the fact that – in a very good approximation – the V_{MPP} is given by

$$V_{\text{MPP}} = k \times V_{\text{oc}}, \tag{19.2}$$

where k is a constant. For crystalline silicon, k usually takes values between 0.7 and 0.8. In general, k of course is dependent on the type of solar cells.

As changes in the open circuit voltage can be easily tracked, changes in the V_{MPP} can be easily estimated just by multiplying by k. This method can thus be implemented easily. However, there are also certain drawbacks.

First, using a constant factor k only allows a rough estimate of the position of the MPP. Therefore, the operating point will not usually be exactly on the MPP but in its proximity. Secondly, every time the system needs to respond to a change in illumination conditions, the V_{OC} must be measured. For this measurement, the PV module needs to be disconnected from the load for a short while, which will lead to a reduced total output of the PV system.

Table 19.1: A summary of the possible options in the P&O algorithm.

Prior Perturbation	Change in Power	Next Perturbation
Positive	Positive	Positive
Positive	Negative	Negative
Negative	Positive	Negative
Negative	Negative	Positive

The more often the V_{OC} is determined, the larger the loss in output will be. This drawback can be overcome by slightly modifying the method. For this modification a pilot PV cell is required, which is highly matched with the rest of the cells in the module. The pilot cell receives the same irradiance as the rest of the PV module, and a measurement of the pilot PV cell's V_{OC} also gives an accurate representation of that of the PV module, hence it can be used for estimating V_{MPP}. Therefore, the operating point of the module can be adjusted without needing to disconnect the PV module.

19.1.2 Direct MPPT

Now we discuss *direct* MPPT, which is more involved than indirect MPPT, because current, voltage or power measurements are required. Further, the system must respond more accurately and faster than in indirect MPPT. We shall look at a couple of the most popular kinds of algorithms.

Perturb and observe (P&O) algorithm

The first algorithm that we discuss is the *perturb and observe* (P&O) algorithm, which is also known as the 'hill climbing' algorithm. In this algorithm, a perturbation is provided to the voltage at which the module is currently driven. This perturbation in voltage will lead to a change in the power output. If an increasing voltage leads to an increasing power, the operating point is at a lower voltage than the MPP, and hence further voltage perturbation towards higher voltages is required to reach the MPP. In contrast, if an increasing voltage leads to a decreasing power, further perturbation towards lower voltages is required in order to reach the MPP. Hence, the algorithm will converge towards the MPP over several perturbations. This principle is summarized in Table 19.1.

A problem with this algorithm is that the operating point is never steady at the MPP but is meandering around the MPP. If very small perturbation steps are used around the MPP, this meandering, however, can be minimized. Additionally, the P&O algorithm struggles from rapidly changing illuminations. For example, if the illumination (and hence the irradiance) changes in-between two sampling instants in the process of convergence, then the algorithm essentially fails in its convergence efforts, as illustrated in Figure 19.3: in the latest perturbation, the algorithm has determined that the MPP lies at a higher voltage than that of point B, and hence the next step is a perturbation to converge towards the MPP accordingly. If the illumination was constant, it would end up at C and the algorithm would conclude that the MPP is a still higher voltage, which is correct. However, as the illumination changes rapidly before the next perturbation, the next perturbation shifts the

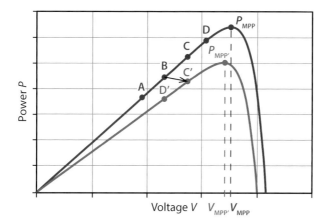

Figure 19.3: The *perturb and observe* algorithm struggles from rapidly changing illumination conditions.

operating point to C' instead of C, such that

$$P_{C'} < P_B. \tag{19.3}$$

While the MPP still lies to the right of C', the P&O algorithm thinks that it is on the left of C' so it moves to point D'. This wrong assumption is detrimental to the speed of convergence of the P&O algorithm, which is one of the critical figures of merit for MPPT techniques. Thus, drastic changes in weather conditions severely affect the efficacy of the P&O algorithms.

Incremental conductance method

Next, we look at the *incremental conductance* method. The *conductance G* of an electrical component is defined as

$$G = \frac{I}{V}. \tag{19.4}$$

At the MPP, the slope of the *P-V* curve is zero, hence

$$\frac{dP}{dV} = 0. \tag{19.5}$$

We can write

$$\frac{dP}{dV} = \frac{d(IV)}{dV} = I + V\frac{dI}{dV}. \tag{19.6}$$

If the sampling steps are small enough, the approximation

$$\frac{dI}{dV} \approx \frac{\Delta I}{\Delta V} \tag{19.7}$$

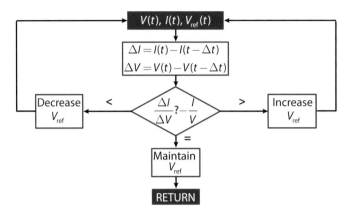

Figure 19.4: A conceptual flowchart of the *incremental conductance* algorithm.

can be used. We call $\Delta I/\Delta V$ the *incremental conductance* and I/V the *instanteneous conductance*. Hence, we have

$$\frac{\Delta I}{\Delta V} = -\frac{I}{V} \quad \text{if} \quad V = V_{\text{MPP}}, \tag{19.8a}$$

$$\frac{\Delta I}{\Delta V} > -\frac{I}{V} \quad \text{if} \quad V < V_{\text{MPP}}, \tag{19.8b}$$

$$\frac{\Delta I}{\Delta V} < -\frac{I}{V} \quad \text{if} \quad V > V_{\text{MPP}}. \tag{19.8c}$$

These relationships are exploited by the incremental conductance algorithm.

Figure 19.4 shows a conceptual flowchart. Note that this flowchart is not exhaustive. While the instantaneous voltage and current are the observable parameters, the instantaneous voltage is also the controllable parameter. V_{ref} is the voltage value forced on the PV module by the MPPT device. It is the latest approximation of the V_{mpp}. For any change of the operating point, the algorithm compares the instantaneous with the incremental conductance values. If the incremental conductance is larger than the negative of the instantaneous conductance, the current operating point is to the left of the MPP; consequently, V_{ref} must be incremented. In contrast, if the incremental conductance is lower than the negative of the instantaneous conductance, the current operating point is to the right of the MPP and V_{ref} is consequently decremented. This process is iterated until the incremental conductance is the same as the negative instantaneous conductance, in which case $V_{\text{ref}} = V_{\text{MPP}}$.

The incremental conductance algorithm can be more efficient than the P&O algorithm as it does not meander around the MPP under steady state conditions. Further, small sampling intervals make it less susceptible to changing illumination conditions. However, under conditions that are strongly varying and under partial shading, the incremental conductance method might also become less efficient. The major drawback of this algorithm is the complexity of its hardware implementation. Not only must currents *and* voltages be measured, but also the instantaneous and incremental conductances must be calculated and compared. How such a hardware design can look like, however, is beyond the scope of this book.

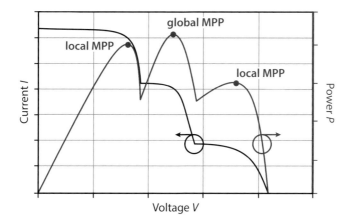

Figure 19.5: The P-V curve of a partially shaded system that exhibits several local maxima.

19.1.3 Some remarks

While MPPT is used to find the MPP by changing the voltage, it does not perform changes of the operating voltage. This is usually done by a DC-DC converter that will be discussed in Section 19.2.2.

In modern PV systems, MPPT is often implemented within other system components like inverters or charge controllers. The list of techniques presented in this section is not exhaustive, we have just discussed the most common ones. The development of more advanced MPPT techniques is progressing rapidly and many scientific papers as well as patents are being published in this area. Furthermore, manufacturers usually use proprietary techniques.

Up to now we have only looked at situations in which the total I-V curve is similar to that of a single cell. Let us now consider a system that is partially shaded, as illustrated in Figure 19.5. In this case, the P-V curve will have different local maxima. Depending on the MPPT algorithm used, it is not guaranteed that the algorithm will find the global maximum. Different companies use proprietary solutions to tackle this issue. Alternatively, each string can be connected to a separate MPPT device. Nowadays, inverters are available, which have connections for several strings (usually two).

19.2 Power electronics

A core technology associated with PV systems is the converter, which is based on power electronics. An ideal PV converter should draw the maximum power from the PV panel and supply it to the load side. It is very important to distinguish between inverters in grid-connected systems and in stand-alone systems.

Note that the term *inverter* can have two different meanings: first, it is used for the actual inverter, which is the electronic building block that performs the DC to AC inversion, as described in Section 19.2.3. Secondly, the term inverter is also used for the total unit produced by manufacturers. Depending on the application, it may contain an MPP tracker, a

DC-DC converter, and/or a DC-AC converter.

In *grid-connected* systems, the inverter is connected directly to the PV array. It converts the DC electricity coming from the PV array into AC electricity. Further, such an inverter usually contains an MPPT system, that we discussed in Section 19.1. As such a inverter is connected to the electricity grid, it must synchronize with the grid, meaning that the phase of the AC signal coming from the inverter must be *in phase* with that of the grid. Further, its signal should have minimal harmonic content. Usually, grid-connected inverters cannot act autonomously, but are switched off when the electricity grid is down. This is to prevent *islanding*, which we discuss in Section 19.2.4.

Inverters used in *stand-alone* systems usually are connected to the batteries. As a PV array is connected to the battery via a charge controller, such an inverter does not require an MPP tracker or a DC-DC converter. Often, these inverters are especially designed for the use with batteries and prevent them from being discharged too far, which would be detrimental to the battery lifetime as discussed in Section 19.3. Sometimes, these inverters can also be used to charge the batteries, for example if an AC generator is connected to the AC side of the inverter. In that case, the inverter also functions as an *AC-DC converter*. As such an inverter is not connected the electricity grid, it must control the AC voltage and frequency.

We now give a short review of different topologies often associated with PV systems. The semiconductor switches in the following are assumed to be ideal.

19.2.1 System architecture

Before going into details about different converter topologies used for power conversion in PV systems, a general overview of different system architectures will be presented. The system architecture determines how PV modules are interconnected and how the interface with the grid is established. Which of these system architectures will be employed in a particular PV plant depends on many factors such as the environment of the plant (whether the plant is situated in an urban environment or in an open area), scalability, costs, etc. In Figure 19.6 an overview of different system architectures is given. The main advantages and disadvantages of the different architectures are discussed below.

In general solar inverters should fulfil the following *requirements* [149]:

- Inverters should be highly efficient because the owner of the solar system requires the absolute maximum possible generated energy to be delivered to the grid/load.

- Special requirements regarding the potential between solar generator and Earth (depending on the solar module type); see potential-induced degradation in Section 19.2.4.

- Special safety features like active islanding detection capability; see Section 19.2.4.

- Low limits for harmonics of the line currents. This requirement is enforced by law in most countries since the harmonic limits of both sources and loads connected to the grid are regulated.

- Special requirements on *electromagnetic interference* (EMI), which are regulated by law in most countries. The goal is to minimize the unwanted influence of EMI on other equipment in the vicinity or connected to the same supply. Think for example of the influence of a mobile phone on an old radio.

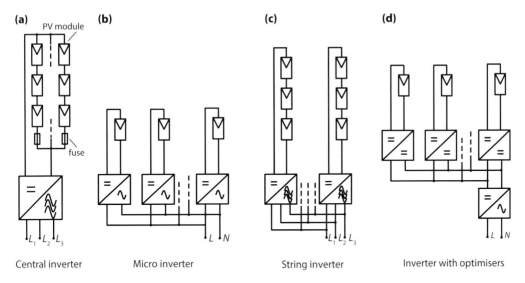

Figure 19.6: Different system architectures employed in PV systems.

- Also the effect of electromagnetic fields to people must be taken into account; the maximally allowed field strength emitted from the inverter is regulated in many countries.
- In many instances the solar system is to be installed outdoors and inverters should adhere to certain specifications regarding temperature and humidity conditions, *e.g.* IP 54.
- Design for high ambient temperatures.
- Design for 10-25 years operation under harsh environmental conditions.
- Silent operation (no audible noise).
- Often, the possibility to monitor the PV system is required by the user.

We have to distinguish between *single-phase* and *three-phase inverters*. For low powers, as they are common in small residential PV systems, single-phase inverters are used. They are connected to one phase of the grid. For higher powers, three-phase inverters are used that are connected to all phases of the grid. If a high power were to be delivered to one phase, the currents flowing across the three phases would become very asymmetric leading to multiple problems in the electricity grid.

Central inverters

This is the simplest architecture employed in PV systems. Here, PV modules are connected in strings leading to an increased system voltage. Many strings are then connected in parallel forming a PV array, which is connected to one central inverter. The inverter performs maximum power point tracking and power conversion as shown in Figure 19.6 (a), where a system with a three-phase inverter is depicted. This configuration is mostly employed in very large-scale PV power plants, with the central inverter usually being DC to three phase [150]. In the case where more than one string is connected in parallel and significant differences in the irradiance from string to string are expected, blocking diodes are required to prevent current circulation inside strings.

Many different inverter topologies are utilized as three-phase inverters. Sometimes they are organized as a single DC to three-phase unit but sometimes as three separate DC to AC single phase units working with a phase displacement of 120 °C each. Having all the PV modules connected in a single array in such a centralized configuration offers the lowest specific cost (cost per kW_p of installed power). Since central inverters only use a few components, they are very reliable which makes them the preferred option in large-scale PV power plants.

In spite of their simplicity and low specific cost, central inverter systems suffer from the following disadvantages:

1. Due to the layout of the system, a large amount of power is carried over considerable distances using DC wiring. This can cause safety issues because fault DC currents are difficult to interrupt. Special precautions must to be taken such as thicker insulation on the DC cabling and special circuit breakers, which can increase costs.

2. All strings operate at the same maximum power point, which leads to mismatch losses in the modules. This is a significant disadvantage. Mismatch losses increase even more with ageing and with partial shading of sections of the array. Mismatches between the different strings may significantly reduce the overall system output.

3. Low flexibility and expandability of the system. Due to the high ratings, a system is normally designed as one unit and hence is difficult to extend. In other words the system design is not very flexible.

4. Power losses in the string diodes (if any), which are put in series with each string to prevent current circulation inside the strings.

Micro inverters

A very different architecture is that of the *micro inverters*, as shown in Figure 19.6 (b). These inverters operate directly at one or several PV modules and have power ratings of several hundreds of watts. Because of the low voltage rating of the PV module, these inverters often require a two-stage power conversion. In the first stage, the DC voltage is boosted to the required value while it is inverted to AC in the second stage. Often, a high frequency transformer is incorporated providing full galvanic isolation, which enhances the system flexibility even further. These inverters are usually placed close to the PV panels; sometimes they are also directly integrated into the PV panels (so called 'AC PV panels'). One of the most distinguishing features of this system is the 'plug and play' characteristic, which allows a complete (and readily expandable) PV system to be built at a low investment cost. Another advantage of these inverters is the minimisation of mismatch losses that can occur because of non-optimal MPPT.

All these advantages come at a price. Because these inverters are mounted on a PV module, they must operate in a harsh environment such as high temperatures and large daily and seasonal temperature variations. Further, these inverters often accept only a very narrow range of input voltages. Therefore it is not possible to use bypass diodes within the modules that bypass one- or two-thirds of the module. Another disadvantage is that the PV module voltage is much lower than the output AC voltage, therefore the DC-DC conversion has to boost the voltage a lot. This has a detrimental effect on the inverter efficiency. Also, the specific costs are the highest of all the inverter topologies. Many topologies for micro inverters have been proposed, with some of them being already implemented in commercially available inverters.

String inverters

String inverters, as illustrated in Figure19.6 (c), combine the advantages of central and module integrated inverter concepts with little trade-offs. A number of PV modules that are connected in series form a PV string with a power rating of up to 5-6 kWp in 1-phase configurations and up to 20-30 kWp in 3-phase configuration.

The open circuit voltages are up to 1 kV, which already makes one disadvantage of this topology apparent: these high DC voltages require special consideration, as was the case for the central inverter architecture. Here, this issue is even more important because string inverters are usually being installed in households or on office buildings without designated support structures or increased safety requirements. The protection of the system also requires special consideration, with emphasis on proper DC cabling.

Although partial shading of the string will influence the overall efficiency of the system, each string can independently be operated at its MPP, if each string has its own MPPT. Also, because no strings are connected in parallel, there is no need for series diodes as in the case of PV arrays with multiple parallel strings. This reduces losses associated with these diodes. However, it is still a risk that within a string, hot-spots will occur because of unequal current and power sharing inside the string.

Central inverter with optimizers

This architecture is a hybrid between a central inverter and micro inverters. An optimizer box is attached to every module, which contains a MPP tracker and a DC-DC converter as illustrated in Figure 19.6 (d). The optimizer boxes of all the modules are connected in series to each other and to the central inverter. The inverter can accept input voltages within a certain range – if the voltage is outside this range, the current is altered such that the voltage falls within the acceptable range again. As a consequence, the output voltage of the optimizers is determined by the input power from the PV module and the current that is enforced by the inverter.

The main advantage of this architecture is that every module can operate at its MPP. This is not only important for the shading of single modules, but also because of the fact that no two modules are the same. Another advantage of this architecture is that all the optimizers can operate at voltages close to the voltage of the PV module. Therefore, the DC-DC conversion is very efficient. Further, the optimizers consume very little power, so there are no problems with heating up, in contrast to what we have seen with micro inverters.

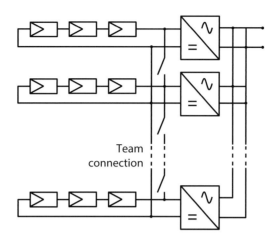

Figure 19.7: The team concept of inverters.

Team concept

Aside from the four system architectures already described, many other concepts are also discussed in the literature. However, these concepts are less widely utilized than the ones already presented. One of the alternative concepts is the so-called *team concept*, which combines string technology with the master-slave concept. A combination of several string inverters working with the team concept is shown in Figure 19.7. At very low irradiation the complete PV array is connected to a single inverter. This reduces the overall losses, as any power electronic converter is designed such that it has maximum efficiency near full load. With increasing solar radiation more inverters are being connected, dividing the PV array into smaller units until every string inverter operates close to its rated power. In this mode every string operates independently with its own MPP tracker. At low solar radiation the inverters are controlled in a master-slave fashion.

For further reading on this subject, we refer for example to [151].

19.2.2 DC-DC converters

DC-DC converters fulfil multiple purposes. In an inverter, DC power is transformed into AC power. The DC input voltage of the inverter is often constant while the output voltage of the modules at MPP is not. Therefore a DC-DC converter is used to transform the variable voltage from the panels into a constant voltage used by the DC-AC inverter. Additionally, as already stated in Section 19.1, the MPP tracker controls the operating point of the modules, but cannot set it. This also is done by the DC-DC converter. Further, in a stand-alone system the MPP voltage of the modules might differ from that required by the batteries and the load. Also, a DC-DC converter is useful here, and hence applied in some high-end charge controllers. Three topologies are used for DC-DC converters: *buck*, *boost*, and *buck-boost* converters. They are described below.

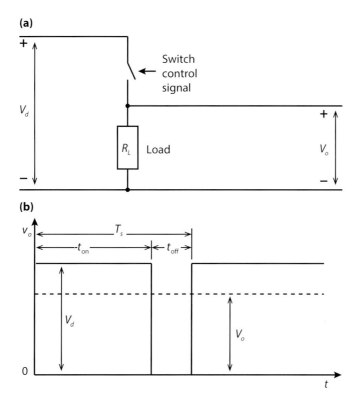

Figure 19.8: (a) A basic buck converter without any filters; and (b) the unfiltered switched waveform generated by this converter.

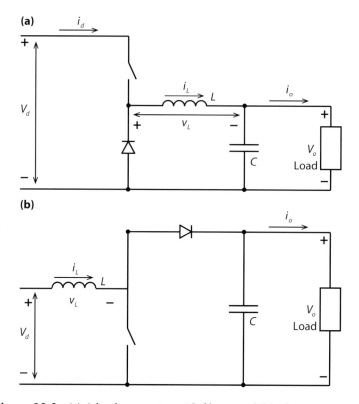

Figure 19.9: (a) A buck converter with filters; and (b) a boost converter.

Step-down (buck) converter

Figure 19.8 (a) illustrates the simplest version of a buck DC-DC converter. The unfiltered output voltage waveform of such a converter operated with pulse width modulation (PWM) is shown in Figure 19.8 (b). If the switch is *on*, the input voltage V_d is applied to the load. When the switch is *off*, the voltage across the load is zero. From the figure we see that the average DC output voltage is denoted as V_o. From the unfiltered voltage, the average output voltage is given as

$$V_o = \frac{1}{T_s} \int_0^{T_s} v_o(t)\,dt = \frac{1}{T_s}(t_{on} V_d + t_{off} \times 0) = \frac{t_{on}}{T_s} V_d. \qquad (19.9)$$

The different variables are defined in Figure 19.8 (b). To simplify the discussion we define a new term, the duty cycle D, as

$$D := \frac{t_{on}}{T_s} \qquad (19.10)$$

and hence

$$V_o = D \times V_d. \qquad (19.11)$$

In general the output voltage with such a high harmonic content is undesirable, and some low-pass filtering is required. Figure 19.9 (a) shows a more complex model of a step-down converter that has output filters included and supplies a purely resistive load. As filter elements, an inductor L and a capacitor C are used. The relationship between the input and the output voltages, as given in Eq. (19.11), is valid in continuous conduction mode, i.e. when the current through the inductor never reaches zero but flows continuously. We can change the ratio between the voltages on the input and output sides by changing the duty cycle D. A detailed discussion about different modes of operation of a buck converter can be found in [152].

In *steady-state operation* the time integral of the voltage across the inductor v_L taken during one switching cycle is equal to zero. If this is not the case, the circuit is not in steady state. Thus, in steady state, we obtain the following *inductor volt-second balance*:

$$\int_0^{T_s} v_L \, \mathrm{d}t = \int_0^{t_{\mathrm{on}}} v_L \, \mathrm{d}t + \int_{t_{\mathrm{on}}}^{T_s} v_L \, \mathrm{d}t = 0. \tag{19.12}$$

Solving this equation leads to

$$V_o = DV_d, \tag{19.13}$$

which is the same result as in Eq. (19.11). A more detailed derivation is given in Appendix F.1.

Step-up (boost) converter

In a boost converter, illustrated in Figure 19.9 (b), an input DC voltage V_d is boosted to a higher DC voltage V_o. By applying the inductor volt-second balance across the inductor as explained in Eq. (19.12), we find

$$V_d t_{\mathrm{on}} + (V_d - V_0) t_{\mathrm{off}} = 0. \tag{19.14}$$

Using the definition for the duty cycle [Eq. (19.10)] we find

$$\frac{V_0}{V_d} = \frac{1}{1-D}, \tag{19.15}$$

which is derived in more detail in Appendix F.2. The above relation is valid in the continuous conduction mode. The principle of operation is that energy stored in the inductor (while the switch is *on*) is later released against higher voltage V_o. In this way the energy is transferred from a lower voltage (solar cell voltage) to a higher voltage (load voltage).

Buck-boost converter

In a *buck-boost converter* the output voltage can be higher or lower than the input voltage. The simplified schematic of a buck-boost converter is depicted in Figure 19.10. Using the inductor volt-second balance as in Eq. (19.12), we find

$$V_d t_{\mathrm{on}} + (-V_o) t_{\mathrm{off}} = 0 \tag{19.16}$$

and hence

$$\frac{V_o}{V_d} = \frac{D}{1-D} \tag{19.17}$$

in the continuous conduction mode. Eq. (19.17) is derived in more detail in Appendix F.3.

The topologies described above are only the most basic DC-DC converter topologies. The interested reader can find more in-depth information in [152].

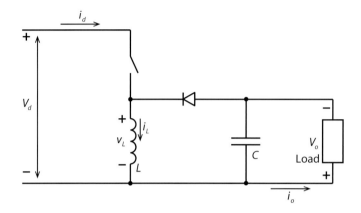

Figure 19.10: A buck-boost converter.

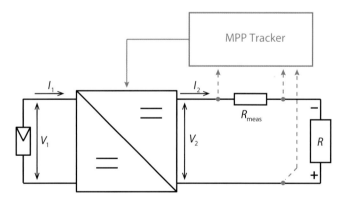

Figure 19.11: A combination of a unit performing an MPPT algorithm and a DC-DC converter (adapted from [153]).

MPP tracking

In section 19.1 we discussed maximum power point tracking extensively. More specifically, we discussed different algorithms that are used for performing MPPT. In these algorithms, the operating point of the module is usually set such that its power output becomes maximal. However, an MPPT algorithm itself cannot actually adjust the voltage or current of the operating point. For this purpose a DC-DC converter is needed. Figure 19.11 shows such a combination of the unit performing the MPPT and a DC-DC converter. As illustrated in the figure, this MPPT unit measures the voltage or the current on the load side and can vary those by adapting the duty cycle of the DC-DC converter. In this illustration, current and voltage on the load side are measured, but they can also be measured on the PV side.

Example

Assume a PV module has its MPP at $V_{PV} = 17$ V and $I_{PV} = 6$ A at a given level of solar irradiance. The module has to power a load with a resistance $R_L = 10\,\Omega$. Calculate the duty cycle of the DC-DC converter, if a buck-boost converter is used.

The maximum power from the module is $P_{MPP} = V_{MPP} \times I_{MPP} = 102$ W. If this power is to be dissipated at the resistor, we have to use the relation

$$P_R = U_R^2 / R$$

and hence find that the voltage at the resistor is

$$V_R = \sqrt{P_{MPP} R_L} = 31.94 \text{ V}.$$

Using Eq. (19.17),

$$\frac{V_o}{V_d} = \frac{D}{1-D}$$

with $V_o = V_R$ and $V_d = V_{PV}$ we find $D = 0.65$.

19.2.3 DC-AC converters (inverters)

Earlier in this section we discussed different architectures that are used for power conversion in PV systems. Further, we looked at DC-DC converters that are mainly used in combination with MPPTs in order to push the variable output from the PV modules to a level of constant voltage.

As nowadays most appliances are designed for the standard AC grids, for most PV systems a DC-AC converter is required. As stated earlier, the term inverter is used for both the DC-AC converter and the combination of all the components that form the actual power converter.

The H-bridge inverter

Figure 19.12 shows a very simple example of a so-called H-bridge or full bridge inverter. On the left, is the DC input. The load (or in our case, the AC output) is situated in-between

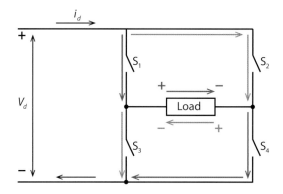

Figure 19.12: A simple representation of an H-bridge.

four switches. During usual operation we can distinguish between three situations:

A All four switches open: No current flows across the load.

B S_1 and S_4 closed, S_2 and S_3 open: Now a current is flowing through the load, where $+$ is connected to the left- and $-$ is connected to the right-hand side of the load.

C S_1 and S_4 open, S_2 and S_3 closed: Now a current is flowing through the load, where $-$ is connected to the left- and $+$ is connected to the right-hand side of the load.

Note It must be assured that S_1 and S_2 are never open at the same time because this would lead to short-circuiting. The same is true for S_3 and S_4.

From this list we see that the H-bridge configuration allows the load to be put on three different levels, which are $+V_d$, 0, and $-V_d$. Continuously switching between positive and negative voltages is exactly what is happening in AC. In the easiest operation mode, the H-bridge switches between situations B and C continuously, which will provide a square wave. Note that there will always be a *dead time* in the order of µs in order to prevent short-circuiting. While such a square wave might be useful for some applications, it is not suited at all for grid-connected installations. The reason for this are harmonic distortions. To understand this we look at a Fourier transform of a square wave, which is given as

$$V_{\text{square}}(t) = \frac{4}{\pi}\left[\sin(2\pi v t) + \frac{1}{3}\sin(6\pi v t) + \frac{1}{5}\sin(10\pi v t) + \cdots\right]. \tag{19.18}$$

Thus, such a square wave contains not only the principal sine function with frequency v, but also all the higher harmonics with frequencies $3v$, $5v$, and so on. These higher harmonics can lead to distortions of the electricity grid and thus must be reduced as much as possible. One method to achieve this is *pulse width modulation* (PWM) that we already discussed in the section on DC-DC conversion (19.2.2). In this configuration, each leg (the one via S_1 and S_4 and the one via S_2 and S_3) in fact acts as a buck converter.

If a low pass filter consisting of capacitors and inductors is used, as in Figure 19.13, the high frequency components are filtered out and hence a very smooth sine curve can be

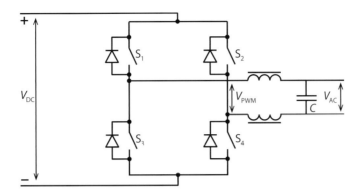

Figure 19.13: Illustration of an H-bridge containing a low pass filter for removing the high-frequency components of the signal.

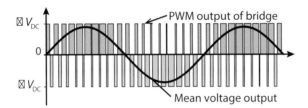

Figure 19.14: The unfiltered PWM signal and the sine signal that are obtained with a low-pass filter.

obtained that complies to the regulations for grid-connected systems. Figure 19.14 shows the unfiltered PWM output and the filtered sine output.

Note that in Figure 19.13 diodes are connected in parallel to the switches. The reason for placing these diodes in parallel is the following: if the switch goes from closed to open very fast, no current can flow through the inductor any more, meaning that the change of current flowing through the inductor is very high. This induces a voltage given by

$$V_\text{induced} = -L\frac{dI}{dt}, \tag{19.19}$$

which increases with faster current changes. As high induced voltages will damage the electric circuitry, they must be prevented. The diode ensures that current can also flow after the switch opens. Hence a current is always flowing and no high induced voltages appear.

Equation (19.19) is an expression of Lenz's law,, named after Heinrich Friedrich Emil Lenz. This law states that the voltage induced due to a change in current is in the direction opposing the current.

Since this configuration is grid-connected, we can easily determine the required DC input voltage, which must at least be equal to the peak of the AC voltage. For an effective AC voltage of 240 V, as used in large parts of the world, the peak voltage is $V_\text{peak} = 1.1\sqrt{2} \times 240$ V $= 373$ V, where we also took a 10% tolerance into account.

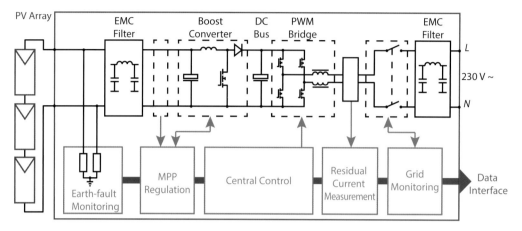

Figure 19.15: An example of a transformer less inverter unit as could be used sold for residential PV systems. As switches, MOSFETs are used (adapted from [153]).

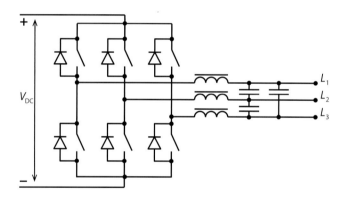

Figure 19.16: A three-phase inverter.

If the PV array delivers a lower voltage, a boost converter will be required prior to the inverter. Alternatively, a transformer can be used that transforms a lower voltage AC signal to the 240 V AC signal. Such a system is sketched in Figure 19.15. This system has the advantage of galvanic isolation with all the advantages discussed below, but on the other hand the transformer reduces the overall efficiency.

Figure 19.16 shows a three-phase inverter. As can be seen, it is very similar to the single-phase inverter discussed above, but contains three legs which each create a sine-wave output with a phase shift of 120° between them.

Half-bridge inverters

A simpler inverter topology, the so-called half-bridge inverter, is shown in Figure 19.17. In contrast to the H-bridge configuration, here two switches are replaced by capacitors and the midpoint between the two capacitors is directly connected to the ground. The half-

19. Components of PV systems

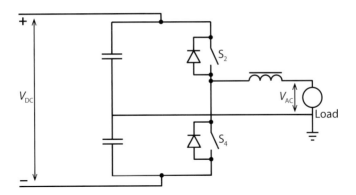

Figure 19.17: A half-bridge inverter.

bridge configuration is much simpler than the H-bridge configuration, but it has some drawbacks.

The main drawback is the requirement for a high DC link voltage, which needs to be two times higher than that of a full bridge inverter. For an effective AC voltage of 240 V this would be 746 V. Because the topology provides two levels in the output voltage (in contrast to three levels for the full bridge configuration, a higher current ripple is present in the output filter inductor. Hence, larger value of the output filters are required.

19.2.4 Some remarks

Switches

All the DC-DC transformers and DC-AC transformers discussed above contain switches. Traditionally, in *line-commutated* inverters, *thyristors* are used as switches. A thyristor is an electronic component consisting of pnpn layers. It thus contains three p-n junctions. One disadvantage is that thyristors cannot be turned off, but only turned on. Thus, one has to wait for the next zero pass of the grid signal [153]. The current flow is thus rectangular which leads to a very high harmonic content requiring additional filters in order to make the output compatible with the electricity grid. Nowadays, thyristors are only used for inverters with a power of 100 kW and above.

All other inverters are *self-commutated* inverters that generate an output with very little harmonic content as described above and in Figure 19.14. The switches there are fully-controllable such that pulse-width modulation becomes possible. As switches, GTOs (gate turn-off thyristors), IGBTs (insulated-gate bipolar transistors) or MOSFETs (metal-oxide-semiconductor field-effect transistors) are used. More information can be found for example in Reference [24].

Overall configuration

Figure 19.15 shows an example of a transformer-less inverter as it could be sold for household systems. Besides the actual DC-AC converter, which is realized as an H-bridge, it also contains a DC-DC boost converter and an MPPT that uses the voltage and current measured on the PV side as input. The system sketched also contains several electronic components for increased safety, such as a residual current measurement sensor to detect leakage currents above a certain threshold and to shut down the inverter if these currents appear. Another example is a grid-monitoring unit to prevent islanding (see below) [153].

Potential-induced degradation (PID)

As mentioned earlier, in PV systems that have a transformer less inverter, no galvanic separation between the DC- and the AC-parts of the system. Because of this lack of galvanic isolation, a potential of -500 V or more between the PV modules and the ground can occur, which can lead to potential-induced degradation (PID) of the PV modules. Many thin-film modules contain a transparent conducting oxide (TCO) front contact that is deposited in superstrate configuration on the glass top plate. Positively charged sodium ions can then travel into the TCO because of this potential. This leads to corrosion and consequently to performance loss of the module. Also, for crystalline silicon modules, PID can be a problem [153]. Therefore, for systems containing thin-film modules the inverter must have a transformer.

Islanding

A potential danger of grid-connected systems is *islanding*. Imagine that a PV system is installed in a street where the electricity grid is shut down in order to do maintenance work on the electricity cables. If it is a sunny day, the PV system will produce power and – without protection – deliver power to the grid. The electricity worker could then be at risk. This dangerous phenomenon is called islanding and must be prevented.

The inverter therefore must be able to detect when the electricity grid is shut-down. If this is the case, the inverter must stop delivering power to the grid.

Efficiency of power converters

To plan a PV system it is very important to know the efficiency of the power converters. This efficiency of DC-DC and DC-AC converters is defined as

$$\eta = \frac{P_o}{P_o + P_d}, \tag{19.20}$$

i.e. the fraction of the output power P_o to the sum of P_o with the dissipated power P_d. To estimate the efficiency we must therefore estimate the dissipated (lost) power, which can be seen as a sum of several components,

$$P_d = P_L + P_{\text{switch}} + P_{\text{other}}, \tag{19.21}$$

i.e. the power lost in the inductor P_L, the power lost in the switch, and other losses. P_L is given as

$$P_L = I_{L,\text{rms}}^2 R_L, \tag{19.22}$$

where $I_{L,\text{rms}}$ is the mean current flowing through the inductor. The losses in the switch strongly depend on the type of switch that is used. Other losses are for example resistive losses in the circuitry in-between the switches.

19. Components of PV systems

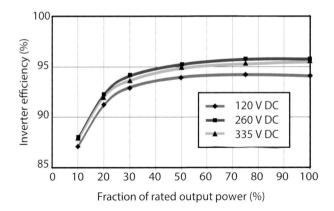

Figure 19.18: The power dependent efficiency for several input voltages of a *Fronius Galvo 1.5-1 208-240* inverter at 240 V AC. The nominal input voltage is 260 V DC. (Data taken from [154] and used with kind permission of the California Energy Commission).

For a complete inverter unit it is convenient to define the efficiency as

$$\eta_{\text{inv}} = \frac{P_{\text{AC}}}{P_{\text{DC}}}, \quad (19.23)$$

which is the ratio of the output AC power to the DC input power. Figure 19.18 shows the efficiency of a commercially available inverter for different input voltages. As we can see, in general the lower the output power, the less efficient the inverter. This is due to the power consumed by the inverter (self consumption) and the power used to control the various semiconductor devices which is relatively high at low output power. This efficiency characteristic must be taken into account when planning a PV system. Further, the efficiency is lower if the input voltage deviates from the nominal value. Indeed up and down conversion losses become relatively large as more energy is stored in the inductors.

19.3 Batteries

In this section we discuss a vital component not only of PV systems but of renewable energy systems in general. *Energy storage* is very important at both small and large scales in order to tackle the intermittency of renewable energy sources. In the case of PV systems, the intermittency of the electricity generation is of three kinds: first, *diurnal* fluctuations, i.e. the difference of irradiance during a 24 hour period. Secondly, fluctuations from one day to another because of changes of weather, especially changes in the cloud coverage. Thirdly, the *seasonal* fluctuations, i.e. the difference of irradiance between the summer and winter months. There are several technological options for realizing storage of energy. Therefore it is important to make an optimal choice.

Figure 19.19 shows a *Ragone* chart, where the *specific energy* is plotted against the *peak power density*. Because it uses a double logarithmic chart, storage technologies with very different storage properties can be compared in one. Solar energy applications require a

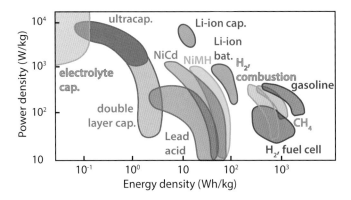

Figure 19.19: A *Ragone* chart of different energy storage methods. Capacitors are indicated with 'cap'.

high energy density and – depending on the application – also a reasonably high power density. For example, we cannot use capacitors because of their very poor energy density. For short to medium-term storage, the most common storage technology of course is the *battery*. Batteries have both the right energy density and power density to meet the daily storage demand in small and medium-sized PV systems. However, the seasonal storage problem at large scales is yet to be solved.

The ease of implementation and efficiency of the batteries are still superior to that of other technologies, such as pumping water to higher levels, compressed air energy storage, conversion to hydrogen, and flywheels. Therefore we will focus on battery technology as a viable storage option for PV systems.

Batteries are electrochemical devices that convert chemical energy into electrical energy. We can distinguish between primary and secondary batteries. Primary batteries convert chemical energy to electrical energy *irreversibly*. For example, *zinc carbon* and *alkaline* batteries are primary batteries.

Secondary batteries or *rechargeable batteries*, as they are more commonly known, convert chemical energy to electrical energy *reversibly*. This means that they can be recharged when an over potential is used. In other words, excess electrical energy is stored in these secondary batteries in the form of chemical energy. Typical examples of rechargeable batteries are *lead acid* or *lithium ion* batteries. For PV systems, only secondary batteries are of interest.

19.3.1 Types of batteries

Lead acid batteries are the oldest and most mature technology available. They will be discussed in detail later in this section.

Other examples are *nickel-metal hydride* (NiMH) and *nickel cadmium* (NiCd) batteries. NiMH batteries have a high energy density, which is comparable to that of lithium-ion batteries (discussed below). However, NiMH batteries suffer from a high rate of self discharge. On the other hand, NiCd batteries have much lower energy density than lithium-ion batteries. Furthermore, because of the toxicity of cadmium, NiCd batteries are widely

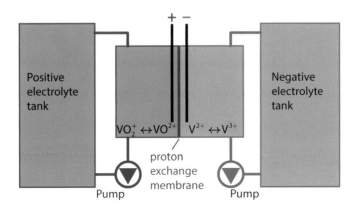

Figure 19.20: Schematic on a *vanadium* redox flow battery that employs vanadium ions.

banned in the European Union for consumer use. Additionally, NiCd batteries suffer from what is called the *memory effect*: the batteries loose their usable energy capacity if they are repeatedly charged after only being partially discharged. These disadvantages make NiMH and NiCd batteries unsuitable candidates for PV storage systems.

Lithium-ion batteries (LIBs) and *lithium-ion polymer* batteries, which are often referred to as lithium polymer (LiPo) batteries, have been heavily investigated in recent years. Their high energy density has already made them the favourite technology for light weight storage applications, for example in mobile telephones. However, these technologies still suffer from high costs and low maturity.

The last and most recent category of batteries that we will discuss in this treatise are *redox flow* batteries. Lead acid batteries and LIBs, the two main storage options for PV systems, are similar in the sense that their electrodes undergo chemical conversion during charging and discharging, which makes their electrodes degenerate with time, leading to inevitable 'ageing' of the battery. In contrast, redox flow batteries combine properties of both batteries and fuel cells, as illustrated in Figure 19.20. Two liquids, a *positive electrolyte* and a *negative electrolyte* are brought together, separated only by a membrane, which is only permeable to protons. The cell can thus be charged and discharged without the reactants being mixed, which in principle prevents the liquids from ageing. The chemical energy in a redox flow battery is stored in its two electrolytes, which are stored in two separate tanks. Since it is easy to make the tanks larger, the maximal energy that can be stored in such a battery is therefore not restricted. Further, the maximal output power can easily be increased by increasing the area of the membrane, for example by using more cells at the same time. The major disadvantage is that such a battery system requires additional components such as pumps, which makes it more complicated than other types of batteries.

Figure 19.21 shows a Ragone chart comparing different battery technologies. In contrast to Figure 19.19, here the gravimetric energy density and the volumetric energy density are plotted against each other. The *gravimetric energy density* is the amount of energy stored per mass of the battery; it typically is measured in Wh/kg. The volumetric energy density is the amount of energy stored per volume of battery; it is given in Wh/l. The higher

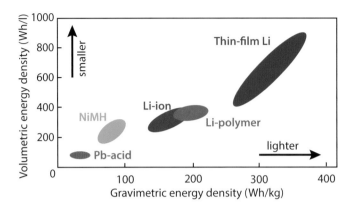

Figure 19.21: A Ragone chart for comparing different secondary battery technologies with each other.

the gravimetric energy density, the lighter the battery can be. The higher the volumetric energy density, the smaller the battery can be.

Figure 19.21 shows that lead-acid batteries have both the lowest volumetric and gravimetric energy densities among the different battery technologies. Lithium-ion batteries show ideal material properties for use as storage devices. Redox flow batteries are very promising. However, both LIB and redox flow batteries are still in the development phase which makes these technologies still very expensive. Thus, because of their unequalled maturity and hence low cost, lead acid batteries are still the storage technology of choice for PV systems.

Figure 19.22 shows a sketch of a lead acid battery. A typical battery is composed of several individual cells, of which each has a nominal cell voltage around 2 V. Different methods of assembly are used. In *block assembly*, the individual cells share the housing and are interconnected internally. For example, to get the typical battery voltage of 12 V, six cells are connected in series. As the name suggests, lead-acid batteries use an acidic electrolyte, namely diluted sulphuric acid H_2SO_4. Two plates of opposite polarity are inserted in the electrolyte solution, which act as the electrodes. The electrodes contain grid-shaped lead carrier and porous active material. This porous active material has a sponge-like structure, which provides sufficient surface area for the electrochemical reaction. The active mass in the negative electrode is lead (Pb), while in the positive electrode lead dioxide (PbO_2) is used. In the figure, the chemistry of charging and discharging the battery is also shown.

When the battery is discharged, electrons flow from the negative to the positive electrode through an external circuit, causing a chemical reaction between the plates and the electrolyte. This forward reaction also depletes the electrolyte, affecting its *state of charge* (SoC). When a source with a voltage higher than the actual battery voltage is connected to the battery, the reverse reaction is enabled. Then, the flow of electrons is reversed and the battery is recharged. In PV systems, this source is nothing but the PV module or array. In grid-connected systems, the inverter operating as an AC-DC converter can be used to charge the battery.

19. Components of PV systems

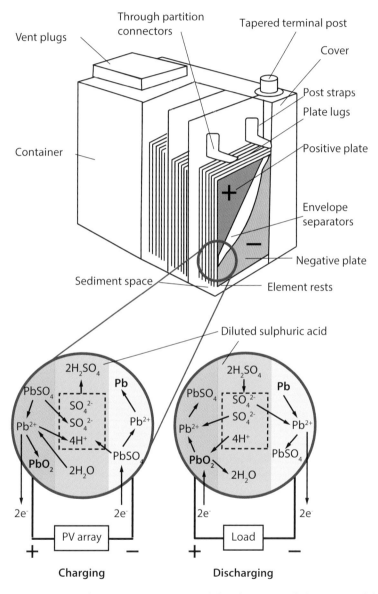

Figure 19.22: Schematic of a lead acid battery and the chemistry of charging and discharging.

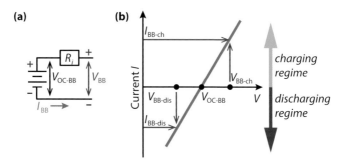

Figure 19.23: A simple model for (a) the equivalent circuit and; (b) the I-V characteristics of a battery.

19.3.2 Equivalent circuit

To understand how a battery works, it is important to look at the equivalent circuit. As we can see from Figure 19.23 (a), a simple representation of the equivalent circuit is given by a voltage source with a constant voltage $V_{\text{OC-BB}}$ and an internal resistance R_i. The subscript BB denotes *battery bank*. In reality, $V_{\text{OC-BB}}$ will not be constant, but a function of the state of charge (SoC, see below), the ambient temperature, and others parameters,

$$V_{\text{OC-BB}} = f(\text{SoC}, T, \ldots). \tag{19.24}$$

For the discussions in this book, we will neglect these dependencies.

Because of R_i, the voltage V_{BB} will differ from $V_{\text{OC-BB}}$. In fact, V_{BB} is also dependent on the current I_{BB} flowing through the battery, as seen by the equation

$$I_{\text{BB}} = \frac{1}{R_i}\left(V_{\text{BB}} - V_{\text{OC-BB}}\right). \tag{19.25}$$

Note that we use the convention that I_{BB} is positive when the battery is charged and negative when it is discharged. The resulting I-V characteristics of the battery bank are illustrated in Figure 19.23 (b). In this figure, the charging regime and the discharging regime are depicted.

The current I_{BB} determines the power loss in the battery,

$$P_{\text{BB}}(\text{loss}) = I_{\text{BB}}^2 R_i. \tag{19.26}$$

This power is always lost, irrespective of the sign of I_{BB}, and hence irrespective of whether the battery is charged or discharged.

19.3.3 Battery parameters

We will now discuss some parameters that are used to characterize batteries.

Voltage

First, we will discuss the *voltage* rating of the battery. The voltage at which the battery is rated and is supposed to operate is the *nominal voltage*. The so-called *solar batteries* or lead acid batteries for PV applications are usually rated at 12 V, 24 V or 48 V. The actual voltage of PV systems may differ from the nominal voltage. This mainly depends on the SoC and the temperature of the battery.

Capacity

When talking about batteries, the term *capacity* refers to the amount of charge that the battery can deliver at the rated voltage. The capacity is directly proportional to the amount of electrode material in the battery. This explains why a small cell has a lower capacity than a large cell that is based on the same chemistry, even though the open circuit voltage across the cell will be the same for both. Thus, the voltage of the cell is more chemistry-based, while the capacity is based more on the quantity of the active materials used.

The capacity C_{bat} is measured in ampere-hours (Ah). Note that charge is usually measured in *coulomb* (C). As the electric current is defined as the rate of flow of electric charge, Ah is another unit of charge. Since 1 C = 1 *ampere-second* (As), 1 Ah = 3,600 C. For batteries, Ah is the more convenient unit, because in the field of electricity the amount of energy is usually measured in watt-hours (Wh). The energy capacity of a battery is simply given by multiplying the rated battery voltage measured in volts by the battery capacity measured in amp-hours,

$$E_{\text{bat}} = C_{\text{bat}} V, \qquad (19.27)$$

which results in the battery energy capacity in watt-hours.

C-rate

A brand new battery with 10 Ah capacity can theoretically deliver a 1 A current for 10 hours at room temperature. Of course, in practice this is seldom the case due to several factors. Therefore, the *C-rate* is used, which is a measure of the rate of discharge of the battery relative to its capacity. It is defined as the maximal capacity that can be drawn from the battery in one hour divided by the battery capacity. For example, a C-rate of 1 for a 10 Ah battery corresponds to a discharge current of 10 A over 1 hour. A C-rate of 2 for the same battery would correspond to a discharge current of 20 A over half an hour. Similarly, a C-rate of 0.5 implies a discharge current of 5 A over 2 hours. In general, it can be said that a C-rate of n corresponds to the battery getting fully discharged in $1/n$ hours, irrespective of the battery capacity.

Battery efficiency

To design PV systems it is very important to know the *efficiency* of the storage system. Usually the *round-trip efficiency* is used, which is given as the ratio of the total storage output to the total storage input,

$$\eta_{\text{bat}} = \frac{E_{\text{out}}}{E_{\text{in}}}. \qquad (19.28)$$

For example, if 10 kWh is pumped into the storage system during charging, but only 8 kWh can be retrieved during discharging, the round-trip efficiency of the storage system is 80%. The round-trip efficiency of batteries can be broken down into two efficiencies: first, the *voltaic efficiency*, which is the ratio of the average discharging voltage to the average charging voltage,

$$\eta_V = \frac{V_{\text{discharge}}}{V_{\text{charge}}}. \tag{19.29}$$

This efficiency covers the fact that the charging voltage is always a little above the rated voltage in order to drive the reverse chemical (charging) reaction in the battery.

Secondly, we have the *coulombic efficiency* (or Faraday efficiency), which is defined as the ratio of the total charge extracted from the battery to the total charge put into the battery over a full charge cycle,

$$\eta_C = \frac{Q_{\text{discharge}}}{Q_{\text{charge}}}. \tag{19.30}$$

The *battery efficiency* then is defined as the product of these two efficiencies,

$$\eta_{\text{bat}} = \eta_V \times \eta_C = \frac{V_{\text{discharge}}}{V_{\text{charge}}} \frac{Q_{\text{discharge}}}{Q_{\text{charge}}}. \tag{19.31}$$

When comparing different storage devices, this round-trip efficiency is usually considered. It includes all the effects of the different chemical and electrical non-idealities occurring in the battery.

State of charge and depth of discharge

Another important battery parameter is the *state of charge* (SoC), which is defined as the percentage of the battery capacity available for discharge,

$$\text{SoC} = \frac{E_{\text{bat}}}{C_{\text{bat}}V}. \tag{19.32}$$

Thus, a 10 Ah rated battery that has been drained by 2 Ah is said to have a SoC of 80%. Also the *depth of discharge* (DoD) is an important parameter. It is defined as the percentage of the battery capacity that has been discharged,

$$\text{DoD} = \frac{C_{\text{bat}}V - E_{\text{bat}}}{C_{\text{bat}}V}. \tag{19.33}$$

For example, a 10 Ah battery that has been drained by 2 Ah has a DoD of 20%. The SoC and the DoD are complementary to each other.

Cycle lifetime

The *cycle lifetime* is defined as the number of charging and discharging cycles after which the battery capacity drops below 80% of the nominal value. Usually, the cycle lifetime is specified by the battery manufacturer as an absolute number. However, stating the battery

19. Components of PV systems

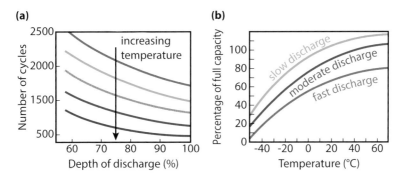

Figure 19.24: Qualitative illustration of (a) the cycle lifetime of a Pb acid battery as a function of the DoD for different temperatures; and (b) effect of the temperature on the battery capacity (with kind permission from S. Bowden, Arizona State University) [155].

lifetime as a single number is an oversimplification because the different battery parameters discussed so far are not only related to each other but are also dependent on the temperature.

Figure 19.24 (a) shows the cycle lifetime as a function of the DoD for different temperatures. Clearly, colder operating temperatures mean longer cycle lifetimes. Furthermore, the cycle lifetime depends strongly on the DoD. The smaller the DoD, the higher the cycle lifetime. So, the battery will last longer if the average DoD can be reduced during the lifetime of the battery. Also, *battery overheating* should be strictly controlled. Overheating can occur because of overcharging and subsequent over voltage of the lead acid battery. To prevent this, charge controllers are used that we address in the next section.

Temperature effects

While the battery lifetime is increased at lower temperatures, another effect needs to be considered. The temperature affects the battery capacity during regular use too.

As seen from Figure 19.24 (b), the lower the temperature, the lower the battery capacity. At higher temperatures, the chemicals in the battery are more active, leading to an increased battery capacity. At high temperatures, it is even possible to reach above-rated battery capacity. However, such high temperatures are severely detrimental to the battery health. In Figure 19.24 (b), it can be seen that the battery capacity increases when the discharge current is lower. This is because the discharge process in the battery is diffusion-limited; if more time is allowed, a better exchange of chemical species between the pores in the plate, and the electrolyte, can take place.

Ageing

The major cause of ageing of the battery is *sulphation*. If the battery is insufficiently recharged after being discharged, sulphate crystals start to grow, which cannot be completely transformed back into lead or lead oxide. Thus the battery slowly loses its active material mass and hence its discharge capacity. *Corrosion* of the lead grid at the electrode is another common ageing mechanism. In the case of lead-acid batteries, *antimony poisoning* is a major cause of accelerated ageing [156]. Corrosion leads to increased grid resistance due to high positive potentials. Further, the electrolyte can *dry out*. At high charging voltages, gassing can occur, which results in the loss of water. Thus, demineralised water should be used to refill the battery from time to time.

19.3.4 Lead acid batteries used in PV applications

One can distinguish lead acid batteries based on the type of plates being used:

- In *flat plate* lead acid batteries the plates are significantly thicker than in the case of starter batteries, implying a cycle life of 1,000 cycles for a daily depth of discharge (DDoD) of 20% at 25 ∘ C.

- The positive plates of *tubular plate* lead-acid batteries are made of porous tubes, which contain the active mass. These tubes prevent the loss of active mass as a result of changes in volume, which occur in cyclic operation. The negative plates in tubular plate batteries are flat plates. The cycle life of this type of battery is considerably longer than for batteries with flat plates described above. For real traction applications, such as in forklift trucks, tubular plates with a fairly high antimony content are used. These batteries have an excellent cycle life, but their self-discharge is far too high for most PV applications. For these, tubular plates are used with an antimony content such that the cycle life is still fairly good, but with a low self-discharge. A typical value for the cycle life is 3,000-4,000 cycles for a daily depth of discharge of 20%.

- In *rod plate* lead acid batteries the positive plate is made up of rods with the active mass between them. The plates have a low antimony content. The construction height of the battery is fairly low, so there is almost no stratification of acid. This battery has superior properties for PV applications, but is rather expensive.

One can also distinguish open (vented) and sealed (maintenance-free) lead acid batteries. In order to make lead acid batteries maintenance-free, all or part of the antimony in the plates is replaced by calcium. A calcium alloy with a very low antimony content (< 1%) has a very low self-discharge (approximately 1% per month at 25 °C) and reduces the splitting of water into oxygen and hydrogen. In most maintenance-free lead acid batteries the electrolyte is contained either in a gel between the plates or in a micro porous separator. Oxygen produced at the positive plate can diffuse through the gel or the micro porous separator and reach the negative plate, where it reacts with hydrogen to form water again. Maintenance-free lead acid batteries frequently used for PV applications are flat plate batteries with a gel or micro porous separator, and tubular plate batteries with a gel.

19.4 Charge controllers

After having discussed different types of batteries and different battery characteristics, it is now time to discuss *charge controllers*, which are used in PV systems that use batteries. As we have seen above, it is very important to charge and discharge batteries at the right voltage and current levels in order to ensure a long battery lifetime. A battery is an electrochemical device that requires a small over potential to be charged. However, batteries have

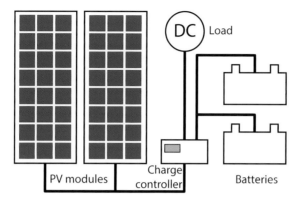

Figure 19.25: Illustrating the position of the *charge controller* in a generic PV system with batteries.

strict voltage limits, which are necessary for their optimal functioning. Further more, the amount of current sent to the battery by the PV array and the current flowing through the battery while being discharged have to be within well defined limits for proper functioning of the battery. We have seen before that lead-acid batteries suffer from both overcharge and over discharge. On the other hand, the PV array responds dynamically to ambient conditions like irradiance, temperature and other factors such as shading. Thus, directly coupling the battery to the PV array and the loads is detrimental to the battery lifetime.

Therefore a device is needed that controls the currents flowing between the battery, the PV array and the load that ensures that the electrical parameters present at the battery are kept within the specifications given by the battery manufacturer. These tasks are done by a *charge controller*, which has several different functionalities, depending on the manufacturer. We will discuss the most important functionalities. A schematic of its location in a PV system is shown in Figure 19.25.

When the sun is shining at peak hours during summer, the generated PV power excesses the load. The excess energy is sent to the battery. When the battery is fully charged and the PV array is still connected to the battery, the battery might overcharge, which can cause several problems like gas formation, capacity loss or overheating. Here, the charge controller plays a vital role by decoupling the PV array from the battery. Similarly, during severe winter days at low irradiance, the load exceeds the power generated by the PV array, such that the battery is heavily discharged. Over discharging the battery has a detrimental effect on the cycle lifetime, as discussed above. The charge controller prevents the battery from being over discharged by disconnecting the battery from the load.

For optimal performance, the battery voltage has to be within specified limits. The charge controller can help maintain an allowed voltage range in order to ensure a healthy operation. Further, the PV array will have its V_{MPP} at different levels, based on the temperature and irradiance conditions. Some charge controllers perform an appropriate voltage regulation to ensure the battery operates in the specified voltage range, while the PV array is operating at the MPP.

However, most simple charge controllers available on the market have no MPP tracker included. In this case the battery will determine the voltage at which the module is op-

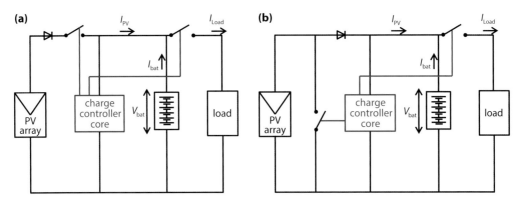

Figure 19.26: Basic wiring scheme of (a) a series; and (b) a shunt charge controller.

erated. In is then important to size the module such that it fits the battery. For example, for systems that are connected to a 12 V battery, crystalline silicon panels with 36 cells in series are often used, while for 24 V batteries modules with 72 cells connected in series are appropriate.

As we have seen above, certain C-rates are used as battery specifications. The higher the charge/discharge rates, the lower the coulombic efficiency of the battery. The optimal charge rates, as specified by the manufacturer, can be reached by manipulating the current flowing into the battery. A charge controller that contains a proper current regulator is also able to control the C-rates. Finally, the charge controller can impose limits on the maximal currents flowing into and out of the battery.

If no blocking diodes are used, it is even possible that the battery can 'load' the PV array, when it is operating at a very low voltage. This means that the battery will impose a forward bias on the PV modules and make them consume the battery power, which leads to the solar cells heating up. Traditionally, blocking diodes are used at the PV panel or string level to prevent this *back discharge* of the battery through the PV array. However, this function is also easily integrated into the charge controller.

We distinguish between *series* and *shunt* controllers, as illustrated in Figure reffig:com-charge-controller. In a series controller, overcharging is prevented by disconnecting the PV array until a particular voltage drop is detected, at which point the array is connected to the battery again. On the other hand, in a parallel or shunt controller, overcharging is prevented by short-circuiting the PV array. This means that the PV modules work under short circuit mode, and that no current flows into the battery. These topologies also ensure over discharge protection using power switches for the load connections which are appropriately controlled by the charge controller algorithms.

As we have seen above, temperature plays a crucial role in the functioning of the battery. Not only does temperature affect the lifespan of the battery, but it also changes its electrical parameters significantly. Thus, modern charge controllers have a temperature sensor included, which is attached to the battery back. This sensor allows the charge controller to adjust the electrical parameters of the battery, such as the operating voltage, to the temperature. The charge controller thus keeps the operating range of the battery within the optimal range of voltages.

19.4.1 Charge controllers for lead acid batteries used for PV applications

The role of the overcharge protection part of controllers will be discussed in more detail now for the case of lead acid batteries. A typical mode of operation of an autonomous PV system is that the battery will be discharged partly during the night and will be charged again during the day. High states of charges will be reached after noon. This implies that the time available subsequently in the afternoon to charge the battery fully is limited, or in other words, the time available to convert all the lead sulphate into lead and lead dioxide is limited. This conversion is crucial because if lead sulphate stays too long in the amorphous state it will change into polycrystalline lead sulphate which cannot then be changed into lead and lead oxide. In order to convert all the amorphous lead sulphite into lead and lead oxide in the short time available, sufficient overvoltage is required. This voltage, however, is not allowed to be too high for too long a period, otherwise voltage-induced corrosion of the plates takes place.

In practice, during charging of the batteries, the battery voltage will rise gradually in the course of a day. A good controller will switch off the array current once a certain voltage level is reached and will switch the array on again at a slightly lower voltage. Keeping the voltage within a narrow voltage band close to the gassing voltage, the optimum charge conditions are realized. In controllers with a boost charge facility, the array current is switched off at a relatively high voltage once (a day) and subsequently the voltage is kept within a narrow band as described above. During boost charging the overvoltage is relatively high during a short period of time, accelerating the conversion of lead sulphate into lead and lead oxide. The precise voltage settings are dependent on the specific type of lead acid battery.

In the case of lead acid batteries, the deep discharge protection acts on the battery voltage, in general. Because the battery terminal voltage depends both on the actual state of charge and the internal voltage drop caused by the battery current flowing through the internal battery resistance (see Section 19.3.2), the controller needs to act on the battery terminal voltage corrected for this internal voltage drop. A deep discharge protection doing so is said to have *current compensation*. If the voltage drop along the battery cables (and fuses if present) cannot be neglected, separate leads (battery voltage sense lines) are used to measure the actual battery terminal voltage.

19.5 Cables

The overall performance of PV systems is also strongly dependent on the correct choice of the cables. We therefore will discuss how to choose suitable cables. But we start our discussion with *colour conventions*.

PV systems usually contain DC and AC parts. To correctly install a PV system, it is important to know the colour conventions. For *DC cables*,

- **red** is used for connecting the + contacts of the different system components with each other, while
- **black** is used for connecting the − *contacts* and for interconnecting the modules with each other.

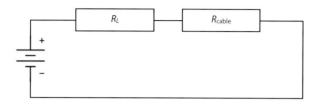

Figure 19.27: A circuit with a load R_L and cable resistance R_{cable}. The connections drawn in between the components are lossless.

For DC string cabling between modules in an array, black cables are often used because of the poor UV stability of coloured cables. Polarity indication is done by the use of dedicated connectors with polarity markings. An example is the famous multi-contact connector, which can be delivered together with the flying cable as one unit.

For *AC wiring*, different colour conventions are used around the world.

- For example, in the *European Union*,
  blue is used for neutral,
  green-yellow is used for the protective earth, and
  brown (or another colour such as **black** or **grey**) is used for the phase.

- In the *United States* and *Canada*,
  silver is used for neutral,
  green-yellow, **green** or a **bare** conductor is used for the protective earth, and
  and **black** (or another colour) is used for the phase.

- In *India* and *Pakistan*, for example
  black is used for neutral,
  green is used for the protective earth and
  blue, **red**, or **yellow** is used for the phase.

Therefore it is very important to check the standards of the country in which the PV system is going to be installed.

The cables have to be chosen such that *resistive losses* are minimal. For estimating these losses, we look at a very simple system that is illustrated in Figure 19.27 The system consists of a power source and a load with resistance R_L, and the cables have a resistance R_{cable}, which is also shown. The power loss at the cables is given as

$$P_{\text{cable}} = I \times \Delta V_{\text{cable}}, \tag{19.34}$$

where ΔV_{cable} is the voltage drop across the cable, which is given as

$$\Delta V_{\text{cable}} = V \frac{R_{\text{cable}}}{R_L + R_{\text{cable}}}. \tag{19.35}$$

Using

$$V = I(R_L + R_{\text{cable}}) \tag{19.36}$$

we find
$$P_{cable} = I^2 R_{cable}. \qquad (19.37)$$
Hence, as the current doubles, four times as much heat will be dissipated at the cables. It is now obvious why modern modules have all cells connected in series.

Let us now calculate the resistance of a cable with length ℓ and cross-section A. It is clear that if ℓ is doubled, R_{cable} also doubles. In contrast, if A doubles, R_{cable} halves. The resistance is thus given by
$$R_{cable} = \rho \frac{\ell}{A} = \frac{1}{\sigma}\frac{\ell}{A}, \qquad (19.38)$$
where ρ is the *specific resistance* or *resistivity* and σ is the *specific conductance* or *conductivity*. If both ℓ and A are given in metres, their units are $[\rho] = \Omega$m and $[\sigma] = S/m$, where S denotes the unit for conductivity, which is *siemens*. Note that the resistivity is also dependent on the temperature.

For electrical cables it is convenient to have ℓ in metres and A in mm^2. When using this convention, we find the following units for ρ and σ:

$$1\,\Omega m = 1\,\Omega \frac{m^2}{m} = 1\,\Omega \frac{10^6\,mm^2}{m} = 10^6\,\Omega \frac{mm^2}{m},$$

$$1\,\frac{S}{m} = 1\,S\frac{m}{m^2} = 1\,\Omega \frac{m}{10^6\,mm^2} = 10^{-6}\,\Omega\frac{m}{mm^2}.$$

The metals most widely used for electrical cables are *copper* and *aluminium*. Their resistances and conductivities are

$$\rho_{Cu} = 1.68 \times 10^{-8}\,\Omega m = 1.68 \cdot 10^{-2}\,\Omega\frac{mm^2}{m},$$

$$\sigma_{Cu} = 5.96 \times 10^{7}\,\frac{S}{m} = 59.6\,S\frac{m}{mm^2},$$

$$\rho_{Al} = 2.82 \times 10^{-8}\,\Omega m = 2.82 \times 10^{-2}\,\Omega\frac{mm^2}{m},$$

$$\sigma_{Al} = 3.55 \times 10^{7}\,\frac{S}{m} = 35.5\,S\frac{m}{mm^2}.$$

Usual cross-sections for cables are 1.5 mm^2, 2.5 mm^2, 4 mm^2, 6 mm^2, 10 mm^2, 16 mm^2, 25 mm^2, 35 mm^2, etc. For series-connected strings of PV modules, 4 mm^2 cables are mostly used. DC cables must usually be certified for voltages up to 1 kV; further they should be coated with a UV-resistive material, as they are often exposed to solar radiation.

19.6 Exercises

19.1 Calculate the maximum power of a solar cell having current and voltage values of $I_{MPP} = 30$ mA and $V_{MPP} = 0.7$ V, respectively, at the maximum power point.

19.2 Consider an ideal solar cell, with open-circuit voltage V_{oc}, short circuit current I_{sc}, maximum power point voltage V_{MPP} and maximum power point current I_{MPP}. Which resistance should the load have for the cell to operate at its maximum?

  (a) V_{OC}/I_{SC}

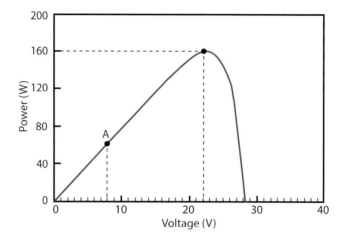

Figure 19.28

(b) V_{OC}/I_{MPP}

(c) V_{MPP}/I_{SC}

(d) V_{MPP}/I_{MPP}.

19.3 Consider a PV module with the following parameters as measured at STC (standard testing conditions): $V_{OC} = 50$ V, $I_{SC} = 4.5$ A, $I_{MPP} = 4.0$ A, $V_{MPP} = 40$ V, $\eta = 18\%$. Suppose that the module is not connected to an MPPT device, and is instead directly connected to a purely resistive, variable load R. The load R can be tuned to give a resistance between 0 and 1 kΩ. At which value of resistance will you keep the load to derive maximum power from the PV module under STC conditions?

19.4 Consider typical I-V and P-V curves of a solar module as shown Figure 19.2. Which of the following statements are true? (More solutions possible.)

(a) For any set of irradiance and temperature conditions, $P_{MPP} > V_{OC} \times I_{SC}$.

(b) With the rest of the conditions remaining the same, if the irradiance falling on the PV module uniformly increases such that the new P-V curve has an MPP at P'_{MPP}, then $P'_{MPP} > P_{MPP}$.

(c) The rest of the conditions remaining the same, if the irradiance falling on the PV module uniformly increases such that the new P-V curve has an MPP at P'_{MPP}, then $P'_{MPP} < P_{MPP}$.

(d) For $V = V_{MPP}$, we have $dP/dV = 0$.

19.5 Consider that an MPPT based on the *perturb and observe* algorithm is connected to a PV module with the P-V curve shown in Figure 19.28. The MPPT has two different modes of operation: coarse and fine. A coarse adjustment corresponds to a voltage change of ± 5 V, while a fine adjustment corresponds to a voltage change of ± 1 V.

If the MPPT has taken a step such that a change in voltage has led to an increase in power, then the next step is a coarse adjustment in the same direction as the previous step.

19. Components of PV systems

On the other hand, if the MPPT has taken a step such that a change in voltage has led to a decrease in power, then the next step is a fine adjustment in the opposite direction to the previous step.

Consider that the module is currently operating at point A, corresponding to a voltage of 8V, and that the MPPT is just about to take a coarse step by increasing the voltage. If the irradiance and temperature conditions are maintained, i.e. the *P-V* curve remains the same, how many steps will the MPPT take before it reaches the MPP?

19.6 Consider the following statements related to the field of MPP tracking:

  (a) The PV output is never steady at the MPP.
  (b) Inclusion of a pilot solar cell can improve the yield of the PV module.
  (c) The hardware complexity is high for implementing this design.
  (d) The output of the PV module is maximum only if load resistance is $R = V_{MPP}/I_{MPP}$.

Consider also the following conditions at the PV output:

  (i) An MPPT device implementing fractional open-circuit voltage method.
  (ii) An MPPT device implementing incremental conductance method.
  (iii) No MPPT device connected; solar module directly connected to the load.
  (iv) An MPPT device implementing perturb and observe method.

Match each of the statements (a) through (d) with one of the statements (i) through (iv).

19.7 Which of the following conditions is **not** required for an ideal solar inverter?

  (a) A very high conversion efficiency.
  (b) Detection and prevention capability against islanding.
  (c) Continuing to supply power to the grid in the event of grid failure.
  (d) A long lifetime.
  (e) Built-in MPP tracking capability.

19.8 Which of the following statements is **not** true regarding the solar inverter topologies?

  (a) Module level (micro)inverters are better suited to get the maximum power out of each PV module compared to central inverters, especially when there are partially shaded arrays.
  (b) String inverters ensure a higher yield than central inverters when the arrays are partially shaded.
  (c) String inverters ensure a higher yield than module level (micro)inverters when the arrays are partially shaded.
  (d) Module level (micro)inverters ensure a higher yield than string inverters when the arrays are partially shaded.

19.9 Which of the following statements is NOT true about solar inverters?

  (a) A stand-alone inverter is supposed to work as an AC voltage source for a specified range of AC loads.
  (b) An ideal grid-tied inverter has its operational voltage and frequency synchronized with those of the grid.

(c) A bimodal inverter can power backup loads even if the grid-connection is disrupted.

(d) An ideal grid-tied inverter works as a perfect voltage source, and changes the grid voltage and frequency to match its own.

19.10 Which of the following is **not** a secondary battery?

(a) Zinc carbon battery

(b) Lead acid battery

(c) Nickel cadmium battery

(d) Lithium polymer battery

19.11 The *Nuon Solar Team* from Delft University of Technology in the Netherlands designed a very fast car powered by solar energy, the *Nuna7*. In 2013, they entered the World Solar Challenge Australia. Directly after coming back to the Netherlands, they started looking at improving their car. For electricity storage, they needed a solution as light as possible. Looking at the Ragone chart in Figure 19.21, which would be the preferred technology for the batteries of the Nuna8 car?

(a) Lead acid batteries.

(b) Nickel or metal hydride batteries.

(c) Lithium batteries.

(d) Lithium-ion batteries.

19.12 Consider a lead acid battery with 100 Ah capacity and a rated voltage of 12 V.

(a) What is the total capacity of energy in watt-hours that can be stored in the battery?

(b) Assume that the battery is at 40% of its rated capacity. The battery is now charging at a C-rate of 2C. What is the charging current that is going into the battery?

(c) How much time will it take for the battery to increase the SoC from 40% to 100%, assuming a constant C-rate of 2C? You may assume a linear rate of charging.

(d) If the voltaic efficiency of the battery is 90%, and the coulombic efficiency of the battery is 90%, how high is the round-trip efficiency of storage?

19.13 If the battery is charged at an average constant voltage of 13 V, and discharged at an average constant voltage of 11.7 V, what is the voltaic efficiency of the battery?

19.14 Below, several aspects are listed that are important in PV systems.

(a) Providing optimum charge to the batteries.

(b) Preventing the batteries from overcharging.

(c) Providing deep discharge protection to the batteries.

(d) Supplying information on the state of charge (SoC) of the batteries.

(e) Maximum power point tracking (MPPT) of the PV module/array.

Which of the above aspects can be performed by a modern day charge controller?

19.15 Consider four identical solar modules connected to a charge controller with two cables having specific length and section. A current flows through the cables. The cables are 10 m in length with a cross-section of 4 mm². The cable material is copper with a conductivity of 59.6 S m mm^{-2}. Assume that the solar modules have a maximum power of 100 Wp.

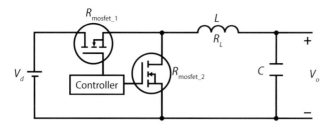

Figure 19.29

(a) How much power is lost as heat in the cables?

(b) Considering the maximum power of the solar modules, what would be the relative power lost in the cables?

(c) Would switching to thicker cables with a cross-section of 10 mm² be helpful in decreasing such relative power loss?

(d) Now repeat questions (a) to (c) for aluminium cables with a conductivity of 35.5 S m mm^{-2}.

19.16 Consider three identical solar modules connected to a charge controller by two copper cables that have lengths of 6 m and conductivity of 59.6 S m mm^{-2}. The average battery charging voltage is 12 V and the maximum power of a single solar module is 100 Wp. Calculate the required cross-section of the cables if the allowed power loss in the cables is 5%.

19.17 350 watt-hours of energy is required daily from a small PV system with a battery. The DC system voltage is 12 V. Calculate the capacity of the battery needed if its depth of discharge is 30%.

19.18 In the schematic shown in Figure 19.29, $V_d = 12$ V, $V_o = 4$ V and $I_o = 3$ A. The values of resistances are $R_{mosfet\_1} = 0.025$ Ω, $R_{mosfet\_2} = 0.02$ Ω and $R_L = 0.1$ Ω. The value of inductance, L is 50 µH. The controller switches *on* exclusively either the MOSFET 1 or the MOSFET 2. Assume that the switching frequency is 300 kHz and consider only the inductor and MOSFET's conduction losses.

(a) What type of converter is this?

(b) How much is the duty cycle D?

(c) How much is the switching time T_s?

(d) What is the efficiency of this converter?

20
PV system design

In Chapter 15, we thoroughly discussed PV modules. Further, in Chapter 18 we discussed how to estimate the irradiance on a PV module in dependence of the position of the Sun and partial absorption and diffusion in the atmosphere. Finally, we introduced all the other components of PV Systems in Chapter 19.

In this chapter, we will combine the knowledge gained so far in order to design complete PV systems. We can design PV systems at different levels of complexity. For a first approximation, the performances of the PV modules and the other components (like the inverter) at standard test conditions (STC) and the number of *equivalent sun hours* (ESH) at the location of the PV system are sufficient. The concept of STC (AM1.5 illumination with a total irradiance of 1,000 W/m^2 and a module temperature of 25 °C) was already introduced in Chapter 9; the notion of ESH will be discussed below. In a more detailed approach, performance changes of the different components due to changing irradiance and weather conditions are taken into account. Since these performance changes can be quite high, they can alter the optimal system design considerably.

There are two main paradigms for designing PV systems. First, the system can be designed such that the generated energy and the loads, i.e. the consumed energy, match. Hence, an *energy balance* must be done Secondly, the design of a PV system can be based on economics. We must distinguish between *grid-connected* and *stand-alone* systems. As we will see, the two have very different demands.

This chapter is organised as follows: First, as an example, we will discuss a design of a simple stand-alone system in Section 20.1. After that we will take a more detailed look at *load profiles* in Section 20.2. In Section 20.3 we discuss how weather and irradiance conditions affect the performance of PV modules and BOS components, mainly inverters. Finally, in Sections 20.4 and 20.5 we learn how to design grid-connected and stand-alone systems, respectively.

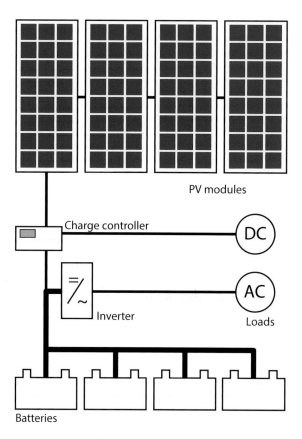

Figure 20.1: Illustrating a simple off-grid PV system with AC and DC loads (see also Figure 17.2).

20. PV system design

20.1 A simple approach for designing stand-alone systems

In this section, we will design a simple stand-alone system, as depicted in Figure 20.1. The design presented here is based on very simple assumptions and does not take any weather-dependent performance changes into account. Nonetheless, we will see the major steps that are necessary for designing a system. Such a simple design can be performed in a six step plan:

1. Determine the total load current[1] and operational time.
2. Add system losses .
3. Determine the solar irradiation in daily equivalent sun hours (ESH).
4. Determine the total solar array current requirements.
5. Determine the optimum module arrangement for solar array.
6. Determine the battery size for recommended reserve time.
7. Choose a suitable charge controller.

1. Determine the total load current and operational time

Before we can determine the current requirements of loads of our PV system, we have to decide on the nominal operational voltage of the PV system. Usual nominal voltages are 12 V, 24 V or 48 V. Once the voltage is known, the next step is to express the daily energy requirements of loads in terms of current and average operational time in ampere-hours (Ah). In the case of DC loads the daily energy (Wh) requirement is calculated by multiplying the power rating [W] of an individual appliance by the average daily operational time (h). Dividing the Wh by the nominal PV system operational voltage, the required Ah of the appliance is obtained.

> **Example**
>
> *A 12 V PV system has two DC appliances A and B requiring 15 and 20 W respectively. The average operational time per day is 6 hours for device A and 3 hours for device B. The daily energy requirements of the devices expressed in Ah are calculated as follows:*
> Device A: $15\,W \times 6\,h = 90\,Wh$
>
> Device B: $20\,W \times 3\,h = 60\,Wh$
> Total: $90\,Wh + 60\,Wh = 150\,Wh$
>
> $150\,Wh/12\,V = 12.5\,Ah$

In the case of AC loads, the energy use has to be expressed as a DC energy requirement since PV modules generate DC electricity. The DC equivalent of the energy use of

---

[1] If in a very simple PV system a charge controller without MPPT is used, we need to determine the current; not the power.

an AC load is determined by dividing the AC load energy use by the efficiency of the inverter, which can be assumed to be 95% for a good inverter. By dividing the DC energy requirement by the nominal PV system voltage, the Ah is determined.

> **Example**
>
> *An AC computer (device C) and TV set (device D) are connected to the PV system. The computer, which has rated power 40 W, runs 2 hours per day and the TV set with rated power 70 W is in operation 3 hours per day. The daily energy requirements of the devices expressed in DC Ah are calculated as follows:*
> *Device C:* $40\,W \times 2\,h = 80\,Wh$
>
> *Device D:* $70\,W \times 3\,h = 210\,Wh$
> *Total:* $80\,Wh + 210\,Wh = 290\,Wh$
>
> *DC requirement:* $290\,Wh/0.95 = 305\,Wh$
> $305\,Wh/12\,V = 25.5\,Ah$

2. Add system losses

Some components of the PV system, such as charge regulators and batteries, require energy to perform their functions. We denote the use of energy by the system components as system energy losses. Therefore, the total energy requirements of loads, which were determined in step 1, are increased by 20 to 30% in order to compensate for the system losses.

> **Example**
>
> *The total DC requirements of loads plus the system losses (20%) are determined as follows:*
> $(12.5\,Ah + 25.5\,Ah) \times 1.2 = 45.6\,Ah$

3. Determine the solar irradiation in daily equivalent sun hours (ESH)

How much energy a PV module delivers depends on several factors, such as local weather conditions, seasonal changes, and installation of modules. As we already discussed in Chapter 18, PV modules should be installed under the optimal *tilt angle* in order to achieve best year-round performance. However, if the PV system is only used during a specific period, the tilt angle needs to be optimized for that specific period. In fact the power output during winter is much less than the annual average, and in the summer months the power output will be above the average. Figure 20.2 shows the average global horizontal irradiance of the world in $kWh/m^2/day$, which is equivalent to the daily equivalent sun hours (ESH). We mainly see that the insolation decreases with latitude. However, other variations are visible because of regional climates.

As we already discussed in Chapter 5, solar cells are usually characterized with the AM1.5 spectrum, which is normalized such that it has a total irradiance of $1{,}000\,W/m^2$.

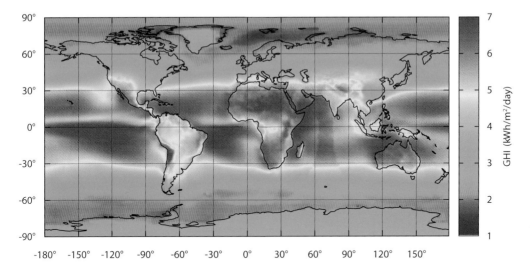

Figure 20.2: The average global horizontal irradiance of the world given in kWh/m^2/day, which is equivalent to the daily equivalent sun hours (ESH). (Data taken from [157].)

Hence 1 equivalent sun means a solar irradiance of 1,000 W/m^2. When solar irradiation data are available for a particular location, the equivalent sun hours can be determined. Figure 20.3 shows the annual irradiation for the Netherlands at an optimally tilted plane with a tilt of 36°. For example, in Delft (Netherlands) the average annual solar irradiation at a horizontal plane is 999 kWh/m^2 [158]. For optimally tilted modules the annual solar irradiation is a bit higher with 1,146 kWh/m^2 [158]. With the value from the AM1.5 spectrum, we hence find 999 equivalent sun hours at a horizontal plane and 1,146 equivalent sun hours at optimal tilt. When we take the length of the year (365.25 days) into account, we calculate the *average number of daily sun hours* to be 2.7 h and 3.1 h for horizontal and optimal tilt, respectively. For the rest of the discussion, we will use a value of 3 h of sun per day.

4. Determine the total solar array current requirements

The current that has to be generated by the solar array is determined by dividing the total DC energy requirement of the PV system including loads and system losses (calculated in step 2 and expressed in Ah) by the daily equivalent sun hours (determined in step 3).

Example

The total DC requirements of loads plus the system losses are 45.6 Ah. The daily ESH for the Netherlands is about 3 hours. The required total current generated by the solar array is 45.6 Ah / 3 h = 15.2 A.

Figure 20.3: Annual insolation at optimal tilt (36°) in the Netherlands in kWh/m^2 (source: PVGIS [158–160]).

5. Determine the optimum module arrangement for the solar array

Usually, PV manufacturers produce modules in a whole series of different output powers. In the optimum arrangement of modules, the required total solar array current (as determined in step 4) is obtained with the minimum number of modules. Modules can either be connected in series or in parallel to form an array. When modules are connected in series, the nominal voltage of the PV system is increased, while the parallel connection of modules results in a higher current.

The output voltage of PV modules should fit the battery voltage for optimal operation. If the modules are based on crystalline silicon technology, they usually contain 36 or 72 cells connected in series for systems based on 12 V or 24 V, respectively. The open circuit voltage (V_{oc}) of the cells used for the module is typically 0.6 V, therefore, the open circuit voltage of the module is 21.6 V or 43.2 V, respectively. As we have seen in Section 19.1, the optimal operation voltage of a solar cell is at around 70%–80% of the V_{oc}, hence 16.2 V or 32.4 V for modules with 36 and 72 cells, respectively. These voltages are very well suited for charging batteries with a nominal voltage of 12 V or 24 V, respectively.

The required number of modules in parallel is calculated by dividing the total current required from the solar array (determined in step 4) by the current generated by the module at maximum power. The number of modules in series is determined by dividing the nominal PV system voltage with the voltage at maximum power. Both this voltage and the current are given in the datasheet. The total number of modules is the product of the number of modules required in parallel and the number required in series.

> **Example**
>
> *The required total current generated by the solar array is* 15.2 A. *We have* Kyocera KD140 *modules with a nominal power of 140 W available, which consist of 36 cells connected in series. At the maximum power point, these modules have a voltage of* $V_{MPP} = 17.7$ V *and a current of* $I_{MPP} = 7.9$ A. *The number of modules in parallel is* 15.2 A / 7.9 A = 1.9 < 2 *modules. The nominal voltage of the PV system is* 12 V. *The required number of modules in series thus is* 12 V / 17.7 V = 0.67 < 1 *module. Therefore, the total number of modules in the array is* 2 × 1 = 2 *modules.*

6. Determine the battery size for recommended reserve time

Batteries are a major component of stand-alone PV systems. The batteries provide load operations at night or in combination with the PV modules during periods of limited sunlight. For a safe operation of the PV system one has to anticipate periods of cloudy weather and plan a reserve energy capacity stored in the batteries. Because of this reserve, the PV system is not dependent on energy generated by PV modules for a certain period of time, called *days of autonomy*. The required days of autonomy depend on the type of loads. For critical loads such as components for telecommunications systems the autonomy can be ten days or more, for residential use it is usually five days or less. This depends also on the weather at the PV system location.

The capacity [Ah] of the batteries is calculated by multiplying the daily total DC energy requirement of the PV system including loads and system losses (calculated in step 2 and

Table 20.1: Worksheet for designing a simple off-grid PV system based on rough assumptions.

Daily DC loads requirements			
DC load	W ×	h =	Wh
Total DC loads energy use:			

Daily AC loads requirements			
DC load	W ×	h =	Wh
Total AC loads energy use:			
/0.85 = **DC energy requirement**			

1	Daily DC energy use (DC loads)			
1	Daily DC energy use (AC loads)	+		
1	Daily DC energy use (all loads)	=		
	PV system nominal voltage	/		
	Daily Ah requirements (all loads)	=		
2	Add PV system losses	×		
	Daily Ah requirements (system)	=		
3	Design EHS	/		
4	Total solar array current	=		
5	**Select module type**			
	Module rated current	/		
	Number of modules in parallel	=		
	PV system nominal voltage			
	Modules nominal voltage	/		
	Number of modules in series	=		
	Number of modules in parallel	×		
	Total number of modules	=		
6	**Determine battery capacity**			
	Daily Ah requirements (system)			
	Recommended reserve time	×		
	Usable battery capacity	/		
	Minimum battery capacity	=		

expressed in Ah) by the number of days of recommended reserve time. In order to prolong the life of lead-acid batteries, which are most commonly used, it is recommended that the battery is discharged maximally by 80%. If this value is decreased, the battery lifetime is prolonged, but the system becomes more expensive. In the end, a cost evaluation has to be made in order to choose the optimal configuration.

> **Example**
>
> *The total DC requirements of loads plus the system losses are* 45.6 Ah. *The recommended reserve time capacity for the installation site in the Netherlands is five days. Battery capacity required by the system is* 45.6 Ah × 5 = 228 Ah. *The minimal battery capacity for a safe operation is therefore* 228 Ah/0.8 = 285 Ah.

Designing a simple PV system as described in this section can be carried out using a worksheet as in Table 20.1, where the PV system design rules are summarized.

Remark

In the example discussed above we sized the PV array according to the daily equivalent sun hours. This approach will work very well in regions that have small changes in irradiation throughout the year, i.e. regions that are in the proximity of the Equator. As we move further away from the equator, differences between the length of the day in summer and winter become larger. As a consequence, the difference between the daily sun hours in winter and summer also becomes larger. To account for this effect, the system can be sized

20. PV system design

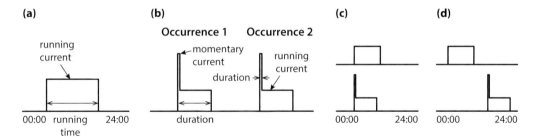

Figure 20.4: Different load profiles.

such that it delivers sufficient energy in the *worst* month, i.e. December or January in the northern hemisphere. However, this may make the system very large and significantly oversized during the summer months, which makes it less economical. In general we can say that designing off-grid systems becomes more difficult the further we are away from the Equator.

20.2 Load profiles

Now we take a look at the load profile. Figure 20.4 illustrates different shapes that loads can have. (a) A simple load draws a constant amount of power for a certain time. (b) However, the consumed power does not need to be constant but can show peaks that correspond to switching electrical appliances on or off. A household of course has several different loads that (c) can be switched on at the same time (coincident) or (d) at different times (non-coincident).

Analyzing load profiles can be performed with increasing complexity and hence accuracy. The simplest method is to determine the loads on a 24-hour basis. To do this, an arbitrary day can be taken and the electricity consumption monitored. However, several loads do not fit in such a scheme. We gave a number of examples of this in Section 20.1. For example, washing machines and dishwashers do not fit in a 24-hour scheme because they are not used every day. Additionally, several loads are seasonal in nature, for example, air conditioning or heating, in case this is performed with a heat pump. Therefore it is advisable to look at load profiles for a whole year.

The total energy consumed in a year is given by

$$E_L^Y = \int_{\text{year}} P_L(t) \, dt, \tag{20.1}$$

where $P_L(t)$ is the power of the load at time t. E_L^Y is expressed in kWh/year.

20.3 Meteorological effects

Standard test conditions (STC) of photovoltaic (PV) modules are not generally representative of the real working conditions of a solar module. For example, high levels of irradiance

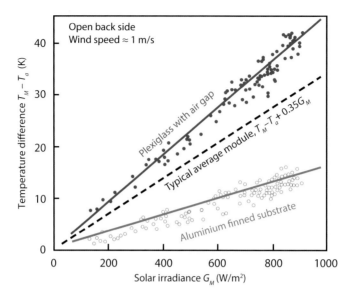

Figure 20.5: Experimentally measured temperature difference $T_M - T_a$ as a function of the irradiance on the PV module G_M for two module types. Linear fits and a linear curve for an average PV module are shown as well (adapted from a figure from NASA [162]).

on the PV module G_M may cause the temperature T_M of a PV module to rise many degrees above the STC temperature of 25 °C. This leads to a lower module voltage and hence output power. On the other hand, in a climate such as the one in the Netherlands, real operating conditions for PV systems correspond to relatively low levels of irradiance combined with a cold and windy weather.

When designing a PV system, it is very important to estimate the effect of the module temperature and the irradiance onto the PV module performance. In this section we first introduce simple models to estimate the solar cell temperature T_M.[2] Then we discuss the effect of temperature and irradiance G_M on the PV module performance. Finally, we develop expressions to estimate the overall performance of PV modules under given temperature and irradiance conditions.

20.3.1 Simplified thermal models for a PV array

The temperature strongly influences the performance of a PV module. While the level of incident irradiation can be easily measured with a pyranometer, the temperature of a solar cell inside a PV module is much harder to evaluate. In order to give an estimate of the average cell temperature, solar manufacturers provide, together with rated performances at STC, the so-called *nominal operating cell temperature* (NOCT). This value corresponds to the temperature of a solar cell under an irradiance level of 800 W/m², ambient temperature of 20 °C and an external wind speed of 1 m/s [161–163].

---

[2]The models in this section estimate the cell temperature, as this is the temperature that affects the performance of the PV module.

Table 20.2: Derivation of the INOCT from the NOCT for various mounting configurations [164].

Rack Mount	INOCT = NOCT − 3°C
Direct Mount	INOCT = NOCT + 18°C
Standoff	INOCT = NOCT + X

where X is given by

W (cm)	X (°C)
2.54	11
7.62	2
15.24	− 1

In a *simplified steady state model*, a linear relationship between the solar irradiance G_M and the difference between the cell and the ambient temperatures $(T_M - T_a)$ is assumed, where the NOCT is used as a reference point [162],

$$T_M = T_a + \frac{T_{\text{NOCT}} - 20°}{800} G_M. \tag{20.2}$$

This model is based on experimental observations showing a linear relationship between $T_M - T_a$ and G_M, as illustrated in Fig. 20.5.

Equation (20.2) only takes the ambient temperature into account but neglects influences from wind speed and mounting configuration, which can lead to significant errors in the predicted cell temperature. In order to take the mounting configuration of the module into account, the *installed nominal operating cell temperature* (INOCT) has been defined [164]. This value is described as the cell temperature of an installed array under NOCT conditions. Its value can therefore be obtained from the NOCT and the mounting configuration. The evaluation of how the INOCT varies with the module mounting configuration has been experimentally determined by measuring the NOCT at various mounting heights; the results are summarized in Table 20.2.

As both the NOCT model and the INOCT approach do not take the wind speed into account, the module temperature might not be accurately predicted, particularly in locations with high wind speeds. The *Duffie–Beckman* (DB) model provides an extension to the NOCT model using an additional empirical term to take the wind speed into account [165]. In this model, the cell temperature T_M is given as

$$T_M = T_a + \frac{T_{\text{NOCT}} - 20°}{800} G_M \left(\frac{9.5}{5.7 + 3.8 \times w} \right) \left(1 - \frac{\eta_{\text{cell}}}{T \times \alpha} \right). \tag{20.3}$$

where w is the wind speed at module height and T is the transmittance of the front layers of the module. α is the absorptivity of the module, hence the product $T\alpha$ gives the fraction of incident light that is absorbed by the solar cells; usually, $T\alpha$ is assumed to be 0.9 [165].

As we can see in Figure 20.6, the NOCT and DB models constitute two extremes that are not always applicable due to meteorological diversity across the globe. To accurately evaluate the influence of external meteorological parameters on the cell temperature, more involved models have to be used. In Appendix G, we present a *fluid-dynamic* (FD) model based on a detailed thermal energy balance between the module and its surroundings. Figure 20.6 also shows results based on the FD model.

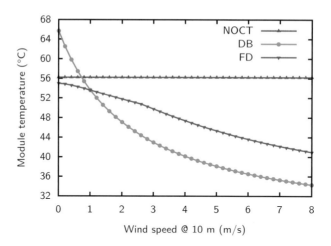

Figure 20.6: Comparison between the NOCT model, the Duffie–Beckman (DB) model, and the fluid-dynamics (FD) model, which is explained in Appendix G. The module temperature of a mono-cSi module was calculated as a function of wind speed.

20.3.2 Effect of temperature on PV module performance

The effect of a solar cell temperature deviating from the 25 °C of STC is expressed by the temperature coefficients that are given on the data sheet provided by the manufacturers. When knowing the temperature coefficient of a certain parameter, such as V_{oc}, I_{sc}, P_{mpp} and the efficiency η, its value at a certain cell temperature T_M can be estimated with

$$V_{oc}(T_M, G_{STC}) = V_{oc} + \frac{\partial V_{oc}}{\partial T}(\text{STC})(T_M - T_{STC}), \tag{20.4}$$

$$I_{sc}(T_M, G_{STC}) = I_{sc} + \frac{\partial I_{sc}}{\partial T}(\text{STC})(T_M - T_{STC}), \tag{20.5}$$

$$P_{mpp}(T_M, G_{STC}) = P_{mpp} + \frac{\partial P_{mpp}}{\partial T}(\text{STC})(T_M - T_{STC}), \tag{20.6}$$

$$\eta(T_M, G_{STC}) = \frac{P_{mpp}(T_M, G_{STC})}{G_{STC} A_M}, \tag{20.7}$$

where A_M is the module area. If the efficiency temperature coefficient $\partial \eta / \partial T$ is not given in the datasheet, it can be obtained by rearranging

$$\eta(T_M, G_{STC}) = \eta(\text{STC}) + \frac{\partial \eta}{\partial T}(\text{STC})(T_M - 25°\text{C}). \tag{20.8}$$

An increased solar cell temperature leads to a shift in the *I-V* curve as shown in Figure 20.7. The slight increase in the short circuit current at higher temperatures is completely outweighed by the decrease in open circuit voltage. Hence, also the efficiency will decrease.

To understand this decrease in V_{oc}, we take another look at Eq. (9.1),

$$V_{oc} \approx \frac{nk_B T}{q} \ln\left(\frac{J_{sc}}{J_0}\right) = \frac{nk_B T}{q} \left(\ln J_{sc} - \ln J_0\right), \tag{20.9}$$

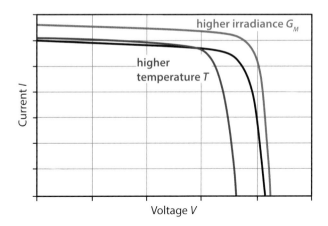

Figure 20.7: Effect of a temperature increase on the I-V solar cell characteristic.

where n is the ideality factor of the solar cell, as introduced in Eq. (9.5). Note that we here and in the following discussion use current densities J instead of currents I. Next to the fraction, where the temperature appears explicitly, it also affects the current densities J_{sc} and J_0. According to Eq. (8.25), J_0 is given as

$$J_0 = q\, n_i^2 \left(\frac{D_N}{L_N N_A} + \frac{D_P}{L_P N_D} \right) = B_J n_i^2, \tag{20.10}$$

where B_J is a constant that is essentially independent of the temperature. Using Eqs. (6.7) and (6.9), we find

$$n_i^2 = 4 \left(\frac{2\pi k_B T}{h^2} \right)^3 \left(m_n^* m_p^* \right)^{\frac{3}{2}} \exp\left(-\frac{E_g}{k_B T}\right) \approx B_i T^\gamma \exp\left(-\frac{E_g}{k_B T}\right), \tag{20.11}$$

where B_i is another constant that is essentially independent of temperature. The exponent 3 was replaced by γ to account for possible other material-dependent temperature dependencies. Hence, we find for the saturation current density

$$J_0 = B_J B_i T^\gamma \exp\left(-\frac{E_g}{k_B T}\right) = B T^\gamma \exp\left(-\frac{E_g}{k_B T}\right) \tag{20.12}$$

with $B := B_J B_i$. Substituting Eq. (20.12) into Eq. (20.9) yields

$$\begin{aligned} V_{oc} &\approx \frac{nk_B T}{q} \left\{ \ln J_{sc} - \ln \left[B T^\gamma \exp\left(-\frac{E_g}{k_B T}\right) \right] \right\} \\ &= \frac{nk_B T}{q} \left(\ln J_{sc} - \ln B - \gamma \ln T + \frac{E_g}{k_B T} \right). \end{aligned} \tag{20.13}$$

As the temperature dependence of J_{sc} and B is very small, we can neglect it when determining the derivative. Hence, we find

$$\frac{\partial V_{oc}}{\partial T} = \frac{V_{oc}}{T} + \frac{nk_B T}{q}\left(-\frac{\gamma}{T} - \frac{E_g}{k_B T^2}\right) = -\left(\frac{nV_g - V_{oc}}{T} + \gamma \frac{nk_B}{q}\right), \tag{20.14}$$

with $V_g = E_g/q$.

The slight increase in the generated current is due to a moderate increase in the photo generated current resulting from an increased number of thermally-generated carriers. The overall reduction of power at high temperatures shows that cold and sunny climates are the best environments for placing PV systems.

Example

The temperature-dependent change of V_{oc} for a c-Si solar cell with $V_{oc}(STC) = 700\,mV$ can be estimated with Eq. (20.14): using $E_g(Si) = 1.12\,eV$, $T(STC) = 298.15\,K$ and $\gamma = 3$, we find

$$\frac{\partial V_{oc}}{\partial T}(Si, STC) \approx 1.7\,mV/K. \tag{20.15}$$

20.3.3 Effect of irradiance on solar cell performance

Intuitively, the power output of a solar cell decreases considerably with decreasing irradiance incident on the PV module. However, the quantitative evaluation of how changing irradiance affects the PV module parameters is less straightforward than for the effect of temperature. This is because solar manufacturers often do not explicitly provide parameters that would allow to derive the PV module parameters at every irradiance level.

By definition the PV module efficiency is given as

$$\eta = \frac{I_{sc} V_{oc} FF}{G_M A_M}. \tag{20.16}$$

where G_M is the irradiance incident on the PV module, as introduced in Section 18.2. The maximum variation of the FF for irradiance values between 1 and 1,000 W/m² is about 2% for CdTe, 5% for a-Si:H, 22% for poly crystalline silicon, and 23% for mono crystalline silicon [166].

The short circuit current of a PV module is directly proportional to the irradiance,

$$I_{sc} = \lambda G_M, \tag{20.17}$$

where λ is proportionality constant. By expressing V_{oc} as in Eq. (20.9), the efficiency is given by

$$\eta = FF \times \lambda \frac{k_B T}{q}(\ln G_M + \ln \lambda - \ln I_0). \tag{20.18}$$

By defining

$$a = FF \times \lambda \frac{k_B T}{q}, \tag{20.19a}$$

$$b = FF \times \lambda \frac{k_B T}{q}(\ln \lambda - \ln I_0), \tag{20.19b}$$

the efficiency can be finally written as

$$\eta(25°C, G_M) = a \ln G_M + b \tag{20.20}$$

Equation (20.20) implies that the PV module efficiency varies linearly with the irradiance [166], provided that a and b are constants. Strictly speaking, Eqs. (20.19a) and (20.19b) are only valid for solar cells, which can be described with a single diode model with a diode quality factor n equal to 1, and where series resistance effects can be neglected. In the case of solar cells like crystalline silicon solar cells, with $n \approx 1.5$ and a non-negligible series resistance, Eq. (20.20) can be used as an approximation, but the relation between a and b and the other parameters is no longer given by Eqs. (20.19). The values of the coefficients a and b are therefore device specific parameters and need to be determined experimentally. These parameters are rarely given by the manufacturer.

From this model the values of I_{sc}, V_{oc}, P_{mpp} and the efficiency at a PV module irradiance level G_M can be determined from the STC values with the expressions

$$V_{oc}(25°C, G_M) = V_{oc}(STC) + \frac{nk_B T}{q} \ln\left(\frac{G_M}{G_{STC}}\right), \quad (20.21)$$

$$I_{sc}(25°C, G_M) = I_{sc}(STC) \frac{G_M}{G_{STC}}, \quad (20.22)$$

$$P_{mpp}(25°C, G_M) = FF \times V_{oc}(25°C, G_M) I_{sc}(25°C, G_M), \quad (20.23)$$

$$\eta(25°C, G_M) = \frac{P_{mpp}(25°C, G_M)}{G_M A_M}, \quad (20.24)$$

where n is the ideality factor and A_M is the module area.

20.3.4 Overall module performance

By combining the two effects of temperature and light intensity, the final efficiency of the module at every level of irradiance and temperature can be determined as [167]

$$\eta(T_M, G_M) = \eta(25°C, G_M)\left[1 + \kappa(T_M - 25°C)\right], \quad (20.25)$$

where

$$\kappa = \frac{1}{\eta(STC)} \frac{\partial \eta}{\partial T}. \quad (20.26)$$

Typical values for κ are $-0.0025/°C$ for CdTe, $-0.0030/°C$ for CIS, and $-0.0035/°C$ for c-Si [167].

All the other parameters such as I_{sc}, V_{oc}, and P_{mpp} can also be evaluated at every level of irradiance and module temperature by simply adapting Eq. (20.25) to the corresponding coefficients and parameters.

At the end of this section, we present a few numerical results that were derived with the fluid-dynamic (FD) model presented in Appendix G. In contrast to Section 20.3.1, where T_M denotes the cell temperature within a PV module, in the FD model the PV module is assumed to be one block with a uniform temperature T_M.

Figure 20.8 shows the effect of changing irradiance on the PV module efficiency. Two curves are shown: first, a curve where the efficiency $\eta(25°C, G_M)$ is shown for varying irradiance while the module temperature is assumed to be constant at $T_M = 25°C$. Secondly, the efficiency $\eta(T_M, G_M)$ for varying irradiance, where changes in T_M due to the varying irradiance are taken into account. The second curve was derived with the fluid-dynamic

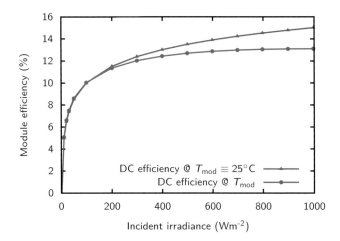

Figure 20.8: Comparison of $\eta(25°C, G_M)$ and $\eta(T_M, G_M)$ with respect to the light irradiance at a wind speed of 1 m/s and ambient temperature of $T_a = 25°C$ for a c-Si solar cell. The curve assuming a variable module temperature was computed using the fluid-dynamic model explained in Appendix G.

model explained in Appendix G: the module temperature T_M was determined in dependance of the irradiance in an iterative process. The wind speed was set to 1 m/s and the ambient temperature was kept constant at 25°C. At low levels of irradiance no effects of the module temperature are observed. At higher level of incident light intensity the difference between the two curves becomes more pronounced.

Figure 20.9 shows the overall efficiency at various light intensities in dependence of the wind speed. The wind has a beneficial effect via turbulent motion, which cools the module. This is reflected by an increase of efficiency.

20.3.5 Summary

In summary, the power output of a PV module before the BOS is given by

$$P_{DC} = \eta\left(T_M, G_M\right) G_M A_M, \tag{20.27}$$

where A_M is the area of the PV module. The power output at STC is given by

$$P_{STC} = \eta\left(25°C, G_{STC}\right) G_{STC} A_M. \tag{20.28}$$

The *energy yield* of the DC side then is defined as

$$Y_{DC} = \frac{P_{DC}}{P_{STC}} \times 100\% = \frac{\eta\left(T_M, G_M\right) G_M}{25°C, \eta\left(G_{STC}\right) G_{STC}} \times 100\%. \tag{20.29}$$

20. PV system design

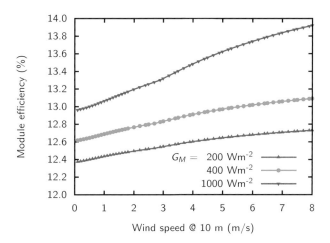

Figure 20.9: The PV module efficiency in dependence of the wind speed at various irradiance levels. The curves were computed using the fluid-dynamic model presented in Appendix G.

20.4 Designing grid-connected PV systems

In this section we learn how to design a PV system, based on the *energy balance* paradigm. This means that we design the system such that the generated energy and the consumed energy match over the year. A visual interactive tool for designing different sorts of grid-connected PV systems in the Netherlands can be found at the *Dutch PV Portal* [168]. Of course, there are also other ways of designing systems for example based on economic arguments.

For the energy balance we first need to calculate the annual load, which was already explained in Section 20.2. The energy yield at the DC side is given by

$$E_{DC}^{Y} = A_{tot} \int_{year} G_M(t)\eta(t)\,dt, \qquad (20.30)$$

where A_{tot} is the total module area. It is related to the area of one module A_M via

$$A_{tot} = N_T \times A_M, \qquad (20.31)$$

where N_T is the number of modules. The energy balance can now be expressed as

$$E_{DC}^{Y} = E_{L}^{Y} \times SF, \qquad (20.32)$$

where SF is a *sizing factor* that is usually assumed to be 1.1. We therefore can calculate the required number of modules,

$$N_T = \left\lceil \frac{E_{L}^{Y} \times SF}{A_M \times \int_{year} G_M(t)\eta(t)\,dt} \right\rceil, \qquad (20.33)$$

where $\lceil x \rceil$ denotes the ceiling function, i.e. the lowest integer that is greater than or equal to x.

Now it is important to decide how many modules are to be connected in *series* (N_S) and in *parallel* (N_P). Of course,

$$N_T = N_S \times N_P. \tag{20.34}$$

Such a PV array hence consists of P strings of S modules each. The N_T determined in Eq. (20.71) does not necessarily need to be a practically divisible number. For example, if $N_T = 11$, one might want to choose $N_T = 12$ panels, because they can be installed as $S \times P = 12 \times 1, 6 \times 2, 4 \times 3, 3 \times 4, 2 \times 6$ or 1×12 strings. In principle, it is preferable to connect as many modules as possible in series since then the currents on the DC side (and hence the cable losses) stay low. Many modern string inverters have two or more independent string inputs, each having its own maximum power point tracker. This can be important if the installation contains two or more areas with different shading and irradiance, for example on two different sides of a roof. An alternative option is to connect different strings to different inverters, which would mean that the system uses several independent inverters.

As for N_S, a maximum exists since each inverter type has a maximum allowed input voltage. The maximum voltage the PV array can generate is the maximum open circuit voltage of the array, occurring when the inverter is not operating. The maximum open circuit voltage is determined by the number of modules in series N_S and the maximum open-circuit voltage of an individual module. Because of the negative value of the temperature coefficient, the maximum open circuit voltage occurs in the coldest period of the year. There is also a minimum value for N_S. Each inverter has an input voltage window, where the maximum power point tracker operates. If the array maximum power point voltage falls below this voltage window, the maximum power point tracker does not operate any more. The lowest array maximum power point voltage occurs in the hottest period of the year.

In a conservative assumption, the power on the DC side at STC now is given as

$$P_{DC}^{STC} = N_T \times P_{MPP}^{STC}. \tag{20.35}$$

The inverter must be chosen such that its maximal power $P_{DC, max}^{inv}$ is above the maximal PV output,

$$P_{DC, max}^{inv} > P_{DC}^{STC}. \tag{20.36}$$

Further, the nominal DC power of the inverter should be approximately equal to the PV power at STC,

$$P_{DC_0} \approx P_{DC}^{STC}. \tag{20.37}$$

In practice, the nominal DC power of the inverter is selected slightly below the PV power at STC, up to 10%, depending on the climate zone, because of the different irradiance distributions. Also, for $P_{DC_0} < 5$ kWp, *single-phase* inverters are used while for $P_{DC_0} > 5$ *three-phase* inverters are advised.

The inverter efficiency is dependent on the input power and voltage. A model discussing the inverter efficiency is presented below.

20.4.1 Inverter efficiency

As we already discussed in Chapter 19, modern *inverters* fulfil two major functions: irst, *maximum power point tracking* (MPPT), and secondly, the actual inverter function, i.e. converting the incoming direct current (DC) to alternating current (AC) that can be fed into the electricity grid.

In order not to waste electricity produced by the PV array, an inverter should always work as close as possible to its maximum achievable efficiency. However, the inverter efficiency is not a constant, but strongly depends on the DC input voltage and the total DC input power. The inverter efficiency η_{inv} with respect to the input DC power at various DC voltage levels is usually given in the data sheet, at least for some values. Sometimes, only a peak value is given.

Weighted efficiencies

A more reliable way of expressing the inverter efficiency in a single number is to use *weighted efficiencies*, which combine the inverter efficiencies over a wide range of solar resource regimes [169]. Two different weighted efficiencies are commonly used. First, the *European efficiency*, which represents a low insolation climate such as in Central Europe, and second, the *California Energy Commission* (CEC) *efficiency*, which represents the PV system performance in high insolation regions such as in the southwest of the United States [169]. They are given by:

$$\eta_{\text{Euro}} = 0.03\,\eta_{5\%} + 0.06\,\eta_{10\%} + 0.13\,\eta_{20\%} + 0.10\,\eta_{30\%} + 0.48\,\eta_{50\%} + 0.20\,\eta_{100\%}, \quad (20.38\text{a})$$

$$\eta_{\text{CEC}} = 0.04\,\eta_{10\%} + 0.05\,\eta_{20\%} + 0.12\,\eta_{30\%} + 0.21\,\eta_{50\%} + 0.53\,\eta_{75\%} + 0.05\,\eta_{100\%}, \quad (20.38\text{b})$$

where $\eta_{x\%}$ denotes the efficiency at $x\%$ of nominal power of the inverter. Note that the CEC efficiency contains a 75% value that is not present in the European efficiency.

Even though the weighted efficiencies represent a more accurate approximation of the effective annual working performance of the inverter compared to the mere peak efficiency, it is still only an approximation of the average performance of a system in the European climate.

If a better estimate of the *real time energy yield* of an extended PV system is needed, a more accurate representation of the instantaneous inverter performance at every level of input power and voltage must be developed, for example the model discussed below.

Sandia National Laboratories (SNL) model

Due to lack of detailed data from inverter manufacturers, many research institutes around the world have published extended data, which are publicly available online. These data present efficiency curves for a large range of inverters as a function of a several characteristic parameters. One useful database is the one provided by the Sandia National Laboratories [171].

Figure 20.10 shows an example of an inverter efficiency curve. For the graph, field test measurements were taken during a period of 13 days with changing weather conditions. While the relationship between the AC and DC power appears to be linear at first, a closer look at the graph reveals that this is not entirely the case, as shown in Figure 20.11. The power consumption of the inverter itself, together with the electrical characteristics of the switching modes and circuits at different power levels, results in a degree of non-linearity between AC and DC power at a given DC voltage level. Assuming that the inverter efficiency is a constant value throughout the whole DC power range is equivalent to assuming a linear relationship between DC and AC power, which has been shown not to be the case.

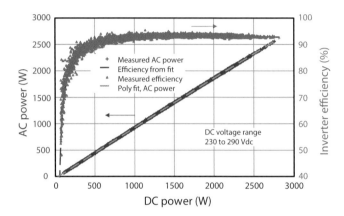

Figure 20.10: The measured AC power and inverter efficiency for a 2.5 kW Solectria PVI2500 inverter recorded during a period of 13 days at Sandia National Laboratories (figure reproduced with kind permission from Sandia National Laboratories) [170].

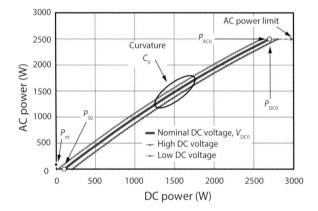

Figure 20.11: A closer look at the relationship between DC input and AC output power for an inverter and the definition of the parameters used in the *Sandia Inverter Perfomance Model* (figure reproduced with kind permission from Sandia National Laboratories) [170].

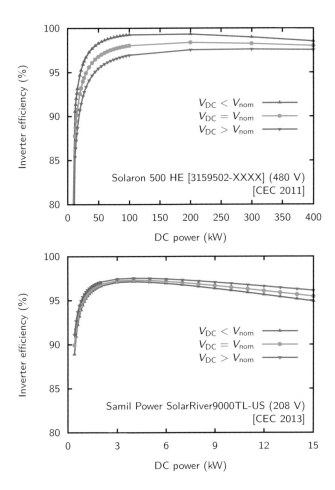

Figure 20.12: Variation in the inverter efficiency with DC input voltage for two different inverters.

The dependency of the inverter efficiency on the DC input voltage is a very complex phenomenon. The differences between the different inverter types can partially be explained by the different types of switches used. Figure 20.12 shows the voltage-dependent inverter efficiency for different inverter types. These curves were determined with the SNL model that is described in the next paragraph.

In the Sandia National Laboratory model the relationship between P_{AC} and P_{DC} is given by [170]

$$P_{AC} = \left[\frac{P_{AC_0}}{A-B} - C(A-B)\right] \times (P_{DC} - B) - C(P_{DC} - B)^2, \tag{20.39}$$

where the coefficients A, B, and C are defined as

$$A = P_{DC_0} \left[1 + C_1 \left(V_{DC} - V_{DC_0}\right)\right], \tag{20.40a}$$
$$B = P_{S_0} \left[1 + C_2 \left(V_{DC} - V_{DC_0}\right)\right], \tag{20.40b}$$
$$C = C_0 \left[1 + C_3 \left(V_{DC} - V_{DC_0}\right)\right]. \tag{20.40c}$$

Several of the parameters are depicted in Figure 20.11. The parameters are [170]:

P_{AC}: AC power output from inverter based on input power and voltage, (W).

P_{DC}: DC power input to inverter, typically assumed to be equal to the PV array maximum power, (W).

V_{DC}: DC voltage input, typically assumed to be equal to the PV array maximum power voltage, (V).

P_{AC_0}: Maximum AC power 'rating' for inverter at reference or nominal operating condition, assumed to be an upper limit value, (W).

P_{DC_0}: DC power level at which the AC power rating is achieved at the reference operating condition, (W).

V_{DC_0}: DC voltage level at which the AC power rating is achieved at the reference operating condition, (V).

P_{S_0}: DC power required to start the inversion process, or self-consumption by inverter, strongly influences inverter efficiency at low power levels, (W).

C_0: Parameter defining the curvature (parabolic) of the relationship between AC power and DC power at the reference operating condition, default value of zero gives a linear relationship, (1/W).

C_i: Empirical coefficient allowing P_{DC_0} to vary linearly with DC-voltage input, default value is zero, ($i = 1, 2, 3, 1/V$).

The model takes the following losses into account [172]:

- Self consumption of the inverter. This value corresponds to the DC power required to start the inversion process.
- Losses proportional to the output power due to fixed voltage drops in semiconductors and switching losses.
- Ohmic losses.

The accuracy of the model depends on the data available for determining the performance parameters. An initial estimate can be made using the little information provided by the manufacturer. Using all the required parameters will provide a model with an error of approximately 0.1% between the modelled and measured inverter efficiency [172].

All the parameters required for handling Eq. (20.39) are given in the SNL database where a full list of a wide range of inverters with nominal powers from 200 W up to 1 MW is given [171]. This model therefore uses the instantaneous value of P_{DC} and V_{DC} produced by the entire PV array to evaluate the AC power output and inverter efficiency.

Maximum power point tracker and additional losses

The efficiency of the MPPT has not been included explicitly in the SNL performance model. This is because the efficiency of most MPPTs used in the inverters ranges between 98% and nearly 100% at every level of input power, and the voltage provided is within the accepted minimum and maximum window for the MPPT to function correctly. A decrease of 1% in the system performance has therefore been used to take the MPPT losses into account.

An additional decrease of 3% in the system performance can be considered to cover losses caused by mismatch between modules (-1.5%), ohmic cable losses (-0.5%) and soiling (-1%) [167], if the cable losses are not determined as described in Section 19.5.

20.4.2 Performance analysis

Now we will put all the concepts together that we have discussed in this section so far. The *instantaneous power output* on the AC side can be described with

$$P_{AC}(t) = A_M G_M(t) \eta_M(t) \eta_{inverter}(t) \eta_{MPPT}(t) \eta_{other}. \tag{20.41}$$

The system efficiency is then given as

$$\eta_{system}(t) = \frac{P_{AC}(t)}{A_{tot} G_M(t)} \times 100\%, \tag{20.42}$$

which leads us to the *instantaneous AC side yield* (also known as *performance ratio*),

$$Y_{AC}(t) = \frac{P_{AC}(t)}{P_{STC}} \times 100\%. \tag{20.43}$$

Then, the *yearly energy yield at the AC side* can be calculated with

$$E_{AC}^Y = \int_{year} P_{AC}(t)\, dt; \tag{20.44}$$

it is given in Wh/year. Another important parameter is the *annual efficiency* of the system

$$\eta_{system}^Y = \frac{E_{AC}^Y}{E_{i,\,sys}^Y} \times 100\%, \tag{20.45}$$

where $E_{i,\,sys}^Y$ is the solar energy incident on the PV system throughout the year. It can be calculated with

$$E_{i,\,sys}^Y = A_{tot} \int_{year} G_M(t)\, dt. \tag{20.46}$$

The last parameter we look at is the *yearly electricity yield*

$$Y_E = \frac{E_{AC}^Y}{N_S N_P \times P_{STC}}, \tag{20.47}$$

which is given by Wh/(year kWp).

At the end of the design phase it is important to check whether the system really fulfils the requirements. If the annual energy yield exceeds the annual load, the system is well designed. Otherwise, another iteration has to be done in order to scale up the system. However, as stated earlier, for a grid-connected system, it can also be a choice to have the whole load covered by PV electricity.

Table 20.3: The specific annual yield and the various losses in a typical Dutch grid-connected PV system with modules made of crystalline silicon solar cells. The irradiation data used are for Delft, The Netherlands. Ohmic losses are according to IEC standard, while MPP tracking losses can retrieved from inverter suppliers.

	Annual loss	Loss factor
Soiling [167]	1.0%	0.990
Reflection [173]	4.0%	0.960
Module mismatch losses [173]	1.0%	0.990
Module temperature losses [174, 175]	6.0%	0.940
Module irradiance losses [174, 175]	4.5%	0.955
Ohmic losses	1.0%	0.990
Inverter conversion losses	5.0%	0.950
Inverter MPP tracking losses	1.0%	0.990
Performance ratio		0.786
Global horizontal irradiation (kWh/m^2)		999
In-plane irradiation (kWh/m^2)		1146
Specific annual energy yield (kWh/kWp)		901

20.4.3 Annual losses in grid-connected PV systems

In Table 20.3 an overview is given of all the system losses that occur in a typical Dutch grid-connected PV system with modules endowed with crystalline silicon solar cells. The modules are roof-integrated with reasonable ventilation and a tilt angle of 36° (the optimal tilt angle for the Netherlands). The losses are given on an annual basis and by a corresponding loss factor. The product of all the loss factors is the performance ratio.

- The reflection losses arise because the peak power of modules is determined at perpendicular incidence whereas the modules are illuminated by the whole hemisphere under operational conditions. For crystalline silicon solar modules the spectral mismatch losses with respect to the AM1.5 spectrum are small (at most 1% on an annual basis).

- The module temperature losses occur because the actual cell temperature is deviating from the 25 °C cell temperature at STC (see Section 20.3.2). The temperature loss given in Table 20.3 is for a roof-integrated system in Delft. In the case of a free-standing array in Delft the temperature loss would be 2-3%, whereas in southern Europe about 6-7%.

- The module irradiance losses occur because the cell efficiency at the actual irradiance deviates from the cell efficiency at the 1000 W/m^2 irradiance at STC (see Section 20.3.3).

- The ohmic losses on an annual basis are always lower than the ohmic losses at STC, because of the irradiance distribution over the various irradiance classes. The product of the performance ratio and the in-plane irradiance gives the specific annual energy yield.

Table 20.4: Recommended Loss of Load Probability (LLP) for some Applications [176].

Application	Recommended LLP
Domestic illumination	10^{-2}
Appliances	10^{-1}
Telecommunications	10^{-4}

20.5 Designing stand alone PV systems

In this final section, we take a closer look at the design of stand alone PV systems (also called off-grid systems). Choosing a good design is more critical for stand alone systems than for grid-connected systems. The reason for this is that stand alone systems cannot fall back on the electricity grid, which increases the requirements on their reliability.

A major component of stand alone systems is the *storage component*, which can store energy in times when the PV modules generate more electricity than required and it can deliver energy to electric appliances when the electricity generated by the PV modules is not sufficient. A major design parameter for stand alone systems is the required number of autonomous days, i.e. the number of days a fully charged storage component must be able to deliver energy to the system until discharged.

The sizing of the PV array and the storage component, usually a *battery bank*, are interrelated. Here, two parameters arise. First, E_{fail}, which is the energy required by the electric load that cannot be delivered by the PV system, for example if the batteries are emptied after several cloudy days. Secondly, E_{dump}, which is the energy produced by the PV array that is neither used for driving a load nor is stored in the batteries, for example, if the batteries are already full after a number of sunny days. We can define the *loss of load probability* (LLP),

$$\text{LLP} = \frac{E_{\text{fail}}}{\int_{\text{year}} P_L(t)\,dt}, \tag{20.48}$$

where P_L is the load power. Note that the lower the LLP, the more stable and reliable the PV system. Table 20.4 shows recommended LLPs for different applications.

Figure 20.13 (a) shows a schematic of an stand alone system with all the required components, i.e. the PV array, a maximum power point tracker (MPPT), a charge controller (CC), a discharge controller (DC), a battery bank (BB), an inverter, and the load. The CC prevents the BB from being overcharged by the PV system, while the DC prevents the battery from being discharged below the minimal allowed SoC. In the simplest case, the (dis)connect the PV system and load are disconnected from the battery by switches. For optimal charging, however, a CC with pulse width modulation (PWM) is often used. Not all stand alone systems contain an MPPT, thus it is represented with a dashed line. However, if an MPPT is present it is usually delivered in one unit together with the CC. Usually, the DC and inverter are combined in one *battery inverter unit*. Especially in larger systems, the inverter currents may become very large. For example, if a 2,400 W load is present for a short time, this means 100 A on the DC side in a 24 V system. Therefore the battery inverter usually is directly connected to the BB with thick cables. In very small systems with maximal powers of several hundred watts, the CC and DC may be combined in one unit.

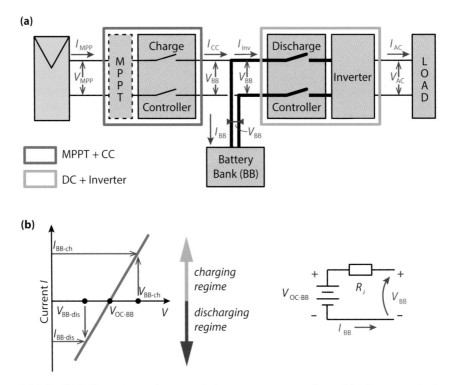

Figure 20.13: (a) PV topology of an stand alone system; note that cables between the batteries and the inverter usually are very thick. (b) Simplified I-V curve and schematic of the battery bank.

20.5.1 A closer look at the battery

The battery bank is the workhorse of any stand alone system, because it is its stable power source. It is thus very important to understand how the battery bank will act in the PV system. When we want to understand how a battery works we have to consider *the net effect of all the components that try to charge or discharge the battery*. In Figure 20.13 (a) the most important currents and voltages that we need for the following discussion are depicted. In Figure 20.13 (b) the equivalent circuit and an I-V curve of an idealized battery bank are shown, similar to what we already discussed in Section 19.3.

As we have seen in Eq. (19.25), the voltage of the battery bank V_{BB} will differ from $V_{OC\text{-}BB}$ and is also dependent on the current I_{BB} flowing through the battery,

$$I_{BB} = \frac{1}{R_i}(V_{BB} - V_{OC\text{-}BB}), \tag{20.49}$$

where R_i is the internal resistance of the battery. Remember that we use the convention that I_{BB} is positive when the battery is charged and negative when it is discharged.

Let us now derive an expression for V_{BB} as a function of the other PV system parameters. We start with the power on the left-hand and right-hand sides of the MPPT,

$$\eta_{MPP} I_{MPP} V_{MPP} = I_{CC} V_{BB} =: \beta, \tag{20.50}$$

and hence

$$I_{CC} = \frac{\eta_{MPP} I_{MPP} V_{MPP}}{V_{BB}} = \frac{\beta}{V_{BB}}. \tag{20.51}$$

Consequently, $\beta = 0$ means that the PV system is not active, for example at night.

In a similar manner, we look at the power on the left-hand and right-hand sides of the inverter:

$$P_L = I_{AC} V_{AC} = \eta_{inv} I_{inv} V_{BB} =: \eta_{Inv}\alpha, \tag{20.52}$$

and consequently

$$I_{inv} = \frac{P_L}{\eta_{inv} V_{BB}} = \frac{\alpha}{V_{BB}}. \tag{20.53}$$

Combining Eqs. (20.51) and (20.53), we find

$$I_{BB} = I_{CC} - I_{inv} = \frac{\beta - \alpha}{V_{BB}}. \tag{20.54}$$

Clearly, $\alpha = 0$ indicates that no load is present.

Combining Eqs. (20.49) and (20.54), multiplying with $R_i V_{BB}$ and rearranging, leads to a quadratic equation,

$$V_{BB}^2 - V_{OC\text{-}BB} V_{BB} - R_i(\beta - \alpha) = 0. \tag{20.55}$$

with the solutions

$$V_{BB}^{\pm} = \frac{V_{OC\text{-}BB}}{2} \pm \sqrt{\left(\frac{V_{OC\text{-}BB}}{2}\right)^2 + R_i(\beta - \alpha)}. \tag{20.56}$$

The correct solution is the '+' solution as we can check with

$$V_{BB}^{+}(\beta - \alpha = 0) = V_{OC\text{-}BB}. \tag{20.57}$$

Table 20.5: Recommended number of autonomous days d_A at several latitudes.

Latitude (°)	Recommended d_A
0-30	5-6
20-50	10-12
50-60	15

Hence, the final solution is

$$V_{BB} = \frac{V_{OC\text{-}BB}}{2} + \sqrt{\left(\frac{V_{OC\text{-}BB}}{2}\right)^2 + R_i(\beta - \alpha)}. \tag{20.58}$$

Let us now take a short look at this solution. If the battery is charged, I_{CC} is higher than I_{inv} and hence $(\beta - \alpha)$ and I_{BB} are positive. From Eq. (20.58) it follows that in this case V_{BB} is higher than $V_{OC\text{-}BB}$. On the other hand, if the battery is discharged, I_{CC} is lower than I_{inv} and hence $(\beta - \alpha)$ and I_{BB} are negative. Therefore, V_{BB} will be lower than $V_{OC\text{-}BB}$.

Note that only the net current $I_{BB} = I_{CC}$ flows in or out of the battery. As mentioned in Section 19.3, only this current determines the power loss in the battery,

$$P_{BB}(\text{loss}) = I_{BB}^2 R_i. \tag{20.59}$$

20.5.2 Designing a PV system with energy balance

Now we discuss how to design a PV system based on the principle of *energy balance*. To determine the adequate components, the analysis for the load side and the PV side is performed separately. Let us begin with the *load side*. We can determine the annual load E_L^Y as discussed in Section 20.2 and Eq. (20.1),

$$E_L^Y = \int_{\text{year}} P_L(t)\, dt. \tag{20.60}$$

From that we can estimate the *average daily load* with

$$E_L^D = \frac{1}{365} E_L^Y. \tag{20.61}$$

Next, we have to choose an adequate number of days of autonomy d_A. Some values are given in Table 20.5. The selection of d_A is not only based on the local irradiance pattern and the year-to-year variations, but also on the specific application. PV-driven buoys at sea always need to work, whereas for a solar home system in rural areas it is acceptable that the system cannot deliver energy for a few days a year.

Now, we can calculate the required energy of the battery bank,

$$E_{BB} = d_A \frac{E_L^D \times SF_{bat}}{DoD_{max}}, \tag{20.62}$$

where SF$_{\text{bat}}$ is the sizing factor of the battery, which is similar to the sizing factor of the PV array already defined in Section 20.4. DoD$_{\text{max}}$ is the maximally allowed depth of discharge of the batteries. The rated energy of the chosen batteries is

$$E_{\text{bat}} = V_{\text{OC-bat}} C_{\text{bat}}, \tag{20.63}$$

where C_{bat} is the battery capacity (in ampere-hours). Hence, the required number of batteries is

$$N_{\text{bat}} = \left\lceil \frac{E_{\text{BB}}}{E_{\text{bat}}} \right\rceil. \tag{20.64}$$

As for grid-connected systems we now need to choose a suitable inverter. Therefore we must consider the maximal load power P_L^{max}. This we can do, for example, by looking at the appliance with the maximal power consumption, or by adding up the power of all the appliances. It may be more beneficial to choose a system design such that not all appliances can be used at the same time. The final decision of course is up to the designer of the system. As for grid-connected systems, the inverter must fulfil several requirements: first, its maximally allowed power output must exceed the maximal power required by the appliances,

$$P_{\text{DC, max}}^{\text{inv}} > P_L^{\text{max}}. \tag{20.65}$$

Secondly, the nominal power of the inverter should be approximately equal to the maximal load power,

$$P_{\text{DC}, 0} \approx P_L^{\text{max}}. \tag{20.66}$$

Thirdly, the nominal inverter input voltage should be approximately equal to the nominal voltage of the battery bank,

$$V_{\text{DC, inv}} \approx V_{\text{OC-BB}}. \tag{20.67}$$

For a more detailed analysis of the inverter performance, the Sandia model that can be used refers to the Section 20.4.

Typical nominal voltages for the battery bank are 12 V, 24 V, 48 V or 96 V. It can be adjusted by the number of batteries that are connected in series,

$$N_{\text{bat}}^S = \frac{V_{\text{OC-BB}}}{V_{\text{OC-bat}}}. \tag{20.68}$$

From that we can also determine the number of batteries that must be connected in parallel,

$$N_{\text{bat}}^P = \left\lceil \frac{N_{\text{bat}}}{N_{\text{bat}}^S} \right\rceil. \tag{20.69}$$

Parallel connection of batteries should be avoided as much as possible, because ageing in one 'string' of battery cells can accelerate the ageing of a whole battery bank. Therefore it is recommended to select a battery with a larger capacity rather than using a number of batteries in parallel.

Now, after designing the load side is completed, having chosen the inverter and batteries, we look at the PV side of the system. Sizing the PV array is very similar to the procedure used for grid-connected systems in section 20.4. The energy balance can now be written as:

$$E_{\text{DC}}^Y = E_L^Y \times \text{SF}, \tag{20.70}$$

where SF is a *sizing factor* that usually is assumed to be 1.1. We therefore can calculate the required number of modules,

$$N_T = \left\lceil \frac{E_L^Y \times \text{SF}}{A_M \times \int_{\text{year}} G_M(t)\eta(t)\,dt} \right\rceil. \tag{20.71}$$

To minimise losses, the MPP voltage of the PV array and the nominal voltages of the inverter and the battery pack should be approximately equal, otherwise the losses of the DC-DC converter that is included in the MPPT-CC unit will be higher. The number of PV modules that are connected in series in the PV array is given by

$$N_S = \left\lfloor \frac{V_{\text{OC-BB}}}{\overline{V}_{\text{mod-MPP}}} \right\rfloor, \tag{20.72}$$

where $\overline{V}_{\text{MPP-mod}}$ denotes the annual average of the MPP voltage of the PV modules. Of course, the maximally allowed input voltage of the MPPT-CC unit must not be exceeded by the PV array,

$$V_{\text{MPP}} \geq N_S \times V_{\text{mod-MPP}}^{\max}. \tag{20.73}$$

The number of required parallel PV strings is given by

$$N_P = \left\lceil \frac{N_T}{N_S} \right\rceil. \tag{20.74}$$

20.5.3 Performance analysis

As for grid-connected systems, a performance analysis should be done to evaluate the chosen design. For this analysis, it is useful to use an algorithm that can simulate the performance of the PV system throughout the year. Conceptually, this algorithm can look as follows:

1. Set a starting state of charge (SoC) of the battery bank.
2. Then calculate the SoC of the battery throughout the year for time steps Δt,
   - depending on the actual load and PV array output, determine the battery current $I_{\text{BB}}(t)$,
   - determine the actualized SoC.

   Of course, the function of the charge controller, i.e. its switching behaviour, must be accurately mimicked by the algorithm.
3. Determine when the system cannot deliver the required load and hence E_{fail}.
4. Now, calculate the loss of load probability of the system (LLP).
5. Finally, determine the annual energy yield on the AC side of the system, E_{AC}^Y.

Mathematically, the annual energy yield on the AC side can be expressed by

$$E_{\text{AC}}^Y = \int_{\text{year}} P_L(t)\,dt - E_{\text{fail}} + \Delta E_{\text{BB}}. \tag{20.75}$$

Table 20.6

Load	Quantity	Power per item (W)	Time of use (h)	Type
Light bulb	10	20	2	DC
TV	1	100	2	AC
DVD	1	40	2	AC
Laptop	1	100	2	AC

In contrast to the expression used for determining the AC yield for grid-connected systems in Eq. (20.44), this equation contains two additional components: E_{fail}, the energy that is required by the load but cannot be delivered and ΔE_{BB} which is the difference in energy stored in the battery bank between the beginning and end of the year,

$$\Delta E_{\text{BB}} = \left(\text{SoC}_{\text{end}} - \text{SoC}_{\text{beginning}}\right) C_{\text{bat}} V_{\text{BB}}. \tag{20.76}$$

If $\Delta E_{\text{BB}} < 0$, the system might not be sustainable. On the other hand, if it is > 0, E_{dump} will increase in the following, if the average meteorological conditions are unchanged.

Because of $E_{\text{fail}} > 0$,

$$E_{\text{AC}}^Y < \int_{\text{year}} P_L(t)\, dt. \tag{20.77}$$

As already stated, the loss of load probability (LLP) is given by

$$\text{LLP} = \frac{E_{\text{fail}}}{\int_{\text{year}} P_L(t)\, dt}. \tag{20.78}$$

T evaluate the design, it is very important to look at the LLP:

- LLP acceptable:
  - E_{dump} low: the system design is OK.
  - E_{dump} high: resize the PV array.
- LLP not acceptable:
  - Increase the size of the PV array.
  - Increase the capacity of the battery bank.

If the LLP is not acceptable, the system must be altered. Whether it is best to resize the PV array, to increase the battery capacity, or to do both, will depend on the specific circumstances.

20.6 Exercises

20.1 The Smith family have a small house in the country side which is not connected to the grid. The place enjoys 3.5 equivalent sun hours. Therefore, Mr Smith has decided to install an off-grid PV system in the house to supply their electricity. He will be using PV modules with the following specifications: $P_{\text{nom}} = 100\, W_p$, $V_{\text{MPP}} = 16$ V, $I_{\text{MPP}} = 6.25$ A, $V_{\text{OC}} = 18$ V, $I_{\text{SC}} = 7$ A. We may assume that the combined efficiency of the cables, the charge controller and the battery system is 90% and that of the inverter is 96%. As the house is only used at the weekend, only two days of autonomy will be needed. In this case, batteries with the following specifications will be used: capacity 100 Ah, voltage 12 V, maximum allowed depth of discharge 50%. The daily house electricity requirements are summarized in Table 20.6.

(a) Calculate the total daily electricity demand in Wh.

(b) What is the total power demanded by the DC loads?

(c) What is the total power demanded by the AC loads?

(d) How much energy in Wh must the panels generate in one day to cover the daily electricity demand of the family?

(e) Assume the panels work under MPP conditions. How many panels will be needed to produce that energy?

(f) The system is designed to use 24 V as its operating voltage. What will be the minimum battery capacity required in Ah?

(g) How many of the specified batteries will be needed?

20.2 Consider a purely grid-connected PV system, i.e. without any battery storage. The PV array efficiency is 20%, while the grid-connected inverter efficiency is 90%. Assuming a 2% loss in the cables, what is the combined efficiency of the PV system from the PV array to the grid?

20.3 Which of the following statements on purely grid-connected systems are false?

(a) The grid-connected system does not necessarily need battery storage, because the grid can uptake and feed power when needed.

(b) The PV system should be designed such that the PV array continues to feed the grid, even in the event of a grid network failure.

(c) The grid-connected PV system can only be designed around the required load that needs to be powered.

(d) The PV array configuration should ensure that the DC current and voltage limits at the input of the inverter are respected.

20.4 Which of the following statements is *not* true about the output of PV cells when the temperature increases, and the other conditions remain the same?

(a) The output power decreases.

(b) The output current decreases.

(c) The output voltage decreases.

(d) The output current increases.

20.5 Consider a PV module with the following external parameters at STC: $P_{max} = 320$ W, $V_{oc} = 45$ V, $I_{sc} = 8$ A, NOCT $= 40$ °C and the temperature coefficient of power is -1 W/°C.

(a) If the ambient temperature rises to 35 °C while the irradiance is 1,000 W/m², what is the cell level temperature, calculated with the NOCT model?

(b) What is the new power output of the PV module under the new ambient temperature of 35 °C and 1,000 W/m² irradiance?

20. PV system design

Table 20.7: Characteristics of PV module A

	Module A
Type	Mono crystalline silicon
P_{MPP} (STC) (W)	327
V_{OC} (STC) (V)	67.6
I_{SC} (STC) (A)	6.07
NOCT (°C)	43
V_{MPP} (STC) (V)	57.3 V
I_{MPP} (STC) (A)	5.71 A
Width (m)	1.56
Height (m)	1.05
Area (m^2)	1.64
Efficiency (STC) (%)	20.3
Temp. coeff of P_{MPP} (%/°C)	-0.30
Temp. coeff of V_{OC} (%/°C)	-0.30
Temp. coeff of I_{SC} (%/°C)	0.06

20.6 A module of 1.65×1.00 m^2 with an efficiency of 18% is placed in the Netherlands. On average, the Netherlands has 1,568 sun hours per year. Assume standard test conditions (STC) which are 1,000 W/m^2, cell temperature of 25 °C and AM1.5 spectrum.

(a) Calculate the power generated by the solar module at standard test conditions.

(b) Calculate the energy generated by the module during a year in the Netherlands.

(c) Calculate the air mass (AM) when the sun is 25°, 40°, and 65° from the zenith.

20.7 A PV installation uses 4,400 modules of module A described by the datasheet specifications as shown in Table 20.7. Assume that the heat transfer coefficients at the back and top of the modules are the same. Top glass emissivity is 0.84 and bottom surface emissivity is 0.893. Reflectivity of the module is 0.1 while the Prandtl number for air is 0.71. The initial module temperature may be taken as 20 °C.

(a) Determine the DC power before BOS components for the plant at the instant when the incident irradiation is 700 W/m^2 and wind speed is 5 m/s. Assume the ambient temperature to be 15 °C, ground temperature of 13 °C and a cloud cover of 4 okta at this instant.

(b) Compare your result with the maximum rated power of the installation.

Hint: You will need to first calculate the module temperature at the instant and then the DC power output.

20.8 Use the NREL website[3] to download the database of inverters. Choose a random inverter and apply the SNL model[4] to calculate the efficiency of the inverter. Plot $P_{AC} = f(P_{DC})$ and $\eta = g(P_{DC})$ for three V_{DC} levels. You may assume:

$$P_{DC} = \left[0, \text{step}, 2 \times P_{DC_0}\right]$$
$$V_{DC} = \left[V_{DC(MPPT_{low})}, V_{DC_0}, V_{DC(MPPT_{high})}\right].$$

---
[3]https://sam.nrel.gov/content/component-databases
[4]http://prod.sandia.gov/techlib/access-control.cgi/2007/075036.pdf

21
PV system economics and ecology

21.1 PV system economics

We conclude our discussions on PV systems by looking at several important topics on the economics of PV systems. The economics of PV can be discussed at several levels, such as the consumer level, the manufacturing level, the level of PV installers, and the technology level where PV is compared to other electricity generation technologies from an economical point of view.

21.1.1 Payback time

We will start this discussion with the definition of the *payback time*, which in finance is defined as the amount of time required to recover the cost of an investment. It can be calculated with

$$\text{payback time} = \frac{\text{initial investment}}{\text{annual return}}. \tag{21.1}$$

Translated to the consumer level, the payback time is the time it takes to recover the initial investment of the PV system as the system continuously reduces the electricity bill. Please note that the *financial* payback time is different from the *energy payback time* that we will discuss in Section 21.2 Let us look at an example:

Example

Let us assume that the Smith family have installed a PV system with a power of 1 kW$_p$ on their rooftop. The initial investment was €2000. Family Smith has an annual electricity bill of €2,000. The installation of the PV system leads to an average annual reduction of the electricity bill of €250.

> *As part of their consumed electricity is provided by their PV system, the electricity bill is therefore constantly reduced. Hence, the average annual return on their PV system is €250. As a consequence, the Smith's have earned the final investment back after 8 years; the* payback time *is 8 years.*

The payback time is strongly influenced by the annual solar radiation on the PV system. As we have seen in Chapter 18, this is dependent on the orientation of the PV modules and the location of the PV systems. In general we can say that the sunnier the location, the greater the PV yield and the shorter the payback time. Another factor that influences the payback time is the grid electricity costs: the higher these costs, the shorter the payback time. Finally, the payback time also is strongly dependent on the initial costs of the PV system.

In practice, more factors must be taken into account than in the simple example above. These will increase the complexity of calculating the payback time. For instance, if we are considering a significant period of time, the changing value of money due to inflation also has to be taken into account. For example € 1,000, today, will have a different *purchasing power* than in ten years' time. Another factor that should be considered are policies regarding renewable energy. For example, subsidies and feed-in tariffs can affect the initial investments and savings.

21.1.2 Compensation schemes

There are different schemes to compensate owners of PV systems for electricity they deliver into the grid. We will discuss two schemes, net metering and feed-in tariffs. In households with PV systems installed, the electricity consumer also becomes an electricity producer, who wants to sell excess electricity to the grid. Hence the consumer is turned into a producer/consumer of electricity, or a *prosumer*.

Net metering

Old-fashioned analogue electricity meters can operate in both directions. If electricity is consumed from the electricity grid, the electricity counter increases. However, when the PV system produces more than is actually consumed in the house, electricity is delivered to the grid. In this case, the electricity counter decreases. In the end, only the net electricity consumption must be paid, for, the energy consumed from the grid minus the energy delivered to the grid. Such a system is interesting for consumers, if the levelized cost of electricity (see below) of the PV system is lower than the price paid for electricity from the grid.

Nowadays, *smart* digital electricity meters are often used. These meters distinguish between electricity consumed from the grid and electricity delivered to the grid. This system allows not only allows the amount of electricity delivered to the grid from the PV system to be monitored, but it also allows the grid utility to adapt its tariff system. For example, the electricity price often contains a certain fee for using the electricity grid. Such a fee could also be imposed on electricity delivered to the grid from the PV system.

21. PV system economics and ecology

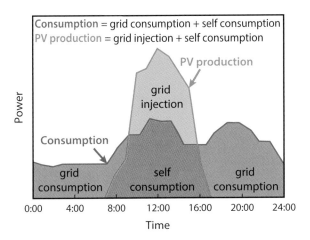

Figure 21.1: Illustrating grid consumption, grid injection and self consumption (reproduced with kind permission from the European Photovoltaic Industry Association (EPIA)) [135].

Feed-in tariffs

With the system of *feed-in tariffs*, electricity generated by the PV system can be sold to the grid utility for a fixed price. For such a system, either two analogue electricity meters (one measuring the power consumed from the grid, the other measuring energy delivered to the grid) or one smart meter are required. We distinguish between two kinds of feed-in tariffs, gross and net. In the system of *gross* feed-in tariffs, all the electricity produced by the system is sold to the grid utility and all the electricity consumed by the household is bought from the grid. In contrast, for *net* feed-in tariffs the actual power consumption is subtracted from the PV power generation and only the surplus electricity is sold to the grid.

Feed-in tariffs stimulate the installation of renewable electricity technologies such as PV, if the feed-in tariffs are above the electricity price. On the other hand, if they are set (slightly) below the electricity price, self consumption can be stimulated, as discussed below.

21.1.3 Self consumption

Self consumption is an important aspect of PV systems. It is defined by the European Photovoltaic Industry Association as the possibility for any kind of electricity consumer to connect a PV system, with a capacity corresponding to their consumption, to their system for on-site consumption, while receiving for non-consumed electricity [135]. In practice this refers to policies that allow a consumer to install a PV system without being charged a premium for connection to the grid.

As PV systems are penetrating the electricity grid power more and more, certain issues start to arise. Figure 21.1 shows how a household with a PV system interacts with the grid. Because of the unpredictable nature of PV electricity generation, the peaks and troughs on this system will be very difficult for utility companies that manage the electricity grids to predict. The grid injection levels may exceed the current demand of a grid well before the PV penetration reaches 100%. This means that there are reasons to encourage prosumers to directly consume the energy they produce in order to avoid grid instability issues. Utilizing the generated PV electricity on site, rather than injecting it into the grid is known as *direct consumption*.

Direct consumption can be achieved in many ways. An unfavourable option would be to limit the PV system size to make sure that the peak power produced is always lower than the peak power consumed. This, of course, would limit the PV capacity of an individual or a community. Other techniques include using storage in various ways, for example batteries. This way, the orange part of the curve in Figure 21.1 will be used to charge the battery, which will be discharged in the 'grid consumption' area.

Different policies can be introduced to encourage direct consumption. Pure net metering, which treats a prosumer equally in consumption and production does nothing to encourage self-consumption as it simply pays the prosumer retail costs of energy back to the grid. However, adjusting feed-in tariffs to be lower than retail energy is a typical scheme used by countries such as Germany and Italy to encourage direct consumption in PV systems.

It is worth repeating that the issue of grid instability can occur well before PV reaches penetrations of 100%. Even if as little as one-third of a grid is being powered by PV, instantaneous power levels can easily rise above demand at peak hours causing energy to be curtailed. For this reason, companies such as Solar City in the United States are already making plans to include batteries with PV systems as early as 2015 [177].

21.1.4 Levelized cost of electricity

Another very important concept is the *levelized cost of electricity* (LCoE), which is defined as the cost per kWh of electricity produced by a power generation facility. It is usually used to compare the lifetime costs of different electricity generation technologies. To be able to estimate the effective price per kWh, the concept of LCoE allocates the costs of an energy plant across its full lifecycle. It is somehow similar to averaging the upfront costs of production over a long period of time. Depending on the number of variables that are to be taken into account, calculating the LCoE can become very complex. In a simple case the LCoE can be determined with

$$\text{LCoE} = \frac{\sum_{t=1}^{n} \frac{I_t + M_t + F_t}{(1+r)^t}}{\sum_{t=1}^{n} \frac{E_t}{(1+r)^t}}. \tag{21.2}$$

The sums expand across the whole lifetime of the system n, and every year accounts for one summand. I_t are the investment expenditures in the year t, M_t are the operational and maintenance expenditures in the year t, and F_t are the fuel expenditures in the year t. Of course, for PV $F_t \equiv 0$. Further, E_t is the electricity yield in the year t. Finally, r is the discount rate which is a factor used to discount future costs and translate them into the present value.

Figure 21.2 shows the estimated LCoE for different methods of electricity generation that enter service in 2019. The values are national averages for the US We see that the LCoE of wind energy is lowest, it is even below the LCoE of fossil fuel and nuclear based

21. PV system economics and ecology

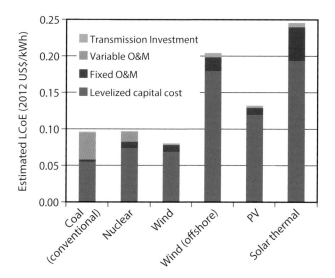

Figure 21.2: The estimated levelized cost of electricity for different electricity generation technologies that enter service in 2019 (data: [178]).

electricity. The LCoE of PV-generated electricity is still slightly above that of the non-renewable technologies.

Depending on the location and the initial investment required for the PV system, the LCoE for PV can vary a lot between different projects. Additionally, the discount rate r used for the calculation will strongly influence the LCoE.

For the electricity supplier, the LCoE is a valuable indicator of the cost competitiveness of a certain energy technology. It is also a good indicator for determining the electricity price: to make a profit this price must be above the LCoE. Surely, the supplier cannot determine the electricity price independently, as it is strongly influenced by policy factors such as feed-in tariffs, subsidies and other incentives.

21.1.5 Grid and socket parity

It is very important to investigate whether electricity generated with PV is competitive when compared to electricity generated by other means. For this purpose, the concepts of grid parity and socket parity are used. We want to warn the reader that many authors use these two concepts interchangeably. However, only a clear distinction between the two concepts allows a well grounded judgement of the economic viability of PV-generated electricity.

The owners of large-scale PV power plants have to compare the LCoE of their system to the cost of electricity production of other sources, ignoring subsidies and other incentives. The point at which the cost of PV electricity is equal to the cost of other electricity generation technologies is called *grid parity*. Of course, if the PV electricity price is below the grid price, the situation becomes even better,

$$\text{LCoE}_{\text{PV}} \leq \text{LCoE}_{\text{conventional}}. \tag{21.3}$$

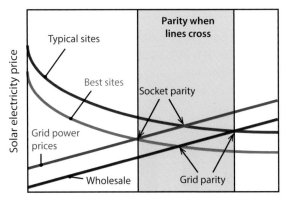

Figure 21.3: Grid and socket parity for PV systems (adapted with kind permission from P.R. Wolfe [179]).

In principle, the concept of grid parity can be generalized to the other renewable technologies as well. However, there is one significant difference between PV and other renewable technologies such as wind and hydro-electricity. Wind and hydro-electricity installations can usually only be financed by companies but are no option for a single consumer. In contrast, PV can be scaled down to the level of a single module, such that a house owner can become an electricity producer with his small scalable PV installation on his roof. The residential electricity price often also includes grid maintenance fees as well as taxes. The point at which the LCoE of a PV system is equal to the price the consumer pays for electricity from the grid is called *socket parity* [180].

Distinguishing between grid parity and socket parity is useful in order to avoid confusion with the cost of energy production. Since residential costs are always higher than production costs, an area will reach socket parity before reaching grid parity, as illustrated in Figure 21.3. The graph shows the volume of installed PV systems versus the price of PV electricity. The installed volume can be directly correlated with time, as the past decade has seen the implemented PV volume rise tremendously. As capital costs decline with increasing volumes, the price of PV-generated electricity is expected to decrease in the future. On the other hand, the price of fossil fuels is expected to rise because of increasing scarcity and costs linked to the right to release CO_2 emissions into the atmosphere. These trends will lead to increasing prices for electricity generated with combusting fossil fuels.

To conclude, grid parity and socket parity are very useful concepts to indicate the feasibility of a renewable energy technology. The closer a technology is to grid parity, the easier it can be integrated into the electricity mix. With the advancements in technology and the maturity of manufacturing processes, grid parity for solar energy systems is expected to be reached at many locations around the world in the next few years.

21.2 PV system ecology

Besides discussing the economics of PV systems, it is also very important to consider their ecological and environmental aspects. The aim of photovoltaics is to generate electricity without any considerable effect on the environment. It is therefore very important to check the ecological aspects of the different PV technologies. In this section we will discuss different concepts to quantify the environmental impact of PV systems.

21.2.1 Carbon footprint

The concept of the *carbon footprint* estimates the CO_2 emissions caused by manufacturing PV modules and compares them with the reduction of CO_2 emissions due to the electricity generated with PV instead of combusting fossil fuels. A more analytical approach is to look at the total energy required to produce either the PV modules or all the components of a PV system. As production processes vary considerably for the different PV technologies, the energy consumption for producing 1 kW_p also varies considerably. If a complete *lifecycle assessment* (LCA) is performed, the energy and carbon footprints of the PV panels are traced where possible, throughout their lifetime. Therefore LCA is also known as *cradle-to-grave analysis*.

21.2.2 Energy yield ratio

We now are going to introduce several indicators that are used to judge the different ecological aspects. The *energy yield ratio* is defined as the ratio of the total energy yield of a PV module or system throughout its lifetime to all the energy that has to be invested in the PV system in that time. This invested energy not only contains the energy for producing the components, transporting them to the location and installing them, but also the energy that is required to recycle the different components at the end of their lifecycle.

As the energy required for producing a PV system depends strongly on the PV technology and also on the quality of the panels, the energy yield ratio for the different technologies varies a lot. While the energy yield ratio for PV modules can be as large as 10 to 15, PV systems usually have a lower ratio because of the energy invested in the components other than the modules.

21.2.3 Energy payback time

A very important concept is the *Energy payback time*, which is defined as the total required energy investment over the lifetime divided by the average annual energy yield of the system,

$$\text{Energy payback time} = \frac{\text{totally invested energy}}{\text{average annual energy yield}}. \quad (21.4)$$

Note that the energy payback time is different from the economic payback time introduced in Section 21.1.

The energy payback time of typical PV systems is between one and seven years and depends on location issues such as the orientation of the PV array as well as the solar irradiance throughout the year.

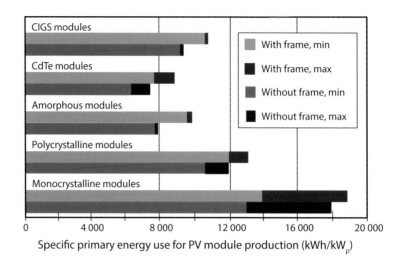

Figure 21.4: Specific primary energy used to produce PV modules of different technologies (data from [181]).

Figure 21.4 shows the specific primary energy required for producing PV modules with different technologies, where the term *specific* refers to the energy required per kW$_p$ of produced modules. As we can see, the differences between the technologies are large. The specific energy required for producing thin-film modules from materials such as amorphous silicon, cadmium telluride and CIGS is significantly below that of modules made from polycrystalline and monocrystalline silicon, where the specific energy can reach values up to 12,000-18,000 kWh/kW$_p$. Because of further improvements in the module efficiency and the manufacturing process, we may expect that the specific energy follows a decreasing trend.

For a PV system it is more difficult to allocate the energy that was used for its production, as all the components constituting the balance of system have to be taken into account. For example, for components like batteries and inverters the technologies and manufacturing processes may vary a lot between the different products available on the market. Nonetheless, studies were carried out that estimated the energy required by whole PV systems. Generally, as we can see in Figure 21.5 the energy required for the BOS is significantly below that used for manufacturing the modules. In the figure, amorphous and crystalline silicon modules are compared. As expected, we see that the energy payback time in regions with high solar irradiance is significantly shorter than in regions with low irradiance. While a-Si:H-based modules have a shorter energy payback time than c-Si-based modules, the energy payback time for the module frame and the BOS of a-Si:H based systems can be significantly higher than that of c-Si-based systems. This can be explained with the lower efficiency of a-Si:H that increases the required framing material per W$_p$.

The irradiance strongly influences the energy payback time and varies between two years (high irradiance) and six years (low irradiance). Roof-mounted systems always have a shorter energy payback time than systems mounted on the ground, mainly because of the BOS that is more energy extensive for ground-mounted systems.

21. PV system economics and ecology

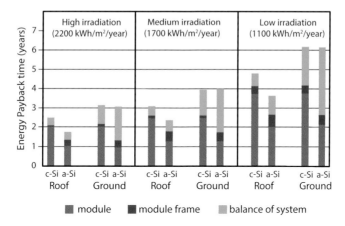

Figure 21.5: The energy payback time for the different components constituting PV systems (adapted with kind permission from E. Alsema) [182].

No matter which PV technology is chosen, the energy payback time is always far below the expected system lifetime, which usually is between 25 and 30 years. For the PV systems discussed in Figure 21.5, the energy yield ratio is between four and ten. Hence, the energy invested in the PV system is paid back several times throughout the lifecycle of the PV system. The urban legend that PV modules require more energy to be manufactured than they will ever produce thus is not backed by any data. In contrast, the net energy produced is much larger than the energy required for PV production.

However, a lot of work still needs to be done. Some studies indicate that the energy required for producing PV modules can be reduced by up to 80%. Further, as the number of installed PV systems increases, recycling of the components at the end of the lifecycle becomes very important. For example, the European Union has already introduced several directives that induce recycling schemes for c-Si-based PV modules.

21.2.4 Pollution

The last environmental issue that we want to mention is pollution caused by the production of PV modules. As many (sometimes toxic) chemicals are required for producing PV modules, this can be a serious threat to the environment. Therefore it is very important to have strong legislation in order to prevent pollution of the surroundings of PV factories. This is especially so in countries with weak environmental legislation.

21.3 Exercises

21.1 Table 21.1 shows the total area, sun hours per day and energy consumption of five different countries.

   (a) If a PV system of 270 W_p is installed in each of these locations, in which country would it be best to place the panels to obtain the maximum output?

Table 21.1

Country	Area (km^2)	Daily sun hours	Energy consumption (MWh/year)
United States	9 826 675	4.0	3 886 400 000
India	3 287 263	5.0	959 070 000
Brazil	8 515 767	4.5	455 700 000
Spain	505 992	4.5	267 500 000
United Kingdom	243 610	2.5	344 700 000

(b) Now let us take a look at the United States. How much area, as a percentage of the total area of the US, would be needed to cover the total annual energy demand of the country with solar panels having an efficiency of 15%?

(c) Which of the five countries would need to cover the highest percentage of their area with solar panels to supply all their electricity demand with solar energy?

(d) Which of the five countries would need to cover the lowest percentage of their area with solar panels to supply all their electricity demand with solar energy?

21.2 The Smith family are interested in buying a solar energy system. After some research on the Internet they have found two different systems that they are considering. System A is a PV system based on multicrystalline silicon solar cells and system B is a PV system based on amorphous silicon solar cells.

System A: The efficiency of the multicrystalline silicon module amounts to 15%. The dimensions of the solar modules are 0.5 m by 1.0 m. Each module has an output of 75 W_p. The modules cost €60 each.

System B: The amorphous silicon solar modules have an efficiency of 6%. The dimensions of the solar modules amount to 0.5 m by 1.0 m. The output of each module is 30 W_p. The modules cost €20 each. The advantage of the amorphous silicon solar modules is that they perform better on cloudy days when there is no direct sunlight. Installed in the Netherlands, this system gives, on an annual basis, 10% more output per installed W_p than the multicrystalline silicon modules.

Both systems are grid-connected using a 900 W_p inverter. The total price of the inverter, the cables, the installation and other costs amounts to €1,000. The solar modules are to be installed on a shed. The roof of the shed can only support 10 m^2 of solar modules.

In the Netherlands a PV system using multicrystalline silicon modules generates on average 850 Wh per W_p in one year. The performance of both types of modules is guaranteed for 20 years. The price of electricity from the grid is €0.23/kWh. Assume that all the power produced by the PV system is completely consumed by the Smith family's.

(a) How much (peak) power can be installed for system A given the peak power of the inverter and the area available on the shed?

(b) How much (peak) power can be installed for system B given the peak power of the inverter and the area available on the shed?

(c) What is the price of a kWh of electricity generated by system A?

(d) What is the price of a kWh of electricity generated by system B?

(e) The Smith family are eligible for a municipal subsidy for sustainable energy that amounts to 15% of the initial costs of the PV system. Using this subsidy, how many years does system A need to be operational, for them to recoup their share of the investment, assuming that the electricity price of €0.23/kWh does not change?

(f) Using the same subsidy of 15% as the previous question, how many years would system B need to be operational, for them to recoup their share of the investment, assuming that the electricity price of €0.23/kWh does not change?

21.3 The Miller family want to install a grid-connected PV system on their rooftop. Until now they have been using the grid electricity at €0.20/kWh, with an average annual consumption of around 3,000 kWh. The PV system will cost them €4,000 in total. The system is expected to have an annual yield of 2,500 kWh.

(a) How high will the average annual savings of the Miller family be if they install the PV system mentioned above? You may assume that any PV-generated power sent to the grid directly offsets the consumption of the household.

(b) Given the initial system costs, what will be the payback period of the PV system in years, assuming the grid costs remain the same? You may neglect inflation.

21.4 Which of the following statements is *not* true?

(a) The levelized cost of electricity depends on the insolation, the cost of PV systems, the discount rate and the cost of electricity coming from the grid at the particular location.

(b) Grid parity is achieved when the levelized cost of electricity is equal to the electricity cost at a particular location.

(c) The levelized cost of electricity can be used to compare the electricity price coming from PV and from solar thermal energy.

(d) The achievement of grid parity depends on the insolation, the cost of PV systems, the discount rate and the cost of electricity coming from the grid at the particular location.

21.5 Which of the following variables does *not* affect the economic payback period of a PV system?

(a) Insolation received by the PV system.

(b) Subsidies and feed-in tariffs.

(c) Cost of electricity.

(d) Overall lifetime of the PV system.

21.6 Consider a 250 W_p rated monocrystalline PV module. It results from a manufacturing process that consumes 10 kWh/W_p of energy. Assume that the PV module is used in a location with an average annual PV yield of 500 kWh. Its lifetime is 20 years.

(a) Determine the total energy consumed during the production of the module.

(b) Calculate the energy yield ratio of the PV module over its lifetime of 20 years.

(c) What is the energy payback period of the PV module?

Part V

Alternative solar energy conversion technologies

22
Solar thermal energy

Solar thermal energy is an application of solar energy that is very different from photovoltaics. In contrast to photovoltaics, where we used electrodynamics and solid state physics to explain the underlying principles, solar thermal energy is mainly based on the laws of thermodynamics. In this chapter we give a brief introduction to that field. After introducing some basics in Section 22.1, we will discuss Solar thermal heating in Section 22.2 and concentrated solar (electric) power (CSP) in Section 22.3.

22.1 Solar thermal basics

We start this section with the definition of *heat*, sometimes called *thermal energy*. The molecules of a body with a temperature different from 0 K exhibit a disordered movement. The kinetic energy of this movement is called *heat*. The average of this kinetic energy is related linearly to the temperature of the body.[1] Usually, we denote heat with the symbol Q. As it is a form of energy, its unit is joule (J).

If two bodies with different temperatures are brought together, heat will flow from the hotter to the cooler body and as a result the cooler body will be heated. Dependent on its physical properties and temperature, this heat can be absorbed into the cooler body in two forms, sensible heat and latent heat.

Sensible heat is the form that results in changes in temperature. It is given as

$$Q = mC_p(T_2 - T_1), \tag{22.1}$$

where Q is the amount of heat that is absorbed by the body, m is its mass, C_p is its *heat capacity* and $(T_2 - T_1)$ is the temperature difference. On the other hand, if a body absorbs

---
[1]The interested reader will find more details in standard textbooks on physics or thermodynamics. This definition is taken from Ref. [183].

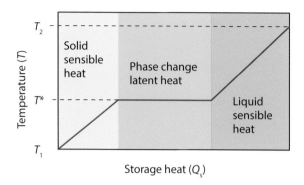

Figure 22.1: Illustrating the difference between sensible and latent heat.

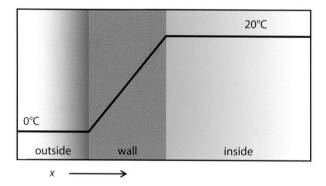

Figure 22.2: Illustrating conductive heat transfer through a wall.

or releases *latent heat*, the temperature stays constant but the phase changes. This happens for example when ice is melting: when its temperature is equivalent to its melting point, heat that is absorbed by the ice will not result in an increase in temperature but in transformation into from the solid to the liquid phase, which is water. Mathematically, this is expressed as

$$Q = mL, \tag{22.2}$$

where L is the *specific latent heat*.

The two forms of heat are illustrated in Figure 22.1, which shows what happens when a body absorbs heat. In the beginning, the body is solid and has temperature T_1. It then heats up and the heat is stored as solid sensible heat. When its melting point T^* is achieved, its temperature will not increase any more but the phase will change from solid to liquid. After everything is molten, the liquid will heat up further, the heat now is stored as liquid sensible heat.

Now we will take a look at the three basic mechanisms of heat transfer: conduction, convection and radiation.

Conduction is the transfer of heat in a medium due to a temperature gradient. For a better understanding we look at the heating of a house during winter, as illustrated in Figure 22.2. The inside of the house is warm at 20 °C, but the outside is cold with a temperature

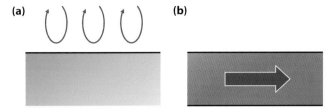

Figure 22.3: Illustrating (a) natural; and (b) forced conductive heat between a solid and a surrounding fluid.

of 0 °C. Because of this heat gradient, heat will be transferred from the inside through the wall to the outside. Mathematically, heat conduction can be described by the *Fourier law*,

$$\frac{dQ_{\text{cond}}}{dt} = -kA\frac{dT}{dx}, \qquad (22.3)$$

where dQ/dt is the heat flow in W, k is the thermal conductivity of the wall (given in W/(Km)), A is the contact area given in m^2 and dT/dx is the temperature gradient in the x direction given in K/m. If we assume the wall to be made from a uniform material, we can assume that dT/dx is constant throughout the wall and we can replace it by $\Delta T/\Delta x$, where Δx is the thickness of the wall.

Convection is the second possible mechanism for heat transfer. It is the transfer of heat by the movement of a fluid. We distinguish between two forms of convection, as sketched in Figure 22.3: In *forced convection* the movement of the fluid is caused by external variables, while in *natural convection* the movement is caused by density differences due to temperature gradients. In both cases, the heat transfer from a medium of temperature T_1 to a fluid of temperature T_2 can be described by *Newton's law*

$$\frac{dQ_{\text{conv}}}{dt} = -hA(T_1 - T_2), \qquad (22.4)$$

where h is the *heat transfer coefficient* which is given in W/(m^{2}K). We will not discuss the calculation of h in detail but we want to mention that it depends on various factors such as the velocity of the fluid, the shape of the surface or the kind of flow that is present, i.e. whether it is laminar or turbulent flow.

The third heat transfer mechanism is *radiative heat transfer*, which is the most important mechanism of heat transfer for solar thermal systems. As we already discussed in Chapter 5, thermal radiation is electromagnetic radiation propagated through space at the speed of light. It is emitted by bodies depending on their temperature and is caused by excited electrons falling back to their ground level and emitting a photon, and hence electromagnetic radiation.

As discussed in Chapter 5, a *blackbody* is an idealized concept of a body, which is a perfect absorber of radiation, independent on the wavelength or direction of the incident light. Further, it is a perfect emitter of thermal radiation. The Stefan-Boltzmann law (Eq. (5.19)) describes the total radiant emittance of a blackbody of temperature T,

$$M_e^{BB}(T) = \sigma T^4, \qquad (22.5)$$

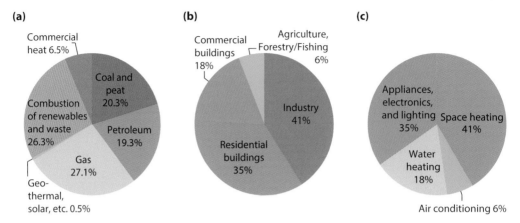

Figure 22.4: (a) Primary energy suppliers for the generation of heat; and (b) the demand of heat by sector [13]. (c) Energy consumption of U.S. homes [184]. All data is for 2009. Percentage of gas in (a) might be significantly higher nowadays due to the strong growth in shale gas usage (©OECD/IEA 2012, Insights Series 2012: Policies for renewable heat, IEA Publishing. Licence: www.iea.org/t&c/termsandconditions).

where M is given in W/m^2 and $\sigma \approx 5.670 \times 10^{-8}$ W/(m^{2}K^4) is the Stefan–Boltzmann constant. In nature, no blackbodies exist. However we can describe them as so-called *greybodies*. The energy emitted by a greybody still can be described by Plank's law [Eq. (5.18a)] when it is multiplied with a wavelength-dependent emission coefficient $\epsilon(\lambda)$. For a blackbody we would have $\epsilon(\lambda) \equiv 1$.

22.2 Solar thermal heating

About half of the world's energy consumption is in the form of *heat*. About two-thirds of the heat demand is supplied by coal, oil, and natural gas, as we can see in Figure 22.4 (a). As shown in Fig. 22.4 (b), heat is mainly used in the industrial sector for facilitating for example chemical processes and in the residential sector for heating and warm water supply.

Figure 22.4 (c) shows the total energy demand of a typical household in the United States. We see that space and water heating represent 59% of the total energy consumption. If the demand for cooling is also taken into account, about two-thirds of the energy consumption is related to the use of heat.

The residential demand for heat can be at least partially covered with a *solar water heater*, which is a combination of a solar collector array, an energy transfer system and a storage tank, as sketched in Figure 22.5. The main part of a solar water heater is the *collector array*, which absorbs solar radiation and converts it into heat. This heat is absorbed by a heat transfer fluid that passes through the collector. The heat can be stored or used directly. The amount of hot water produced by a solar water heater throughout the year depends on the type and size of the solar collector array, the size of the water storage, the amount of sunshine available at the site and the seasonal hot water demand pattern.

22. Solar thermal energy

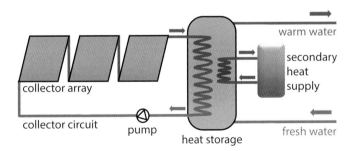

Figure 22.5: The main components of a solar water heating system.

There are several ways to classify solar water heating systems. One way is by the fluid heated in the collector. When the fluid used in the application is the same as that is heated in the collector, it is called a direct or open loop. In contrast, when the fluid heated in the collector goes to a heat exchanger to heat up the utility fluid, it is called an indirect or closed loop.

Another way to classify the systems is by the way the heat transfer fluid is transported. This can either be passive, where no pumps are required, or by forced circulation, using a pump. Passive solar water heating systems use natural convection to transport the fluid from the collector to the storage tank. This happens because the density of the fluid drops when the temperature increases, such that the fluid rises from the bottom to the top of the collector – this is the same as natural convection that we discussed in Section 22.1. The advantage of passive systems is that they do not require any pumps or controllers, which make them very reliable and durable. However, depending on the quality of the used water, pipes can get clogged, which considerably reduces the flow rate.

On the other hand, active systems like the one sketched in Figure 22.5 require pumps that force the fluid to circulate from the collector to the storage tank and the rest of the circuit. These systems are usually more expensive than passive systems. However, they have the advantage that the flow rates can be tuned more easily.

22.2.1 Solar thermal collectors

Now we will take a closer look at the solar collector, in which the working fluid is heated by the solar radiation. The collector determines how efficiently the incident light is used. It usually consists of a black surface, called the absorber, and a transparent cover. The absorber is able to absorb most of the incident energy from the sun, Q_{sun}, raising its temperature and transferring that heat to a working fluid. Hence, the absorber can be cooled and the heat transferred elsewhere.

Not all of Q_{sun} can be used, as there are some losses, as illustrated in Figure 22.6. A part Q_{refl} is lost as reflection either in the encapsulation or in the absorber itself. Other losses are related to the heat exchanged with the surrounding air by the convection mechanism, Q_{conv} and radiation from the hot absorber, Q_{rad}. When we combine all these energies to form an energy balance, we find for the heat Q_{col} that can be collected by the collector:

$$Q_{col} = Q_{sun} - Q_{refl} - Q_{conv} - Q_{rad}. \tag{22.6}$$

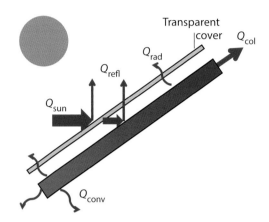

Figure 22.6: The major energy fluxes in a covered solar collector.

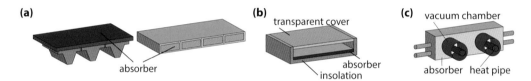

Figure 22.7: Illustrating (a) an uncovered, (b) a covered and (c) a vacuum collector.

The efficiency of the collector depends mainly on two factors: the extent to which the sunlight is converted into heat by the absorber, and the heat losses to the surroundings. It will therefore depend on the weather conditions and the characteristics of the collector itself. To reduce losses, insulation from the surroundings is important, especially when the temperature difference between the absorber and the ambient is high.

Usually, collectors are classified into three categories: uncovered, covered and vacuum, as shown in Figure 22.7. *Uncovered collectors* do not have a transparent cover, so the sun strikes the absorber surface directly, hence the reflection losses are minimized. This collector type only is used for applications where the temperature differences between the absorber and the surroundings are small, for example for swimming pools. *Covered collectors* are covered by a transparent material, providing extra insulation but increasing reflection losses. These collectors are used for absorber temperatures of up to 100 °C. Finally, in *vacuum collectors* the absorber is confined in vacuum tubes, and little heat is lost to the surroundings. The manufacturing process of these collectors is more complicated and expensive, but the collector can be used for high temperature applications since convection losses to the surroundings are much lower than for the other types.

Modern solar collectors often use light management techniques just as solar cells. For example, transparent conducting oxide layers at the front window, that we discussed in Section 13.1, can be used. If they have the plasma frequency in the infrared, almost all the solar radiation can enter the window, but the long wavelength infrared radiation originating from the hot parts in the collector cannot leave the collector as it is reflected back at the window.

Collectors can also be classified by their shape: we can distinguish between *flat plate collectors* and *concentrating* collectors. *Flat plate collectors* consist of flat absorbers that are oriented towards the sun. They can deliver moderate temperatures, up to around 100 °C. They use both direct and diffuse solar radiation and no tracking systems are required. Their main applications are solar water heating, heating of buildings, air conditioning and heat for industrial processes. In contrast, *concentrating collectors* are suited for systems that require a higher temperature than is achievable with flat collectors. The performance of concentrating collectors can be optimized by decreasing the area of heat loss. This is done by placing an optical device between the source of radiation and the energy-absorbing surface. Because of this optical device the absorber will be smaller and hence will have a lower heat loss compared to a flat plate collector at the same absorber temperature. One disadvantage of concentrator systems is that they require a tracking system to maximise the incident radiation at all times. This increases the cost and leads to additional maintenance requirements.

Just as for PV systems,[2] for solar heat collector arrays it is important to decide whether the collectors should be connected in series or in parallel. Connecting collectors in parallel means that all collectors have the same input temperature, while for those connected in series the outlet temperature of one collector is the input temperature of the next. Most commercial and industrial systems require a large number of collectors to satisfy the heating demand. Therefore most of the time, a combination of collectors in series and in parallel is used. Parallel flow is used more frequently because it is inherently balanced and minimizes the pressure drop. In the end, the choice of series or parallel arrangement will ultimately depend on the temperature required by the application.

22.2.2 Heat storage

Now we will discuss the heat storage, which is an extremely important component as it has an enormous influence on the overall system cost, performance and reliability. Its design affects the other basic elements such as the collector or the thermal distribution system. The task of the storage is twofold. First, it improves the utilization of the collected solar energy. Secondly, it improves the system efficiency by preventing the fluid flowing through the collectors from quickly reaching high temperatures.

Several storage technologies are available; some of them can even be combined to cover daily and seasonal fluctuations. In general, heat can be stored in liquids, solids or phase change materials, abbreviated as PCM. Water is the most frequently used storage medium for liquid systems, because it is inexpensive, non-toxic, and it has a high specific heat capacity. In addition, the energy can be transported by the storage water itself, without the need for additional heat exchangers.

The usable energy stored in a water tank can be calculated using:

$$Q_{\text{stored}} = V \rho C_p \Delta T, \quad (22.7)$$

where V is the volume of the tank, ρ is the density of water, C_p is the specific heat capacity of water and ΔT is the temperature range of operation. The lower temperature limit is often set by external boundaries, such as the temperature of the cold water, or by specific process

---

[2]See Chapter 15.

requirements. The upper limit may be determined by the process, the vapour pressure of the liquid or the heat loss of the water storage. For example, for residential water heating systems the maximally allowed temperature is set to 80 °C because at higher temperatures calcium carbonate will be released from the water, clogging the warm water tubes [185].

The heat loss of the tank, $\dot{Q}_{\text{loss}}$, can be determined using:

$$\dot{Q}_{\text{loss}} = UA\Delta T, \tag{22.8}$$

where A is the area of the heat storage tank. U is the global heat exchange coefficient and is a measure of the quality of the insulation. Usually it varies between 2 and 10 W/K. Further, U is also a function of the different media between which the heat exchange takes place.

The same principles can be applied to small and big storage systems. Small water energy storage systems can cover daily fluctuations and are usually in the form of water tanks with volumes from several hundreds up to several thousands of litres. Large storage systems can be used for seasonal storage. Often they are realized as underground reservoirs.

Another type of energy storage are so-called *packed beds*, which are based on heat storage in solids. They use the heat capacity of a bed of loosely packed particulate material to store the heat. A fluid, usually air, is circulated through the bed to add or remove energy. A variety of solids can be used, rock being the most common. In operation, flow is maintained through the bed in one direction during the addition of heat and in the opposite direction during removal. The bed is in general heated during the day with hot air from the collector. In the evening and during night, energy is removed with air temperatures around 20 °C flowing upward.

Another way to store heat in buildings is to use their solid walls and roofs as a storage material. A case of particular interest is the *collector storage wall*, which is arranged such that the solar radiation is transmitted through a glazing and absorbed in one side of the wall. As a result, the temperature of the wall increases and the energy can be transferred from the wall to the room by radiation and convection. Some of these walls are vented to transfer more energy to the room via forced convection.

The last method that we will discuss to store heat is to use phase change materials (PCM). While in all the storage methods discussed so far energy is stored as sensible heat, in PCM storage takes place as *latent heat*, which is used for a phase change without any change in the material temperature. PCM used for energy storage must have a high specific latent heat L, such that a large amount of energy can be stored. In addition, the phase change must be reversible, and survive cycles without significant degradation. The stored amount of heat can be calculated using:

$$Q_{\text{stored}} = m\left[C_s(T^* - T_1) + L + C_l(T_2 - T^*)\right], \tag{22.9}$$

where the different temperatures are also shown in Figure 22.1. We consider the specific heat capacity in the solid phase, C_s, to be constant from the initial temperature T_1 up to the phase change temperature T^*. The latent heat of the material is L, and the specific heat capacity in the liquid phase C_l, rises from T^* up to the final temperature T_2. Materials that are commonly used as PCM are molten salts, such as sodium sulphate (Na_2SO_4), calcium chloride ($CaCl_2$) or magnesium chloride ($MgCl_2$). PCM storage in general is used for high temperature applications.

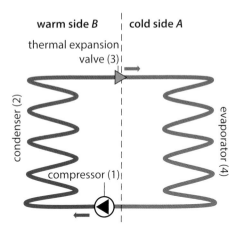

Figure 22.8: A compressor-driven cooling circuit.

22.2.3 System design

Often, at least at higher latitudes with large differences between summer and winter, solar thermal systems include a boiler as a backup. Its main function is to provide the necessary energy when the solar power is not sufficient. It is basically a normal heater that adds the remaining heat needed to achieve the desired temperature in the storage tank. As an energy source, usually natural gas, oil, or biomass is used. Sometimes, the additional heat is also provided with electricity or a heat pump.

Up to now we have discussed how to collect and store the energy, but have not considered how to transport it from the collector to the storage system. The transport of heat is done with a *collector circuit*, usually using either a liquid or gas. If a liquid is used, it is important that it neither freezes nor boils, even at the most extreme operating conditions. Further, the medium should have a large specific heat capacity, a low viscosity, and it should be non-toxic, cheap and abundant. The most common fluids are water, oils or air.

As mentioned above, the flux can be caused either naturally by the temperature gradients, forced by a pump, or by a heat pipe in which the fluid is allowed to boil and condense again. The optimal choice will depend on the specific system design. Further, heat losses in the collector circuit must be taken into account, especially when the pipes are very long. During the planning phase it is therefore important to minimize the circuit length.

In systems with a pump, often a *controller* is present that regulates the fluxes of fluid through collector, storage, and boiler, such that the heat transport from collector to storage and boiler is maximized. This can for example be done by calculating the optimal flux with respect to the fluid temperatures at the collectors and in the storage.

22.2.4 Solar cooling

Another interesting application is *solar cooling* (also called *solar air conditioning*), which seems a bit contradictory first sight. Before we start with the actual discussion on solar air conditioning, we briefly recap the principle of an air conditioning system based on the

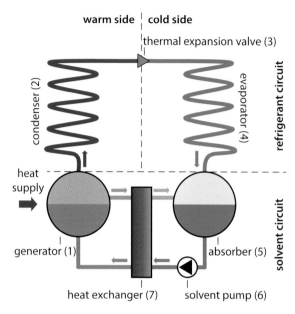

Figure 22.9: An absorption cooling circuit. The numbers are explained in Section 22.2.4.

vapour compression cycle, which is sketched in Figure 22.8. Similar to every *heat pump*, the task of this cycle is to transport heat from a cool reservoir B to a warmer reservoir A. Such a system typically has four components: a compressor, a condenser, a thermal expansion valve (throttle) and an evaporator.

In the compressor (1), vapourised coolant is compressed to a higher pressure. On release of its latent heat to the surroundings, it is condensed to its liquid state in the condenser (2). Then the coolant passes a thermal expansion valve (3), such that its pressure on the other side of this valve is so low that it can evaporate again. For the evaporation, it must absorb latent heat from the cool reservoir A. Then, the coolant containing the latent heat from A is compressed again. We see that this cooling circuit utilizes the latent heat stored and released during phase changes.

Naturally, heat only flows from warm to cold reservoirs. To reverse the heat flow, as happens in a heat engine, additional energy must be added to the system. This happens in the compressor and traditionally is supplied as electric energy. Probably the simplest option to realize solar cooling is to generate this electricity by a conventional PV system or a solar thermal power system as discussed in Section 22.3. It is also possible to drive the compressor with mechanic energy obtained directly from solar energy with a Rankine cycle, which is discussed in Section 22.3.

The choice of *coolant* is based on the temperature regime in which the cooling system will operate. However, its effect on the environment has to be taken into account. For example, *chlorofluorocarbons*, which were widely used as coolants, were heavily regulated in the 1980s because of their destructive effect on the ozone layer [186].

Another concept that can be driven directly by solar heat is that of an *absorption cooling machine*. In contrast to compressor-driven cooling, no mechanical energy is required for

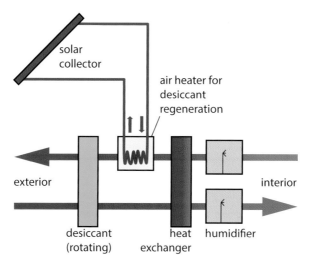

Figure 22.10: Desiccant cooling.

absorption cooling; the cooling process is directly driven by heat that can be supplied by solar collectors. Instead of one coolant, here *two substances* are used: a *refrigerant* and a *solvent*. Often, ammonia (NH_3) is used as refrigerant, while water is used as solvent. Under atmospheric pressure, ammonia has a boiling point of -33 °C. Further, it is very soluble in water, a property that is utilized in the cooling process.

Figure 22.9 illustrates an absorption cooler. In the generator (1), a mixture of both substances is heated with an external heat supply (for example from a solar collector). Because of its low boiling point the refrigerant will leave the mixture in the gas phase. The solvent still present in this gas is separated and brought back into the boiler (not drawn). Then, the refrigerant is condensed in a condenser (2) and brought to a thermal expansion valve (3). Similar to the compressor circuit, latent heat is used to evaporate the expanded refrigerant (4), such that cooling takes place. Next, the refrigerant is absorbed by the solvent (5) and the refrigerant-rich solvent is pumped back into the generator using a solvent pump (6). The solvent circuit is closed with the solvent flowing from the generator (1) to the absorber (5). A heat exchanger (7) between the two branches of the solvent cycle prevents heat from flowing to the generator side, hence reducing the heat supply necessary.

Instead of using a throttle valve (3) to expand the refrigerant, it can also be mixed with a third gas, for example hydrogen (H_2), such that the partial pressure of the refrigerant is reduced allowing it to evaporate. In this case, the whole system operates at one pressure.

The last option of cooling that we discuss is that of solar *desiccant cooling*, illustrated in Figure 22.10. In such a system, air is dried when passing through a desiccant, like silica. Next, the air passes a heat exchanger where a large fraction of the heat is taken out of the air stream and transported to the outgoing air stream. Then, water is sprayed into the air. As it evaporates, the air is cooled and humidified, such that it guarantees optimal interior conditions. Air that leaves the interior is first preheated passing the heat exchanger and then heated up further using a solar collector. The hot air streams through the desiccant. Hence, the desiccant is dried and can be reused for adsorbing humidity.

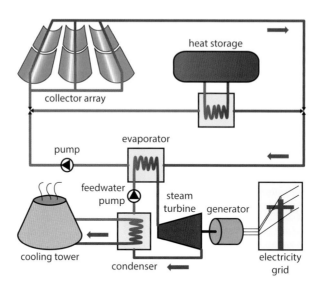

Figure 22.11: A solar thermal electric power system.

22.3 Concentrated solar power (CSP)

In this section we take a closer look at *concentrated solar power* (CSP), where high temperature fluids are used in steam turbines to produce electricity. Much of the early attention on CSP systems was on small-scale applications, mainly for pumping water. However, since about 1985 several large-scale power systems with a power output of up to 80 MW have been built.

The CSP technology is especially interesting for desert regions, where almost all the solar radiation is incident as direct radiation. As illustrated in Figure 22.11, these systems consist basically of a collector, where the solar energy is absorbed, a storage system, usually water or phase change storage, a boiler that acts as a heat exchanger between the operational fluids of the collector and the heat engine, and the heat engine itself, which converts the thermal energy into mechanical energy. In large CSP plants, the heat engine is a *steam turbine*, where the energy stored in the hot steam is partially converted to rotational (*mechanical*) energy as the steam expands along the turbine. This thermodynamic process is called the *Rankine cycle*, named after the Scottish engineer William John Macquorn Rankine (1820–1872).

The mechanical energy can be further used to drive an electrical generator. The collectors are built as concentrator systems in order to reach the high temperatures of several hundreds of degrees Celsius, which are required to operate the heat engine.

One problem of CSP systems is that the efficiency of the collector diminishes as its operating temperature rises, while the efficiency of the engine increases with temperature as we have already seen when discussing the thermodynamic efficiency limit in Section 10.1. Hence, a compromise between the two has to be found when choosing the operating temperature. Current systems have efficiencies up to about 30% [187].

Now we will take a closer look at the solar concentrator system, illustrated in Figure

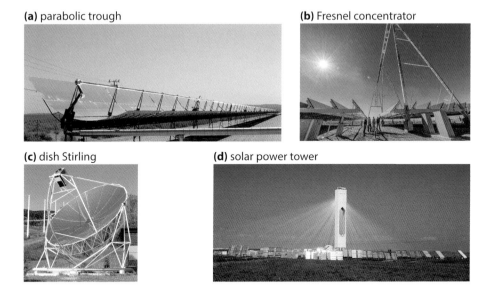

Figure 22.12: Different types of concentrators used in CSP systems [188–191].

22.12. First concepts were developed by the ancient scientist Archimedes (287 BC–212 BC); Leonardo da Vinci (1452–1519) later designed some concentrators. Different types of concentrators produce different peak temperatures and hence thermodynamic efficiencies, due to the different ways of tracking the sun and focussing light. Innovations in this field are leading to more energy efficient and cost effective systems.

The most common concentrator type with about 96% of all installed CSP installations is the *parabolic trough* [192], which consists of a linear parabolic reflector that concentrates the light onto an absorber tube located in the middle of the parabolic mirror, in which the working fluid is located. The fluid is heated to 150 to 350 °C, and then used in a heat engine. This technology is highly developed; well known examples are the Nevada Solar One and the Plataforma Solar de Almeria (PSA), Spain, power plants.

Another concentrator concept is that of *Fresnel concentrators*, where flat mirrors are used, as illustrated in Figure 22.12. Flat mirrors allow more reflection in the same amount of space as parabolic mirrors, reflect more sunlight, and are much cheaper.

The *dish Stirling* or dish engine system consists of a parabolic reflector that concentrates light to the focal point, where the working fluid absorbs the energy and is heated up to about 500 °C. The heat is then converted to mechanical energy using a *Stirling engine*. With about 31.25% efficiency demonstrated at Sandia National Laboratories, this design currently has the highest demonstrated efficiency [187].

The last concentrator type is that of *solar power tower plants*. Here, an array of dual axis tracking reflectors, commonly named *heliostats*, are arranged around a tower. The concentrated sunlight hits a central receiver, which contains the working fluid, which can then be heated to 500 or even 1,000 °C. Examples of solar power towers are Solar One and Solar Two in the Unites States and the Eureka project in Spain [192].

CSP systems can be combined with heat storage, such as phase change materials, discussed in Section 22.2. Hence, CSP systems can operate during day *and* night.

22.4 Exercises

22.1 During the winter, the inside of an average house is maintained at 20 °C, while the outside temperature is 0 °C. Assuming that the only mechanism of heat transfer is conduction, the walls are 10 cm thick and the heat conductivity of the walls is 0.5 W/(Km).

(a) Calculate the heat flux from the room to the surroundings in W/m^2.

(b) To reduce the heat loss through the walls, the material should be changed to an insulator material. The new overall conductivity will be 0.1 W/(Km); the thickness of the walls is maintained. Calculate the reduction of the heat flux through the walls compared to the initial case.

22.2 What is the most important heat transfer mechanism in domestic solar water heating systems?

(a) Conduction.

(b) Convection.

(c) Radiation.

(d) They are all equally important.

22.3 What does it mean when the water heating system is in an open loop?

(a) The solar water heating system is used for heating instead of power production.

(b) The fluid that is heated in the collector is directly used to cover the heating demand.

(c) The flow of the collector liquid is caused by natural convection.

(d) The fluid of the collector liquid is caused by forced convection.

22.4 A solar collector with an area of 1.5 m^2 is installed on the rooftop of a house. Assume that the radiative energy arriving from the sun is 1,000 W/m^2. The collector reflects 10% of the energy arriving on its surface. Also, the collector is not perfectly insulated, and losses occur. The collector has a heat transfer coefficient h of 2 $Wm^{-2}K^{-1}$. The side areas of the collector are assumed to be negligible. The ambient temperature is 20 °C and the collector is assumed to be at a temperature of 50 °C. Consider that this temperature is constant throughout the whole collector. The collector is assumed to behave like a black body.

(a) What is the power output of the collector?

(b) What percentage of the total losses is caused by radiation?

22.5 The Smith family have already installed PV in their house. Now, they also want to cover their needs for warm water with solar energy. For this, they are considering having a solar thermal water heating system. They need 100 L/day of warm water and the water has to be heated from 10 °C to 60 °C. The specific heat capacity of water is 4.18 J/gK. Assume a solar irradiance of 1,000 W/m^2 for three equivalent sun hours and an efficiency of 70% for the solar thermal installation.

(a) How much heat does the system need to generate per day to meet the warm water demand?

(b) How much collector area will be needed to cover the demand?

(c) If only half of the required hot water has to be stored, what would the minimum size of the storage tank be?

(d) The cost of the solar collector is estimated at €120/m^2, and the extra costs for the water tank and piping are €6 per L of storage. How much will the whole system cost?

(e) Mr Smith has read a newspaper article in which it was claimed that at the moment it is more expensive to install a solar thermal system for water heating than to use the PV electricity to heat up the water directly. He wants to assess whether this is true. The price of a PV panel is €1/W$_p$, the external costs are €400, and the efficiency of converting electricity to heat is assumed to be 85%. How much will it cost to generate the same amount of energy with PV technology?

(f) Assume that the lifetime of both systems is 20 years and the maintenance costs are equal for both cases. Which will be the better choice for water heating?

22.6 Which one of the following methods cannot be used as a cooling method using solar thermal techniques?

(a) Solar desiccant.

(b) Heat pump.

(c) Solar absorption.

(d) Phase change materials (PCM).

22.7 What are the main advantages of a solar tower power plant? (More then one correct answer possible.)

(a) Easy storage of energy.

(b) High temperature fluid, which results in higher efficiencies.

(c) Compact and modular.

(d) Low investment costs.

22.8 The world electricity consumption is roughly 20,300 TWh per year. We want to cover the total electricity demand of the world by installing solar thermal power in the Sahara desert, where the average daily solar insolation is 6.3 kWh/m^2. The Sahara has an area of 9,400,000 km^2.

(a) Assuming that the overall efficiency of a solar thermal power plant is 20%, how much area will be needed to cover the world electricity demand?

(b) What percentage of the area of the Sahara would be covered?

23
Solar fuels

In this chapter we discuss how to store solar energy in the form of chemical energy in so-called *solar fuels*. As we have seen in previous chapters, we are able to convert solar energy to electricity with efficiencies of up to 44%. Also, we are able to convert solar energy easily into heat, which we can use for preparing warm water, heating and even cooling.

The biggest obstacle towards an economy that is driven by renewable energy is not the generation of sustainable energy sources but the storage of the renewable energy. As we have seen in Chapter 19, storing electricity with batteries is very difficult. While day-night storage that makes solar electricity generated during daytime available at night is easily achieved, seasonal storage that allows electricity generated in the summer to be stored until winter would require very large battery systems that are not usually feasible. This problem becomes more severe in regions that are far away from the Equator where the differences in solar irradiance between summer and winter are significant. The same is also true for solar heat that was discussed in Chapter 22: while a hot water storage tank with a volume of several hundreds of litres is sufficient as day-night storage, seasonal storage that would allow solar-heated water to be used throughout the year would require storage tanks with a volume of tens of thousands of litres. This tank also would need to be heavily insulated. If we manage to store solar energy as chemical energy, we do not have this problem as chemicals can be easily stored.

Figure 23.1 shows the Ragone chart that we analyzed in Section 19.3. The chart shows the amount of energy per mass stored in a certain storage technology versus the specific power provided by the technology. As we know, by far the most used form of stored energy is fossil fuels. As we can see from the graph, fossil fuels have high energy density properties, and are stable and reliable. But, fossil fuels are depleting and their use is detrimental to the environment. Hydrogen (H_2) has an even better gravimetric energy density than fossil fuels, but it is a very light gas. Hence, the volumetric energy density, i.e. the energy per unit volume, is much lower. This is one of the reasons it has not been widely used

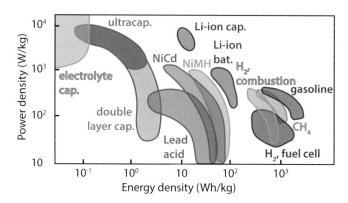

Figure 23.1: *Ragone* chart of different energy storage methods. Capacitors indicated with 'cap'.

until now. When we compare batteries and hydrogen as energy storage for solar electricity, we see that batteries have a lower energy density. Further, they require a much higher initial investment than hydrogen. Hence, it seems an interesting option to use hydrogen as a storage material, i.e. as a solar fuel.

In this chapter we discuss how to produce hydrogen using solar energy by utilizing electrochemistry. In nature, *photosynthesis* takes place, where sunlight is used to convert carbon dioxide and water into oxygen and sugars, i.e. is a form of chemical energy. Here, we try to mimic nature with inorganic semiconductor materials that are able to split a water molecule into oxygen and hydrogen using the energy of sunlight. This process is sometimes referred to as *artificial photosynthesis*.

For storing solar energy as chemical energy in the form of hydrogen, *water splitting* can be used,

$$2H_2O \rightarrow 2H_2 + O_2. \tag{23.1}$$

The energy required for this reaction is given by the *Gibbs free energy* and it has a value of $G = 237.2$ kJ/mol. In *solar* water splitting this energy is provided by the Sun.

Another product that seems very promising for energy storage is *methane* (CH_4), as it is easier to store and has fewer hazardous problems than H_2. It can be produced from hydrogen and carbon dioxide (CO_2) with the *Fischer–Tropsch process*. In this process, first the hydrogen is combined with carbon dioxide in a *water-gas shift reaction* to obtain carbon monoxide and water in the form of steam,

$$H_2 + CO_2 \rightarrow 2H_2O + CO. \tag{23.2}$$

This gas mixture, known as *synthesis gas*, can then be refined to finally obtain methane,

$$CO + 3H_2 \rightarrow CH_4 + H_2O. \tag{23.3}$$

When the stored energy is required, the methane can be burned in a combustion reaction,

$$CH_4 + 2O_2 \rightarrow CO_2 + 2H_2O \tag{23.4}$$

that will give water and carbon dioxide as by-products. The water can be reused for the electrolysis and the carbon dioxide can be used for the water gas shift reaction, hence, the

23. Solar fuels

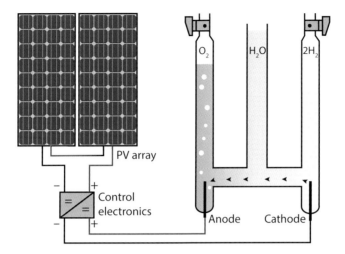

Figure 23.2: Illustrating a lab-scale Hofmann voltameter that is used to split water. In the diagram, the voltmeter is connected to a PV system via some control electronics.

cycle is closed. Each reaction in the cycle has a certain efficiency; the overall efficiency can be obtained by combining all those efficiencies.

Many different methods can be used for water splitting. In this chapter we will discuss two methods: *electrolysis of water* and *photoelectrochemical water splitting*.

23.1 Electrolysis of water

Figure 23.2 shows a typical lab-scale setup for the hydrolysis of water. It is called a *Hofmann voltameter* after the German chemist August Wilhelm von Hofmann (1818–1892). It contains three upright cylinders that are joined at the bottom. While electrodes are placed in the left and right cylinders, the central cylinder is used to refill the device with water. For starting the electrolysis, a voltage has to be applied between the two electrodes.

Water splitting is a *reduction-oxidation* reaction, which is commonly abbreviated as *redox reaction*. In such redox reactions, the reaction happens due to the exchange of electrons between atoms or molecules. They can be divided into two half reactions, the *oxidation* and the *reduction*. For water splitting these half reactions look as follows. During the oxidation, which happens at the *anode*, water is split giving oxygen and protons to the solutions and electrons to the anode,

$$H_2O \rightarrow \tfrac{1}{2}O_2 + 2H^+ + 2e^-. \tag{23.5}$$

In the reduction, electrons originating from the *cathode* react with the protons to hydrogen,

$$2H^+ + 2e^- \rightarrow H_2. \tag{23.6}$$

The total reaction therefore can be written as

$$H_2O \rightarrow \tfrac{1}{2}O_2 + H_2. \tag{23.7}$$

Each half reaction has an associated potential and the sum of the potentials of each half reaction is the potential for the whole redox reaction. Potentials are always defined with respect to a reference, i.e. a zero point. For redox reactions, the zero is defined as the hydrogen half reaction that happens at the cathode. For calculating the potential at the anode, i.e. of the oxygen production reaction, we can apply the equation

$$V^0 = \frac{G}{nF}, \qquad (23.8)$$

where G is the Gibbs free energy, n is the number of electrons transferred between the anode and the cathode for the production of one H_2 molecule, and $F = 96,485$ C/mol is the *Faraday constant* that tells us the amount of charge per mole of electrons. Using $G = 237.2$ kJ/mol and $n = 2$, as two electrons are transferred per H_2 molecule, we find $V^0 = 1.23$ V.

In reality, an applied voltage of 1.23 V is not enough to drive the redox reaction because of some voltage losses. These voltage losses are mainly because of the activation energy at the electrodes as well as ohmic losses in the cables, electrodes, and also in the solution. To increase the conductivity of the water, a base or an acid can be added. For the classic Hofmann voltameter, a small amount of sulphuric acid (H_2SO_4) is used.

The potential difference between V_0 and the potential used in the real device is called *overpotential* ΔV. The overpotential due to the activation energy required at the electrodes is strongly related to the electrode material and the gas that is produced at that electrode. For hydrogen production at the cathode, *platinized (black) platinum* is one of the best materials with an overpotential of -0.07 V. For the oxygen production at the anode, *nickel* with an overpotential of $+0.56$ V is very well suited. Combining these two voltages and the overpotentials we can understand that the typical overpotential is usually around 0.8 V.

It is important to realize that all the energy consumed by the hydrolysis process related to the overpotential is lost and will lead to heating of the hydrolysis device. Only a fraction

$$\eta_{WS} = \frac{V^0}{V^0 + \Delta V} \qquad (23.9)$$

of the supplied energy is actually stored in the hydrogen. It is clear that it is a very important task for science and industry to reduce the required overpotential. Further, the current requirement for platinum electrodes makes hydrolysis unfeasible from an economic point of view.

For an overpotential of 0.8 V, we would get an efficiency of around 60%. If we were to use a good PV system with an overall efficiency of $\eta_{PV} = 18\%$ we would have a total maximal *solar-to-hydrogen efficiency* of

$$\eta_{STH} = \eta_{PV}\eta_{WS} = 10.9\%. \qquad (23.10)$$

23.2 Photoelectrochemical (PEC) water splitting

Hydrogen can also be produced utilizing solar energy by using a *photoelectrode*, which uses light to produce an electrochemical reaction, in which water is split into oxygen and hydrogen. In this process, the photons reach the surface of the photoelectrode, which is made of a photoactive semiconductor. As in any other semiconductor, photons with an energy equal to or larger than the semiconductor bandgap energy will create an electron-hole pair. The electrons and holes will be separated by an electric field, and both will be used in the two half reactions involved in the overall water splitting process. To generate the required electrical field, we need a voltage source, for example a solar cell. The photoelectrode can be either an anode or a cathode.

If the solar-splitting device is made of a solar cell that is placed behind the photoanode, the solar cell will receive the light transmitted through the photoanode. This light creates another electron-hole pair and an electric field that brings the electrons to the photoanode and the holes to the photocathode with enough potential to drive the redox reaction in the electrolyte. As a result, the water molecule is split into oxygen and hydrogen.

When the photoelectrode is made from an *n*-type semiconductor it acts as an electron donor; if it is *p* type it acts as an electron acceptor. As an acceptor, the material will attract more holes to the interface, which will enhance the reduction half reaction. Hence it acts as the photocathode and produces hydrogen. If the semiconductor is *n* type, it acts as the photoanode. Electrons are moved to the interface by the internal electric field, and those electrons are involved in the oxidation half reaction, such that oxygen is produced.

The semiconductor material has to fulfil several requirements:

1. It has to absorb light incident on its surface.

2. Charge carrier transport inside the material and separation into the two electrodes must be efficient.

3. A bandgap of 1.23 eV is not enough to drive the reaction because of the overpotential already discussed above. It has been estimated that materials with an energy bandgap close to 2.1 eV have the potential to split water.

4. The energy levels of the material have to be adequate to couple with the energy needed for the reactions. In particular, the energy levels of the reactions have to be located in the energy bandgap of the semiconductor, which is called the favourable position. To further enhance the reaction, a catalyst may be added to the semiconductor surface.

5. On the practical side, it is important that the materials used are photochemically stable and relatively cheap. From these criteria we can conclude that the main technical challenges to be addressed are light absorption, the separation of charges and the catalysis of the reaction.

The absorption and catalysis problems can be tackled by carefully choosing the semiconductor material and its corresponding catalyst. Several materials can be considered for solar water splitting. Some of the most popular materials studied for this application are titanium dioxide (TiO_2), tungsten oxide (WO_3), bismuth vanadate ($BiVO_4$), hematite (Fe_2O_3), and amorphous silicon carbide (a-SiC). As shown in Figure 23.3, $BiVO_4$, Fe_2O_3, and a-SiC have the most promising potential solar-to-hydrogen efficiency. Hematite and silicon carbide are the best choices for the optimal absorption of light, since they have bandgap energies closest to the optimal bandgap of 2.1 eV. However, if other factors such as the band position or stability are considered, materials such as bismuth vanadate may also be a viable option. Because of their large bandgap energies, the other materials are not considered.

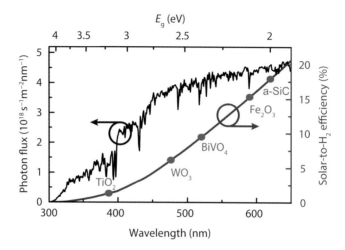

Figure 23.3: The solar-to-hydrogen efficiency and bandgap of different potential photocathode materials. The photon flux is also shown.

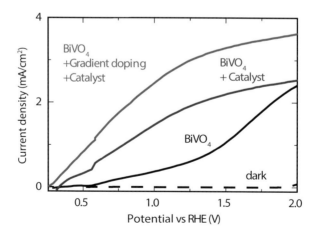

Figure 23.4: The effect of using a catalyst and gradient doping on the performance of a $BiVO_4$ photoanode [193].

The overall efficiency of water splitting depends on the catalytic efficiency and the separation efficiency of the photoelectrode. The *catalytic efficiency* can be improved by placing a catalyst on the semiconductor surface. For example, for a $BiVO_4$ photoanode the inclusion of a cobalt phosphate $(Co_3(PO_4)_2)$ catalyst on the surface will ease the oxidation reaction and hence increase the water splitting efficiency. The *separation efficiency* can be improved by introducing an electric field inside the material. One way to do this is to introduce gradient doping, starting from no doping at the surface to 1% doping close to the back of the electrode. With gradient doping, a depletion region is created in-between the semiconductor and the electrolyte. As a result, electrons will move more easily from the electrolyte into the semiconductor. Combining both effects can greatly improve the efficiency of the overall device. The effects of both the catalyst and the gradient doping are shown in Figure 23.4, where the current density of the illuminated photoanode is shown with respect to the applied voltage with respect to the *reversible hydrogen electrode* (RHE). The RHE is a special electrode with the property that the measured potential does not change when the pH value of the solution is changed.

As we already mentioned above, for the photoelectrochemical (PEC) water splitting process, a potential difference of at least 1.23 V plus the overpotential ΔV must be present. The value of ΔV will depend on the materials and electrolyte used, but it is usually around 0.8V. The sum of these potentials will result in the total potential difference required to drive the redox reaction. This voltage will be partially covered by the potential difference created within the photoelectrode when light shines on it. However, depending on the material, the PEC only creates 0.6 V of the required voltage. For the extra potential that is required to allow water splitting, the photoelectrode can be combined with solar cells. The combination of a photoelectrode and a solar cell forms a photoelectrochemical device.

Figure 23.5 (a) shows such a PEC device. The photoelectrode (in this case a photoanode) is connected to a solar cell in series. The photoanode is connected to the positive contact of the solar cell. The negative contact is connected to another electrode via an external circuit. This counter electrode may or may not be photoactive. The electric circuit is closed by the electrolyte. The band diagram of this device is shown in Figure 23.5 (b).

Since the photoelectrode and the solar cell are connected in series, the same current will go through both devices, similarly to a multi-junction solar cell. In this device, the light is utilized better than it would be in the solar cell alone. In the photoelectrode, the photons with energies exceeding that of the photoelectrode bandgap will be absorbed. The fraction of the light that has not been absorbed or reflected, called the *transmitted spectrum*, reaches the solar cell where it can be absorbed to generate the extra potential difference required for water splitting. The solar cell must be optimized for the transmitted spectrum, which is different from the standard AM1.5 spectrum that is usually used for solar cell optimization.

As for all semiconductor devices, the photoelectrode has its own characteristic *J-V* curve. When a solar cell and a photoelectrode are combined, the conditions at which they will work can be estimated by studying their *J-V* curve characteristics. Figure 23.6 shows the *J-V* curves of the solar cell and the photoanode in this case. Since both elements are connected in series, the current of both the solar cell and the photoelectrode must be the same. Hence, the operational point will be where the two *J-V* curves cross.

The *solar-to-hydrogen efficiency* η_{STH} is directly calculated from the current density measured at the operating point, which we can understand by realizing that current is the rate of flow of electric charge. When we assume that all generated charges produced are in-

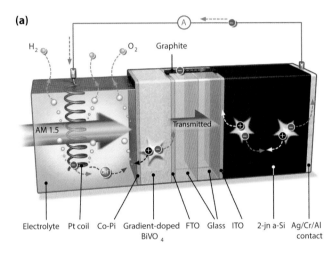

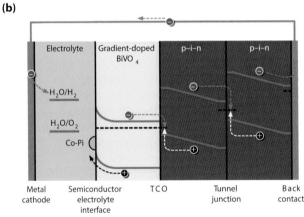

Figure 23.5: (a) A photoelectrochemical (PEC) device consisting of a $BiVO_4$ photoanode and a tandem a-Si:H/a-Si:H solar cell. (b) The band diagram of this PEC device (reprinted by permission from Macmillan Publishers Ltd: F.F. Abdi, L.Han, A.H.M. Smets, M. Zeman, B. Dam, and R. van de Krol, Nature Communications, vol. 4, 2195, copyright (2013)) [193].

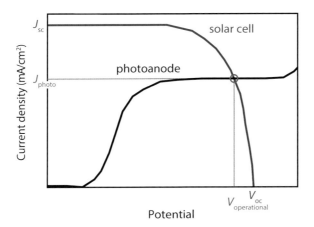

Figure 23.6: The J-V characteristics of the photoelectrode and the solar cell. The operating point is at the crossing of the two.

volved in the production of hydrogen, the overall solar-to-hydrogen conversion efficiency is described by

$$\eta_{STH} = \frac{J_{ph} \times 1.23\,V}{P_{in}}, \qquad (23.11)$$

where J_{ph} is the operational current density, P_{in} is the irradiance arriving at the PEC device, usually it will be 1,000 W/m² from the AM1.5 spectrum.

The only free variable in Eq. (23.11) is the current density: the higher the operational current, the more hydrogen is produced and the higher the device efficiency. The major focus of current academic research therefore is to improve the current density. Besides utilizing gradient doping water-oxidation catalysts and other improvements can be made to increase the performance of PEC devices. For example, optimizing the materials and layer thicknesses in the solar cell, and texturing of the photoanode for better light management.

For a device in which a bismuth vanadate photoanode was combined with a double junction amorphous silicon solar cell and a platinum cathode, a solar-to-hydrogen efficiency of $\eta_{STH} = 4.9\%$ was reported [193].

23.3 Exercises

23.1 We assess the efficiency of two methods of chemical energy storage.

(a) First, we store the energy as natural gas (methane, CH_4) and utilize it via combustion. We assume the following efficiencies: electrolysis 70%, water-gas shift reaction 75%, natural gas storage and circulation 80%, Fischer–Tropsch reaction 80%, combustion 75%. Based on these efficiencies, calculate the efficiency of the total process.

(b) Secondly, we want to store the energy as hydrogen and utilize it via a fuel cell. We assume the following efficiencies: electrolysis 70%, hydrogen storage and circulation 70%, fuel cell 50%. Calculate the efficiency of the total process.

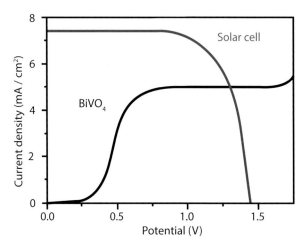

Figure 23.7

   (c) Which of the two methods would you choose to use the hydrogen produced during electrolysis?

23.2 We want to build a photoelectrochemical device by joining a $BiVO_4$ photoanode and a double-junction amorphous silicon solar cell. The J-V curves of both the photoanode and the solar cell at STC are shown in Figure 23.7.

   (a) Calculate the solar-to-hydrogen efficiency of the device.

   (b) Considering the J-V curves of the solar cell and the photoanode, which one is the limiting factor in this case for achieving higher efficiencies, the solar cell or the photoanode?

   (c) Which of the effects listed below are considered to be the main limiting factors for increasing the current density of the photoelectrodes? (More than one answer possible.)

   i. Bandgap of the semiconductor forming the photoelectrode.
   ii. Reflection in the electrolyte between the light source and the photoelectrode.
   iii. Separation of charges in the photoelectrode semiconductor material.
   iv. Catalytic effects in the photoelectrode surface.
   v. Mass transport of ions within the electrolyte.

23.3 Figure 23.8 shows the band diagram of a semiconductor material. Can this material be used as a photoanode or a photocathode?

23.4 In this chapter, we talked about combining a solar cell with a photoelectrode to form a photoelectrochemical device. Why do we add a solar cell to the system?

   (a) Because the photoelectrode is not able to produce any voltage difference to drive the reaction.

   (b) Because the photoelectrode is not able to produce any current to drive the reaction.

   (c) Because the voltage difference created in the photoelectrode is not able to produce enough voltage to drive the reaction.

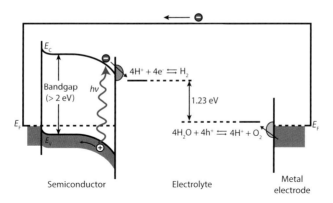

Figure 23.8

(d) Because the current created in the photoelectrode is not enough to drive the reaction.

23.5 What is the approximate voltage that the solar cell needs to deliver for the photoelectrochemical device to work?

(a) 0.60 V; (b) 1.23 V; (c) 1.43 V; (d) 0.80 V.

Appendices

A

Derivations in electrodynamics

A.1 The Maxwell equations

The four Maxwell equations couple the electric and magnetic fields to their sources, i.e. electric charges and current densities, and to each other. They are given by

$$\nabla \cdot \mathbf{D}(\mathbf{r},t) = \rho_F(\mathbf{r}), \tag{A.1a}$$

$$\nabla \times \mathbf{E}(\mathbf{r},t) = -\frac{\partial \mathbf{B}(\mathbf{r},t)}{\partial t}, \tag{A.1b}$$

$$\nabla \cdot \mathbf{B}(\mathbf{r},t) = 0, \tag{A.1c}$$

$$\nabla \times \mathbf{H}(\mathbf{r},t) = +\frac{\partial \mathbf{D}(\mathbf{r},t)}{\partial t} + \mathbf{J}_F(\mathbf{r}), \tag{A.1d}$$

where $\mathbf{r}$ and t denote location and time, respectively. $\mathbf{D}$ is the *electric displacement*, $\mathbf{E}$ the *electric field*, $\mathbf{B}$ the *magnetic induction* and $\mathbf{H}$ the *magnetic field*.[1] ρ_F is the *free charge density* and $\mathbf{J}_F$ is the *free current density*.

The electric displacement and field are related to each other via

$$\mathbf{D} = \epsilon \epsilon_0 \mathbf{E}, \tag{A.2a}$$

where ϵ is the relative permittivity of the medium in which the fields are observed and $\epsilon_0 = 8.854 \times 10^{-12}$ As/(Vm) is the permittivity *in vacuo*. Similarly, the magnetic field and induction are related to each other via

$$\mathbf{B} = \mu \mu_0 \mathbf{H}, \tag{A.2b}$$

---

[1] In this appendix we use $\mathbf{E}$ instead of $\boldsymbol{\xi}$ for the electric field, and $\mathbf{H}$ instead of $\boldsymbol{\zeta}$ for the magnetic field.

where μ is the relative permeability of the medium in which the fields are observed and $\mu_0 = 4\pi \times 10^{-7}$ Vs/(Am) is the permeability of vacuum. Equations (A.2) are only valid if the medium is *isotropic*, i.e. ϵ and μ are independent of the direction. We may assume all the important materials for solar cells to be *non-magnetic*, i.e. $\mu \equiv 1$.

A.2 Derivation of the electromagnetic wave equation

We now derive the electromagnetic wave equations in source-free space, $\rho_F \equiv 0$ and $\mathbf{j}_F \equiv 0$. For the derivation of the electromagnetic wave equations we start by applying the rotation operator $\nabla \times$ to the second Maxwell equation, Eq. (A.1b),

$$\nabla \times (\nabla \times \mathbf{E}) = -\nabla \times \left(\frac{\partial \mathbf{B}}{\partial t}\right). \tag{A.3}$$

Now we take the fourth Maxwell equation, Eq. (A.1d), with $\mathbf{j}_F = 0$ and Eqs. (A.2),

$$\frac{1}{\mu\mu_0} \nabla \times \mathbf{B} = \epsilon\epsilon_0 \frac{\partial \mathbf{E}}{\partial t},$$

and substitute it into Eq. (A.3),

$$\nabla \times (\nabla \times \mathbf{E}) = -\epsilon\epsilon_0\mu\mu_0 \left(\frac{\partial^2 \mathbf{E}}{\partial t^2}\right). \tag{A.4}$$

By using the relation

$$\nabla \times (\nabla \times \mathbf{E}) = \nabla(\nabla \times \mathbf{E}) - \Delta \mathbf{E} = -\Delta \mathbf{E}, \tag{A.5}$$

we find

$$\Delta \mathbf{E} = \epsilon\epsilon_0\mu\mu_0 \left(\frac{\partial^2 \mathbf{E}}{\partial t^2}\right). \tag{A.6}$$

In Eq. (A.5) we used the fact that we are in source-free space, i.e. $\nabla \mathbf{E} = 0$. Equation (A.6) is the *wave equation* for the *electric field*. Note that the factor

$$\frac{1}{\epsilon\epsilon_0\mu\mu_0}$$

has the unit of $(m/s)^2$, i.e. a speed to the square. In easy terms, it is the squared propagation speed of the wave.[1] We now set

$$c_0^2 := \frac{1}{\epsilon_0\mu_0} \tag{A.7}$$

and

$$n^2 = \epsilon, \tag{A.8}$$

where c_0 is the speed of light in *vacuo* and n is the *refractive index* of the medium. Since we also assume $\mu \equiv 1$, we finally obtain for the wave equation for the electric field

$$\Delta \mathbf{E} - \frac{n^2}{c_0^2}\left(\frac{\partial^2 \mathbf{E}}{\partial t^2}\right) = 0. \tag{A.9a}$$

---

[1] In reality, if the medium is absorbing, things are getting much more complex.

A. Derivations in electrodynamics

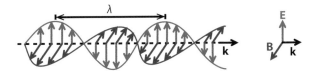

Figure A.1: Illustrating the **E** and **B** fields and the **k** vector of a plane electromagnetic wave.

In a similar manner we can derive the wave equation for the *magnetic field*,

$$\Delta \mathbf{H} - \frac{n^2}{c_0^2}\left(\frac{\partial^2 \mathbf{H}}{\partial t^2}\right) = 0. \quad \text{(A.9b)}$$

A.3 Properties of electromagnetic waves

In Section A.2, we found that plane waves can be described by

$$\mathbf{E}(\mathbf{x}, t) = \mathbf{E}_0 \times e^{ik_z z - i\omega t}, \quad \text{(A.10a)}$$

$$\mathbf{H}(\mathbf{x}, t) = \mathbf{H}_0 \times e^{ik_z z - i\omega t}. \quad \text{(A.10b)}$$

In this section we study some general properties of plane electromagnetic waves, illustrated in Figure A.1. Substituting Eq. (A.10a) into the first Maxwell equation, Eq. (A.1a), yields

$$\nabla \mathbf{D} = \epsilon \epsilon_0 \nabla \mathbf{E} = 0,$$
$$\nabla \mathbf{E} = 0,$$
$$\nabla \left(\mathbf{E}_0 e^{ik_z z - i\omega t}\right) = 0,$$
$$\underbrace{E_{x,0} \frac{\partial}{\partial x} e^{ik_z z - i\omega t}}_{=0} + \underbrace{E_{y,0} \frac{\partial}{\partial y} e^{ik_z z - i\omega t}}_{=0} + \quad \text{(A.11a)}$$
$$+ E_{z,0} \frac{\partial}{\partial z} e^{ik_z z - i\omega t} = 0,$$
$$iE_{z,0} k_z e^{ik_z z - i\omega t} = 0,$$
$$E_{z,0} = 0,$$

where we used the notation $\mathbf{E}_0 = (E_{x,0}, E_{y,0}, E_{z,0})$. In a similar manner, by substituting Eq. (A.10b) into the first Maxwell equation, Eq. (A.1c), we obtain

$$H_{z,0} = 0. \quad \text{(A.11b)}$$

Thus, neither the electric nor the magnetic fields have components in the propagation direction but only components perpendicular to the propagation direction (the x and y directions in our case).

Without loss of generality we now assume that the electric field only has an x component, $\mathbf{E}_0 = (E_{x,0}, 0\, 0)$. Substituting this electric field into the left-hand side of the second

Maxwell equation, Eq. (A.1b), yields

$$\nabla \times \mathbf{E} = \begin{pmatrix} \frac{\partial}{\partial x} \\ \frac{\partial}{\partial x} \\ \frac{\partial}{\partial x} \end{pmatrix} \times \begin{pmatrix} E_x \\ E_y \\ E_z \end{pmatrix} = \begin{pmatrix} \frac{\partial}{\partial y} E_z - \frac{\partial}{\partial z} E_y \\ \frac{\partial}{\partial z} E_x - \frac{\partial}{\partial x} E_z \\ \frac{\partial}{\partial x} E_y - \frac{\partial}{\partial y} E_x \end{pmatrix} \quad (A.12)$$

$$= \begin{pmatrix} 0 \\ \frac{\partial}{\partial z} E_x \\ 0 \end{pmatrix} = \begin{pmatrix} 0 \\ E_{x,0} \frac{\partial}{\partial z} e^{ik_z z - i\omega t} \\ 0 \end{pmatrix} = \begin{pmatrix} 0 \\ i E_{x,0} k_z e^{ik_z z - i\omega t} \\ 0 \end{pmatrix}. \quad (A.13)$$

The right-hand side of Eq. (A.1b) yields

$$\frac{\partial \mathbf{B}(\mathbf{r},t)}{\partial t} = \mu_0 \frac{\partial \mathbf{H}(\mathbf{r},t)}{\partial t} = \quad (A.14)$$

$$\mu_0 \begin{pmatrix} \frac{\partial}{\partial t} H_x \\ \frac{\partial}{\partial t} H_y \\ \frac{\partial}{\partial t} H_z \end{pmatrix} = \mu_0 \begin{pmatrix} H_{x,0} \frac{\partial}{\partial t} e^{ik_z z - i\omega t} \\ H_{y,0} \frac{\partial}{\partial t} e^{ik_z z - i\omega t} \\ 0 \end{pmatrix} = \mu_0 \begin{pmatrix} -i\omega H_{x,0} e^{ik_z z - i\omega t} \\ -i\omega H_{y,0} e^{ik_z z - i\omega t} \\ 0 \end{pmatrix}. \quad (A.15)$$

Thus, we obtain for Eq. (A.1b)

$$\begin{pmatrix} 0 \\ iE_{x,0} k_z e^{ik_z z - i\omega t} \\ 0 \end{pmatrix} = -\mu_0 \begin{pmatrix} -i\omega H_{x,0} e^{ik_z z - i\omega t} \\ -i\omega H_{y,0} e^{ik_z z - i\omega t} \\ 0 \end{pmatrix}, \quad (A.16)$$

i.e. only the y component of the magnetic field survives. The electric and magnetic components are related to each other via

$$E_{x,0} k_z = \mu_0 \omega H_{y,0}. \quad (A.17)$$

By substituting Eq. (4.4) into Eq. (A.17), we find

$$H_{y,0} = \frac{n}{c\mu_0} E_{x,0} = \frac{n}{Z_0} E_{x,0}, \quad (A.18)$$

where

$$Z_0 = c\mu_0 = \sqrt{\frac{\mu_0}{\epsilon_0}} = 376.7\,\Omega \quad (A.19)$$

is the *impedance of free space*.

In summary, we found the following properties of the electromagnetic field:

- The electric and magnetic field vectors are perpendicular to each other and also perpendicular to the propagation vector,

$$\mathbf{k} \times \mathbf{E}_0 = \mathbf{k} \times \mathbf{H}_0 = \mathbf{H}_0 \times \mathbf{E}_0 = 0. \quad (A.20)$$

- The electric and magnetic fields are proportional to the propagation direction, hence electromagnetic waves are *transverse waves*.

A. Derivations in electrodynamics

- The electric and magnetic vectors have a constant, material-dependent ratio. If the electric field is along the x direction and the magnetic field is along the y direction, this ratio is given by

$$H_{y,0} = \frac{n}{c\mu_0} E_{x,0} = \frac{n}{Z_0} E_{x,0}, \quad (A.21)$$

where

$$Z_0 = c\mu_0 = \sqrt{\frac{\mu_0}{\epsilon_0}} = 376.7\,\Omega \quad (A.22)$$

is the *impedance of free space*.

The derivations in this section were done for plane waves. However, it can be shown that the properties of electromagnetic waves summarized in the bullet points above are valid for electromagnetic waves in general.

B
Derivation of homojunction *J-V* curves

In this appendix we present the derivations for the *J-V* curves in the dark and under illumination for *p-n* homojunctions, as introduced in Section 8.1.

B.1 The *J-V* characteristic in the dark

When an external voltage, V_a, is applied to a *p-n* junction the potential difference between the *n*-type and *p*-type regions will change and the electrostatic potential across the space-charge region will become $(\psi_0 - V_a)$. Under the forward-bias condition an applied external voltage decreases the electrostatic-potential difference across the *p-n* junction. The concentration of the minority carriers at the edge of the space-charge region increases exponentially with the applied forward-bias voltage but it is still much lower than the concentration of the majority carriers (low-level injection conditions). The concentration of the majority carriers in the quasi-neutral regions do not change significantly under forward bias. The concentration of charge carriers in a *p-n* junction under forward bias is schematically presented in Fig. B.1

The concentrations of the minority carriers at the edges of the space-charge region, electrons in the *p*-type semiconductor and holes in the *n*-type semiconductor, after applying forward-bias voltage are described by

$$n_p(a) = n_{p0}\, e^{\frac{qV_a}{k_B T}} = \frac{n_i^2}{N_A}\, e^{\frac{qV_a}{k_B T}}, \tag{B.1a}$$

$$p_n(b) = p_{n0}\, e^{\frac{qV_a}{k_B T}} = \frac{n_i^2}{N_D}\, e^{\frac{qV_a}{k_B T}}. \tag{B.1b}$$

Since we assume that there is no electric field in the quasi-neutral region, the current density equations of carriers reduce to only diffusion terms and are not coupled by the

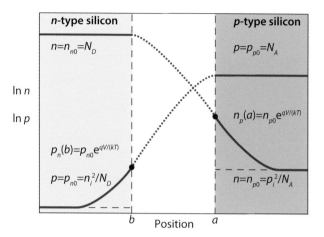

Figure B.1: Concentration profiles of mobile charge carriers in a p-n junction under forward bias (red lines). Concentration profiles of carriers under thermal equilibrium are shown for comparison (black lines).

electric field. The current is based on the diffusive flows of carriers in the quasi-neutral regions and is determined by the diffusion of the minority carriers. The minority-carrier concentration can be calculated separately for both quasi-neutral regions. The electron current density in the quasi-neutral region of the p-type semiconductor and the hole current density in the quasi-neutral region of the n-type semiconductor are described by

$$J_n = qD_N \frac{dn}{dx}, \tag{B.2a}$$

$$J_p = -qD_P \frac{dp}{dx}. \tag{B.2b}$$

The continuity equations (4.34) for electrons and holes in steady-state ($\partial n/\partial t = 0$ and $\partial p/\partial t = 0$) can be written as

$$\frac{1}{q}\frac{dJ_N}{dx} - R_N + G_N = 0, \tag{B.3a}$$

$$-\frac{1}{q}\frac{dJ_P}{dx} - R_P + G_P = 0. \tag{B.3b}$$

Under *low level injection conditions*, a change in the concentration of the majority carriers due to generation and recombination can be neglected. However, the recombination-generation rate of minority carriers depends strongly on the injection and is proportional to the excess of minority carriers at the edges of the depletion region. The recombination-generation rate of electrons, R_n, in the p-type semiconductor, and holes, R_p, in the n-type

B. Derivation of homojunction J-V curves

semiconductor, is described by Eqs. (7.18) and (7.23),

$$R_n = \frac{\Delta n}{\tau_n}, \tag{B.4a}$$

$$R_p = \frac{\Delta p}{\tau_p}, \tag{B.4b}$$

where Δn is the excess concentration of electrons in the p-type semiconductor with respect to the equilibrium concentration n_{p0}, and τ_n is the electron's (minority-carrier) lifetime. Δp is the excess concentration of holes in the n-type semiconductor with respect to the equilibrium concentration p_{n0} and τ_p is the hole's (minority-carrier) lifetime. Δn and Δp are given by

$$\Delta n = n_p(x) - n_{p0}, \tag{B.5a}$$

$$\Delta p = p_n(x) - p_{n0}. \tag{B.5b}$$

By combining Eqs. (B.2a), (B.3a) and (B.4a) we obtain

$$D_N \frac{d^2 n_p(x)}{dx^2} = \frac{\Delta n}{\tau_n} - G_N. \tag{B.6a}$$

This equation describes the diffusion of electrons in the p-type semiconductor. Similarly, by combining Eqs. (B.2b), (B.3b) and (B.4b) we obtain

$$D_P \frac{d^2 p_n(x)}{dx^2} = \frac{\Delta p}{\tau_p} - G_P, \tag{B.6b}$$

which describes the diffusion of holes in the n-type semiconductor.

Now we substitute $n_p(x)$ from Eq. (B.5a) and $p_n(x)$ from Eq. (B.5b) into Eqs. (B.6a) and (B.6b), respectively. By using that $d^2 n_{p0}/dx^2 = d^2 p_{n0}/dx^2 = 0$, and in dark $G_N = G_P = 0$ we obtain

$$\frac{d^2 \Delta n}{dx^2} = \frac{\Delta n}{D_N \tau_n}, \tag{B.7a}$$

$$\frac{d^2 \Delta p}{dx^2} = \frac{\Delta p}{D_P \tau_p}. \tag{B.7b}$$

The electron concentration profile in the quasi-neutral region of the p-type semiconductor is given by the general solution to Eq. (B.7a),

$$\Delta n(x) = A \exp\left(\frac{x}{L_N}\right) + B \exp\left(-\frac{x}{L_N}\right),$$

where $L_N = \sqrt{D_N \tau_n}$ (see Eq. (7.28a)) is the electron minority-carrier diffusion length. The origin of the x axis is set to the edge of the depletion region in the p-type semiconductor and denoted as a in Fig. B.1. For the p-type semiconductor, infinite thickness is assumed (*approximation of the infinite thickness*). The constants A and B can be determined from the boundary conditions:

1. At $x = 0$, $n_p(a) = n_{p0} \exp(qV_a/k_BT)$.
2. n_p is finite at $x \to \infty$, therefore $A = 0$.

With these boundary conditions the solution for the concentration profile of electrons in the p-type quasi-neutral region is found to be

$$n_p(x) = n_{p0} + n_{p0}\left(e^{\frac{qV_a}{k_BT}} - 1\right)e^{-\frac{x}{L_N}}. \tag{B.8a}$$

The hole concentration profile in the quasi-neutral region of the n-type semiconductor is given by the general solution to Eq. (B.7a),

$$\Delta p(x') = A' \exp\left(\frac{x'}{L_P}\right) + B' \exp\left(-\frac{x'}{L_P}\right),$$

where $L_P = \sqrt{D_P \tau_p}$ (see Eq. (7.28b)) is the hole minority-carrier diffusion length. The origin of the x' axis ($x' := -x$) is set at the edge of the depletion region in the n-type semiconductor and denoted as b in Fig. B.1. Here, we use the approximation of infinite thickness for the n-type semiconductor. The constants A' and B' can be determined from the boundary conditions

1. At $x' = 0$, $p_n(b) = p_{n0} \exp(qV_a/k_BT)$.
2. $p - n$ is finite at $x' \to \infty$, therefore $A' = 0$.

We thus find the concentration profile of holes in the quasi-neutral region of the n-type semiconductor to be

$$p_n(x') = p_{n0} + p_{n0}\left(e^{\frac{qV_a}{k_BT}} - 1\right)e^{-\frac{x'}{L_P}}. \tag{B.8b}$$

When substituting the corresponding concentration profiles of minority carriers Eqs.(B.8) into Eqs. (B.2) we obtain for the current densities

$$J_N(x) = \frac{qD_N n_{p0}}{L_N}\left(e^{\frac{qV_a}{k_BT}} - 1\right)e^{-\frac{x}{L_N}}, \tag{B.9a}$$

$$J_P(x') = \frac{qD_P p_{n0}}{L_P}\left(e^{\frac{qV_a}{k_BT}} - 1\right)e^{-\frac{x'}{L_P}}. \tag{B.9b}$$

Under the assumptions that the effect of recombination and thermal generation of carriers in the depletion region can be neglected, which means that the electron and hole current densities are essentially constant across the depletion region, one can write for the current densities at the edges of the depletion region

$$J_N|_{x=0} = J_N|_{x'=0} = \frac{qD_N n_{p0}}{L_N}\left(e^{\frac{qV_a}{k_BT}} - 1\right), \tag{B.10a}$$

$$J_P|_{x'=0} = J_P|_{x=0} = \frac{qD_P p_{n0}}{L_P}\left(e^{\frac{qV_a}{k_BT}} - 1\right). \tag{B.10b}$$

The total current density flowing through the p-n junction at the steady state is constant across the device therefore we can determine the total current density as the sum of the

B. Derivation of homojunction J-V curves

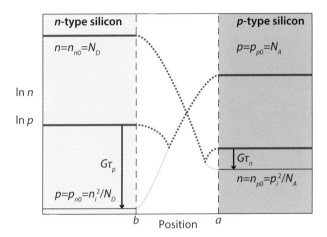

Figure B.2: Concentration profiles of mobile charge carriers in an illuminated p-n junction with uniform generation rate G (red line). Concentration profiles of charge carriers under equilibrium conditions are shown for comparison (black line).

electron and hole current densities at the edges of the depletion region,

$$J(V_a) = J_N |_{x=0} + J_P |_{x=0} \tag{B.11}$$
$$= \left(\frac{q D_N n_{p0}}{L_N} + \frac{q D_P p_{n0}}{L_P} \right) \left(e^{\frac{qV_a}{k_B T}} - 1 \right).$$

Using Eqs. (8.1b) and (8.2b), Eq. (B.11) can be rewritten as

$$J(V_a) = J_0 \left[\exp\left(\frac{qV_a}{k_B T} \right) - 1 \right], \tag{B.12}$$

where J_0 is the saturation current density of the p-n junction which is given by

$$J_0 = \left(\frac{q D_N n_i^2}{L_N N_A} + \frac{q D_P n_i^2}{L_P N_D} \right). \tag{B.13}$$

Equation (B.12) is known as the *Shockley equation* that describes the current-voltage behaviour of an ideal p-n diode. It is a fundamental equation for microelectronics device physics.

B.2 J-V characteristic under illumination

When a p-n junction is illuminated the additional electron-hole pairs are generated through the junction. In the case of moderate illumination the concentration of majority carriers does not change significantly while the concentration of minority carriers (electrons in the p-type region and holes in the n-type region) will strongly increase. In the following section it is assumed that the photo generation rate, G, is uniform throughout the p-n junction, this is called the *uniform generation-rate approximation*. This assumption reflects the situation where the device is illuminated with a long wavelength light, which is weakly absorbed by the semiconductor. The concentration of charge carriers in a *p-n* junction with uniform photo generation rate is schematically presented in Figure B.2.

Eqs. (B.6a) described the steady-state situation for minority carriers. In Section B.1 we then proceeded with assuming that the generation rate is zero. Now we analyse the case where the junction is illuminated, i.e. the generation rate is not zero. The equations thus can be rewritten:

$$\frac{d^2 \Delta n}{dx^2} = \frac{\Delta n}{D_N \tau_n} - \frac{G}{D_N}, \tag{B.14a}$$

$$\frac{d^2 \Delta p}{dx^2} = \frac{\Delta p}{D_P \tau_p} - \frac{G}{D_P}. \tag{B.14b}$$

Under the assumption that G/D_N and G/D_P are constant, the general solution to Eq. (B.14) is

$$\Delta n(x) = G \tau_n + C e^{\frac{x}{L_N}} + D e^{-\frac{x}{L_N}}, \tag{B.15a}$$

$$\Delta p(x') = G \tau_p + C' e^{\frac{x'}{L_P}} + D' e^{-\frac{x'}{L_P}}. \tag{B.15b}$$

The constants in Eqs. (B.15) can be determined from the same boundary conditions as were used in the analysis of the *p-n* junction in the dark. The particular solution for the concentration profile of electrons in the quasi-neutral region of the *p*-type semiconductor and holes in the quasi-neutral region of the *n*-type semiconductor is described by

$$n_p(x) = n_{p0} + G \tau_n + \left[n_{p0} \left(e^{\frac{qV}{k_B T}} - 1 \right) - G \tau_n \right] e^{-\frac{x}{L_N}}, \tag{B.16a}$$

$$p_n(x') = p_{n0} + G \tau_p + \left[p_{n0} \left(e^{\frac{qV}{k_B T}} - 1 \right) - G \tau_p \right] e^{-\frac{x'}{L_P}}. \tag{B.16b}$$

By substituting these concentration profiles of minority carriers into Eqs. (B.2), we obtain for the current densities

$$J_N(x) = \frac{q D_N n_{p0}}{L_N} \left(e^{\frac{qV}{k_B T}} - 1 \right) e^{-\frac{x}{L_N}} - q G L_N e^{-\frac{x}{L_N}}, \tag{B.17a}$$

$$J_P(x') = \frac{q D_P p_{n0}}{L_P} \left(e^{\frac{qV}{k_B T}} - 1 \right) e^{-\frac{x'}{L_P}} - q G L_P e^{-\frac{x'}{L_P}}. \tag{B.17b}$$

In the case of an ideal *p-n* junction the effect of recombination in the depletion region was neglected. However, the contribution of photo generated charge carriers to the current in the depletion region has to be taken into account. The contribution of photo generation from the depletion region to the current density is given by

$$J_N |_{x=0} = q \int_{-W}^{0} -G \, dx = -q G W, \tag{B.18a}$$

$$J_P |_{x'=0} = q \int_{-W}^{0} -G \, dx' = -q G W. \tag{B.18b}$$

The total current density flowing through the p-n junction in the steady state is constant across the junction. Under the *superposition approximation* we thus can determine the total current density as the sum of the electron and hole current densities at the edges of the depletion region,

$$J(V_a) = J_N|_{x=0} + J_P|_{x=0}$$
$$= \left(\frac{qD_N n_{p0}}{L_N} + \frac{qD_P p_{n0}}{L_P} \right) \left(e^{\frac{qV_a}{k_BT}} - 1 \right) \quad \text{(B.19)}$$
$$- eG(L_N + L_P + W.)$$

Eq. (B.19) can be rewritten as

$$J(V_a) = J_0 \left[\exp\left(\frac{qV_a}{k_BT} \right) - 1 \right] - J_{\text{ph}}, \quad \text{(B.20)}$$

where J_{ph} is the photo current given by

$$J_{\text{ph}} = qG(L_N + L_P + W). \quad \text{(B.21)}$$

A number of approximations have been made in order to derive the analytical expressions for the current–voltage characteristics of an ideal p-n junction in the dark and under illumination. These approximations are summarized below:

- the depletion-region approximation;
- the Boltzmann approximation;
- low level injection conditions;
- the superposition principle;
- infinite thickness of doped regions;
- uniform generation rate.

The derived expressions describe the behaviour of an ideal p-n junction and help us understand the basic processes behind the operation of this junction. However, they do not fully and correctly describe real p-n junctions. For example, the thickness of a real p-n junction is limited, which means that the recombination at the surface of the doped regions has to be taken into account. The thinner the p-n junction, the more important surface recombination becomes. The surface recombination modifies the value of the saturation current density. Further it was assumed that there are no recombination-generation processes in the depletion region. However, in real p-n junctions, the recombination in the depletion region represents a substantial loss mechanism. These and other losses in a solar cell are discussed in Chapter 10.

C

Some aspects of surface recombination

In this appendix we look at several aspects of surface recombination for a p-type substrate with dopant concentration N_B. In Section C.1 we look at the case when the surface recombination velocity is $S \to \infty$. The opposite case, $S = 0$, is studied in Section C.2. Finally, in Section C.3 we look at how the open circuit voltage of solar cells is dependent on the absorber thickness, if $S = 0$.

C.1 Infinite surface recombination velocity S

In order to calculate the current density, first the minority-carrier distribution should be worked out. For this we start with the continuity equation,

$$D\frac{d^2 \delta n(x)}{dx^2} + \mu E \frac{d\delta n(x)}{dx} + G - \frac{\delta n(x)}{\tau} = \frac{\partial \delta n(x)}{\partial t}. \tag{C.1}$$

Taking into account that in the quasi-neutral region the electric field is zero, that there is no generation, and that we are considering the steady state condition, this equation reduces to

$$D\frac{d^2 \delta n(x)}{dx^2} - \frac{\partial \delta n(x)}{\partial t} = 0. \tag{C.2}$$

The general solution for this reduced continuity equation is:

$$\delta n(x) = A \exp(-x/L) + B \exp(x/L) \quad \text{with} \quad L = \sqrt{D\tau}. \tag{C.3}$$

To solve this problem, we need to apply the right boundary conditions:

1. $n(x=0) = n_{B0} \exp\left(\dfrac{qV}{k_B T}\right) \Rightarrow$

$$\delta n(x=0) = n(x=0) - n_{B0} = n_{B0}\left[\exp\left(\dfrac{qV}{k_B T}\right) - 1\right],$$

with n_{B0} the thermal equilibrium electron concentration in the p-type Si.

2. $\delta n(x=W) = 0$ expressing the fact that for $x = W$ the recombination velocity S is infinite.

We then find the following set of equations

$$A \qquad\qquad + B \qquad\qquad = n_{B0}\left[\exp\left(\dfrac{qV}{k_B T}\right) - 1\right] \tag{C.4a}$$

$$A\times \exp(-W/L) + B\times \exp(W/L) = 0. \tag{C.4b}$$

We find for the constants A and B:

$$A = \dfrac{\exp(W/L)}{2\sinh(W/L)} \times n_{B0}\left[\exp\left(\dfrac{qV}{k_B T}\right) - 1\right], \tag{C.5a}$$

$$B = -\dfrac{\exp(-W/L)}{2\sinh(W/L)} \times n_{B0}\left[\exp\left(\dfrac{qV}{k_B T}\right) - 1\right]. \tag{C.5b}$$

The full solution for the excess electron concentration is therefore

$$\begin{aligned}\delta n(x) &= \dfrac{n_{B0}}{2\sinh(W/L)} \times \left[\exp\left(\dfrac{qV}{k_B T}\right) - 1\right]\left[e^{-\frac{x-W}{L}} - e^{\frac{x-W}{L}}\right] \\ &= -n_{B0}\left[\exp\left(\dfrac{qV}{k_B T}\right) - 1\right] \times \dfrac{\sinh[(x-W)/L]}{\sinh(W/L)}.\end{aligned} \tag{C.6}$$

We find for the electron current density

$$J_n(x) = qD\dfrac{\mathrm{d}\delta n(x)}{\mathrm{d}x} = -\dfrac{qDn_{B0}}{L}\left[\exp\left(\dfrac{qV}{k_B T}\right) - 1\right] \times \dfrac{\cosh[(x-W)/L]}{\sinh(W/L)}. \tag{C.7}$$

If the emitter is much more highly doped than the base, the current is virtually only carried by the electrons. This current should be constant anywhere in the device, so also for $x = 0$, we find

$$J = \dfrac{qDn_{B0}}{L} \times \dfrac{1}{\tanh(W/L)} \times \left[\exp\left(\dfrac{qV}{k_B T}\right) - 1\right]. \tag{C.8}$$

Using $L = \sqrt{D\tau}$ and $n_{B0} = n_i^2/N_B$, we finally obtain for the saturation current density

$$J_0 = \dfrac{qn_i^2}{N_B} \times \sqrt{\dfrac{D}{\tau}} \cdot \dfrac{1}{\tanh(W/L)}. \tag{C.9}$$

C.2 Surface recombination velocity $S = 0$

Also in this case, we first need to work out the minority carrier distribution. Again, we start with the continuity equation

$$D\dfrac{\mathrm{d}^2\delta n(x)}{\mathrm{d}x^2} + \mu E\dfrac{\mathrm{d}\delta n(x)}{\mathrm{d}x} + G - \dfrac{\delta n(x)}{\tau} = \dfrac{\partial \delta n(x)}{\partial t}. \tag{C.10}$$

C. Some aspects of surface recombination

Taking into account that in the quasi-neutral region the electric field is zero, that there is no generation, and that we are considering the steady-state condition, this equation reduces to

$$D\frac{d^2 \delta n(x)}{dx^2} - \frac{\partial \delta n(x)}{\partial t} = 0. \tag{C.11}$$

The general solution for this reduced continuity equation is

$$\delta n(x) = A \exp(-x/L) + B \exp(x/L) \quad \text{with} \quad L = \sqrt{D\tau}. \tag{C.12}$$

We need to apply the boundary conditions for this problem:

1. $n(x=0) = n_{B0} \exp\left(\frac{qV}{k_B T}\right) \Rightarrow$

$$\delta n(x=0) = n(x=0) - n_{B0} = n_{B0}\left[\exp\left(\frac{qV}{k_B T}\right) - 1\right],$$

where n_{B0} is the thermal equilibrium electron concentration in the p-type Si.

2. As the recombination velocity S is zero, we find in this case

$$-D\left[\hat{\mathbf{n}} \cdot \frac{d\delta n(x)}{dx}\right]_{x=W} = S\,\delta n(x)|_{x=W} = 0 \Rightarrow$$

$$-\frac{A}{L}\exp(-W/L) + \frac{B}{L}\exp(W/L) = 0.$$

With these boundary conditions we find the two equations

$$A + B = n_{B0}\left[\exp\left(\frac{qV}{k_B T}\right) - 1\right] \tag{C.13a}$$

$$-\frac{A}{L}e^{-W/L} + \frac{B}{L}e^{W/L} = 0. \tag{C.13b}$$

We find for the constants A and B

$$A = \frac{\exp(W/L)}{2\cosh(W/L)} \times n_{B0}\left[\exp\left(\frac{qV}{k_B T}\right) - 1\right], \tag{C.14a}$$

$$B = \frac{\exp(-W/L)}{2\cosh(W/L)} \times n_{B0}\left[\exp\left(\frac{qV}{k_B T}\right) - 1\right]. \tag{C.14b}$$

The full solution for the excess electron concentration is then

$$\delta n(x) = \frac{n_{B0}}{2\cosh(W/L)} \times \left[\exp\left(\frac{qV}{k_B T}\right) - 1\right]\left[e^{-(x-W)/L} + e^{(x-W)/L}\right]$$
$$= n_{B0}\left[\exp\left(\frac{qV}{k_B T}\right) - 1\right] \times \frac{\cosh[(x-W)/L]}{\cosh(W/L)}. \tag{C.15}$$

We find for the electron current density

$$J_n(x) = qD\frac{d\delta n(x)}{dx} = -\frac{qDn_{B0}}{L}\left[\exp\left(\frac{qV}{k_B T}\right) - 1\right] \times \frac{\sinh[(x-W)/L]}{\cosh(W/L)} \tag{C.16}$$

When the emitter is much more highly doped than the base, the current is virtually only carried by the electrons. This current should be constant anywhere in the device, so also for $x = 0$, we find

$$J = \frac{qDn_{B0}}{L} \times \tanh(W/L) \times \left[\exp\left(\frac{qV}{k_BT}\right) - 1\right]. \tag{C.17}$$

Using $L = \sqrt{D\tau}$ and $n_{B0} = n_i^2/N_B$ we finally find the saturation current density J_0 to be

$$J_0 = \frac{qn_i^2}{N_B} \times \sqrt{\frac{D}{\tau}} \times \tanh(W/L). \tag{C.18}$$

C.3 Open circuit voltage for *p-n* junction solar cells

The V_{oc} for a *p-n* junction solar cell generating a photocurrent density J_L is given by

$$V_{oc} = \frac{k_BT}{q} \times \ln\left(1 + \frac{J_L}{J_0}\right) \approx \frac{k_BT}{q} \times \ln\left(\frac{J_L}{J_0}\right). \tag{C.19}$$

With the expression for the saturation current density from Eq. (C.18), this becomes

$$\begin{aligned}V_{oc} &= \frac{k_BT}{q} \times \ln\left(J_L \times \frac{N}{en_i^2} \times \sqrt{\frac{\tau}{D}} \times \frac{1}{\tanh(W/L)}\right) \\ &= \frac{k_BT}{q} \times \left[\ln\left(J_L \times \frac{N}{en_i^2} \times \sqrt{\frac{\tau}{D}}\right) + \ln\left(\frac{1}{\tanh(W/L)}\right)\right].\end{aligned} \tag{C.20}$$

For $W \ll L$ we can use the approximation $\tanh(W/L) \approx W/L$. Combined with Eq. (C.20), we find

$$V_{oc} \approx C + \frac{k_BT}{q} \times \ln\left(\frac{L}{W}\right). \tag{C.21}$$

From this result we can draw an important conclusion: the thinner the base, the larger the V_{oc}!

D
The morphology of selected TCO samples

As we have mentioned in Chapter 10, *light management* is crucial for designing solar cells. Very often, the TCO layers in thin-film solar cells have nano textured surfaces. The incident light is scattered at these interfaces such that the average photon path length in the absorber layer is enhanced. Thus, more light can be absorbed and the photocurrent density can be increased. However, the texture may also influence the growth of layers that are deposited onto the TCO and thus alter the electrical properties, which may lead to a reduction in V_{oc} and the fill factor. Further, the increased interface area due to the texture may lead to more surface recombination and also reduce the voltage.

D.1 Surface parameters

To compare the morphology of different samples, many surface parameters can be used. Here, we discuss two parameters, the *root mean squared (rms) roughness* σ_r and the *correlation length* ℓ_c.

The rms roughness in principle is the standard deviation of the height profile. It is defined as

$$\sigma_r = \sqrt{\frac{1}{N-1} \sum_{i=1}^{N} (z_i - \bar{z})^2}, \tag{D.1}$$

where N is the number of data points, z_i is the height of the i_{th} datapoint and $\bar{z}$ is the average height. These data can be obtained with *atomic force microscopy* (AFM). Note that σ_r is insensitive to the lateral feature sizes. Samples with very small or large lateral features can both have the same rms roughness.

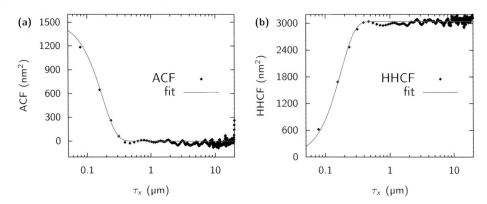

Figure D.1: The ACF, HHCF, and fitted Gaussian functions for SnO_2 : F.

The correlation length gives an indication of the lateral feature sizes. Its derivation is less straightforward than that of σ_r: it has to be extracted from the autocorrelation function (ACF) and/or the height-height correlation function (HHCF) [194].

For a discrete set of data, the two-dimensional ACF is given by

$$\text{ACF}(\tau_x, \tau_y) = \frac{1}{(N-n)(M-m)} \times \sum_{l=1}^{N-n} \sum_{k=1}^{M-m} z(k\delta + \tau_x, l\delta + \tau_y) \, z(k\delta, l\delta), \quad \text{(D.2)}$$

where δ is the distance between two data points, $m = \tau_x/\delta$ and $n = \tau_y/\delta$. For AFM scans the one-dimensional ACF along the fast scanning axis (x) is usually used:

$$\text{ACF}_x(\tau_x) = \text{ACF}(\tau_x, 0) = \frac{1}{N(M-m)} \sum_{l=1}^{N} \sum_{k=1}^{M-m} z(k\delta + \tau_x, l\delta) \, z(k\delta, l\delta). \quad \text{(D.3)}$$

The one-dimensional HHCF is given by

$$\text{HHCF}_x(\tau_x) = \frac{1}{N(M-m)} \sum_{l=1}^{N} \sum_{k=1}^{M-m} [z(k\delta + \tau_x, l\delta) - z(k\delta, l\delta)]^2. \quad \text{(D.4)}$$

To determine ℓ_c, Gaussian functions can be fitted to the ACF and the HHCF. They are given by

$$\text{ACF}_x^{\text{fit}}(\tau_x) = \sigma_r^2 \exp\left(-\frac{\tau_x^2}{\ell_c^2}\right), \quad \text{(D.5)}$$

$$\text{HHCF}_x^{\text{fit}}(\tau_x) = 2\sigma_r^2 \left[1 - \exp\left(-\frac{\tau_x^2}{\ell_c^2}\right)\right], \quad \text{(D.6)}$$

respectively. The correlation length is then the length at which the (fitted) ACF has decayed to $1/e$ of its highest value. Instead of applying Eq. (D.1) directly, the rms roughness can also be obtained by fitting Gaussian functions to the ACF and/or the HHCF.

Figure D.1 shows the ACF, the HHCF, and fitted Gaussian functions for the SnO_2:F sample depicted in Fig. D.2 (a). Instead of fitting to Gaussian functions, some authors use exponential functions for fitting. For this sample, Gaussians lead to better fits.

D.2 Examples

Figure D.2 shows the surface morphology, rms roughness σ_r and correlation length ℓ_c of some nano textured TCO samples made from three different materials. The data was obtained with atomic force microscopy. Fluorine-doped tin oxide (SnO_2:F) of Asahi U-type [196] was grown using atmospheric-pressure chemical vapour deposition (AP-CVD). Boron-doped zinc-oxide (ZnO:B) of so-called 'B-type' from PV-LAB of the École polytechnique fédérale de Lausanne (EPFL), Switzerland [197], was prepared using low-pressure chemical vapour deposition (LP-CVD). Both materials obtain their nano structure during the deposition process. The nano structures are due to the crystal growth of the TCO layers and have a *pyramid-like* shape.

In contrast, RF-sputtered, aluminium-doped zinc oxide (ZnO:Al) is flat after deposition ($\sigma_r \approx 3$ nm, depending on the deposition condition). To obtain a nano texture, ZnO:Al is etched in a 0.5% HCl solution [198, 199]. This etching process leads to *crater-like* features. The lateral size of these craters is influenced by the grain size of the zinc oxide crystals. Figure D.2 (c) and (d) shows etched ZnO:Al with broad craters. Figure D.2 (e) and (f) shows etched ZnO:Al with narrow craters.

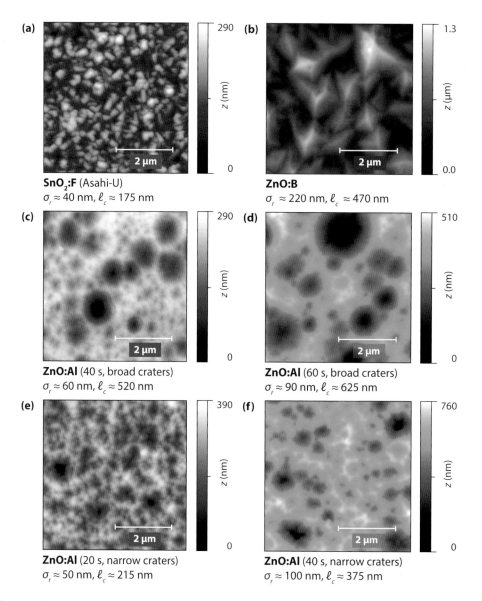

Figure D.2: The morphology, rms roughness σ_r, and correlation length ℓ_c for some selected TCO samples [195]. The statistical parameters σ_r and ℓ_c were obtained with AFM scans of 256×256 points over an area of 20×20 μm^2.

E

Some aspects on location issues

In Section E.1 we will derive the equations, which are used to calculate the *position of the Sun* in the horizontal coordinate system that is illustrated in Figure 18.1. In Section E.2, we will discuss the equation of time. The last two sections are devoted to calculating the cosine between PV modules and the Sun: in Section E.3 this is done for modules installed on a horizontal plane while in Section E.4 the more complex case of PV modules on a roof is treated.

E.1 The position of the Sun

Earth orbits the Sun in an elliptic orbit at an average distance of about 150 million kilometres. Due to the elliptic orbit the speed of Earth is not constant, which is described by Kepler's second law, named after Johannes Kepler (1571-1630). It states that 'A line joining a planet and the Sun sweeps out equal areas during equal intervals of time.' On the celestial sphere the Sun seems to move on a circular path with one revolution per year. This path is called the *ecliptic* and illustrated in Figure E.1. To describe the apparent movement of the Sun on the celestial sphere it is convenient to use coordinates in which the ecliptic lies in the fundamental plane. This coordinate system is called the *ecliptic coordinate system*; it is shown in Figure E.2 (a). The principal direction of this coordinate system is the position of the Sun at the spring (*vernal*) *equinox* (thus around 21 March) is used, which is indicated by the sign of Aries, ♈. As is obvious from Figure E.1, ♈ lies in both the ecliptic and the equatorial planes. The two angular coordinates are called the *ecliptic longitude* λ and the *ecliptic latitude* β. Note, that in this coordinate system the rotation of the Earth around its axis is not taken into account.

In ecliptic coordinates the position of the Sun can be expressed easily. The approxim-

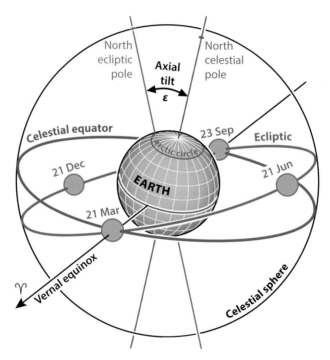

Figure E.1: Illustrating the *ecliptic*, i.e. the apparent movement of the Sun around the Earth. Further, the *celestial equator* and the direction of the *vernal equinox* are indicated. The axial tilt has a value of $\epsilon \approx 23.4°$. Sun and Earth are not drawn to scale.

ation presented here has an accuracy of about 1 arcminute[1] within two centuries around the year 2,000 and is published by the *Astronomical Applications Department* of the *US Naval Observatory* [200].

To express the position of the Sun we first have to express the time D elapsed since Greenwich noon, terrestrial time, on 1 January 2000, in days. For astronomical purposes it may be convenient to relate D to the Julian date JD via

$$D = \text{JD} - 2451545.0. \tag{E.1}$$

The Julian date is defined as the number of days since 1 January 4713 BC in a proleptic[2] Julian calendar or since 24 November 4717 BC in a proleptic Gregorian calendar.

Now, the *mean longitude* of the Sun corrected to the aberration of the light is given by

$$q = 280.459° + 0.98564736° D. \tag{E.2}$$

Because of the elliptic orbit of the Earth and hence a varying speed throughout the year, we have to correct with the so-called *mean anomaly* of the Sun,

$$g = 357.529° + 0.98560028° D. \tag{E.3}$$

It may be convenient to normalise q and g to the range $[0°, 360°)$ by adding or subtracting multiples of $360°$.

Now the ecliptic longitude of the Sun is given by

$$\lambda_S = q + 1.915° \sin g + 0.020° \sin 2g. \tag{E.4}$$

The ecliptic latitude can be approximated by

$$\beta_S = 0. \tag{E.5}$$

To estimate the radiation it might also be convenient to approximate the distance of the Sun from the Earth. In astronomical units (AU) this is given by

$$R = 1.00014 - 0.01671 \cos g - 0.00014 \cos 2g. \tag{E.6}$$

As stated in Chapter 18, for PV applications it is convenient to use horizontal coordinates. We therefore have to transform from ecliptic to horizontal coordinates. This is done via *three rotations* that are to be performed consecutively.

First, we have to transform from ecliptic to equatorial coordinates. As illustrated in Fig. E.1, the fundamental plane of these coordinates is tilted to the ecliptic with an angle ϵ,

$$\epsilon = 23.429° - 0.00000036° D. \tag{E.7}$$

The principal direction is again given by the vernal equinox Υ. Figure E.2 (b) shows the equatorial coordinate system. The two coordinates are called the *right ascension* α and the *declination* δ. The transformation from ecliptic to equatorial coordinates is a rotation by the angle ϵ about the vernal equinox as rotational axis. Mathematically this is expressed by

$$\begin{pmatrix} \cos \delta \cos \alpha \\ \cos \delta \sin \alpha \\ \sin \delta \end{pmatrix} = \begin{pmatrix} 1 & 0 & 0 \\ 0 & \cos \epsilon & -\sin \epsilon \\ 0 & \sin \epsilon & \cos \epsilon \end{pmatrix} \begin{pmatrix} \cos \beta \cos \lambda \\ \cos \beta \sin \lambda \\ \sin \beta \end{pmatrix}. \tag{E.8}$$

---
[1] One arcminute is 1/60 of a degree.
[2] *Proleptic* means that a calendar is applied to dates before its introduction.

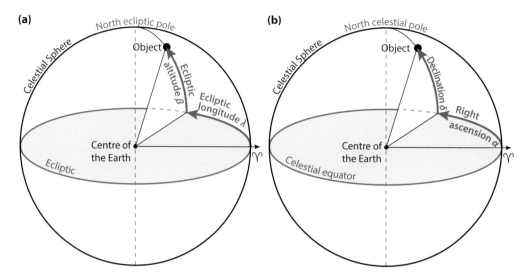

Figure E.2: Illustrating (a) the *ecliptic coordinate system* and (b) the *equatorial coordinate system*.

Secondly, we have to take the rotation of the Earth around its axis into account. We do this by using the so-called *hour angle h* instead of the right ascension α. Those two angles are connected to each other via

$$h = \theta_L - \alpha, \tag{E.9}$$

where θ_L is the *local mean sidereal time*, i.e. the angle between the vernal equinox and the meridian. All these angles are illustrated in Figure E.3. A sidereal day is the duration between two passes of the vernal equinox through the meridian and it is slightly shorter than a solar day. We can understand this by realizing that the Earth has to rotate by 360° and approximately 360°/365.25 between two passes through the meridian. The duration of *mean* sidereal day is approximately 23 h, 56 m and 4 s.

To calculate θ_L we first have to determine the Greenwich mean sidereal time (GMST), which is (approximately) given by

$$\text{GMST} = 18.697374558\,\text{h} + 24.06570982441908\,\text{h} \times D + 0.000026\,\text{h} \times T^2, \tag{E.10}$$

where D is as defined above and T is the number of centuries past since Greenwich noon, Terrestrial Time, on 1 January 2000,

$$T = \frac{D}{36525}. \tag{E.11}$$

For many applications, the quadratic term may be omitted. GMST is given in hours and has to be normalized to the range [0 h, 24 h). We can then obtain the *local* mean sidereal time in degrees with

$$\theta_L = \text{GMST} \frac{15°}{\text{hour}} + \lambda_0, \tag{E.12}$$

where λ_0 is the longitude of the observer.

E. Some aspects on location issues

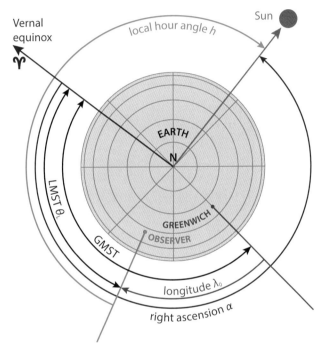

Figure E.3: Illustrating the right ascension α, the local hour angle h, the Greenwich mean sidereal time GMST and the local mean sidereal time θ_L.

We thus have the following transform about the rotational axis of the earth,

$$\begin{pmatrix} \cos\delta\cos h \\ \cos\delta\sin h \\ \sin\delta \end{pmatrix} = \begin{pmatrix} \cos\theta_L & \sin\theta_L & 0 \\ \sin\theta_L & -\cos\theta_L & 0 \\ 0 & 0 & 1 \end{pmatrix} \begin{pmatrix} \cos\delta\cos\alpha \\ \cos\delta\sin\alpha \\ \sin\delta \end{pmatrix}. \quad \text{(E.13)}$$

Note that this transform is no rotation but a reflection at an angle of $\theta_L/2$. We have thus transformed the principal direction of the coordinate system from vernal equinox to the local mean sidereal time.

Thirdly we transform to the horizontal coordinate system by rotating with the latitude angle ϕ_0 of the observer. While the ecliptic and equatorial coordinate systems use the centre of the Earth as origin, the horizontal coordinate system uses the actual position on the surface of Earth as origin. However, because of the distance of celestial objects in general and the Sun in particular being much larger than the radius of the Earth, we may neglect this translational shift of the origin of the coordinate systems.

The rotational axis of the third transform is the axis that is normal to both the principal direction θ_L and the rotational axis of the Earth ($\delta = 90°$).

$$\begin{pmatrix} \xi' \\ v' \\ \zeta \end{pmatrix} = \begin{pmatrix} \sin\phi_0 & 0 & -\cos\phi_0 \\ 0 & 1 & 0 \\ \cos\phi_0 & 0 & \sin\phi_0 \end{pmatrix} \begin{pmatrix} \cos\delta\cos h \\ \cos\delta\sin h \\ \sin\delta \end{pmatrix}. \quad \text{(E.14)}$$

However, here the directions ξ' and v' point due South and West, respectively. We thus must apply $\xi' \to -\xi$ and $v' \to -v$ in order to get the directions as they were defined in Figure 18.1 (ξ and v pointing due North and East, respectively). We do this with the matrix transform

$$\begin{pmatrix} \xi \\ v \\ \zeta \end{pmatrix} = \begin{pmatrix} -1 & 0 & 0 \\ 0 & -1 & 0 \\ 0 & 0 & 1 \end{pmatrix} \begin{pmatrix} \xi' \\ v' \\ \zeta \end{pmatrix}. \quad \text{(E.15)}$$

By combining Eqs. (E.8), (E.13)–(E.15) we can directly transform from ecliptic to horizontal coordinates,

$$\begin{pmatrix} \xi \\ v \\ \zeta \end{pmatrix} = \begin{pmatrix} -1 & 0 & 0 \\ 0 & -1 & 0 \\ 0 & 0 & 1 \end{pmatrix} \begin{pmatrix} \sin\phi_0 & 0 & -\cos\phi_0 \\ 0 & 1 & 0 \\ \cos\phi_0 & 0 & \sin\phi_0 \end{pmatrix}$$
$$\times \begin{pmatrix} \cos\theta_L & \sin\theta_L & 0 \\ \sin\theta_L & -\cos\theta_L & 0 \\ 0 & 0 & 1 \end{pmatrix} \begin{pmatrix} 1 & 0 & 0 \\ 0 & \cos\epsilon & -\sin\epsilon \\ 0 & \sin\epsilon & \cos\epsilon \end{pmatrix} \begin{pmatrix} \cos\beta\cos\lambda \\ \cos\beta\sin\lambda \\ \sin\beta \end{pmatrix}. \quad \text{(E.16)}$$

Please note that matrix multiplications do not commute, i.e. the order in which the rotations are applied must not be altered. Now, we apply the ecliptic latitude of the Sun $\beta_S = 0$. By calculating Eq. (E.16) we find

$$\xi_S = \cos a_S \cos A_S = -\sin\phi_0 \cos\theta_L \cos\lambda_S \quad \text{(E.17a)}$$
$$- (\sin\phi_0 \sin\theta_L \cos\epsilon - \cos\phi_0 \sin\epsilon) \sin\lambda_S,$$
$$v_S = \cos a_S \sin A_S = -\sin\theta_L \cos\lambda_S + \cos\theta_L \cos\epsilon \sin\lambda_S, \quad \text{(E.17b)}$$
$$\zeta_S = \sin a_S \quad = \cos\phi_0 \cos\theta_L \cos\lambda_S$$
$$+ (\cos\phi_0 \sin\theta_L \cos\epsilon + \sin\phi_0 \sin\epsilon) \sin\lambda_S, \quad \text{(E.17c)}$$

E. Some aspects on location issues

where we also used the relationship between Cartesian and spherical horizontal coordinates from Eq. (18.1). Dividing Eq. (E.17b) by Eq. (E.17a) and leaving Eq. (E.17c) unchanged leads to the final expressions for the solar position,

$$\tan A_S = \frac{v_S}{\zeta_S} = \frac{-\sin\theta_L \cos\lambda_S + \cos\theta_L \cos\epsilon \sin\lambda_S}{-\sin\phi_0 \cos\theta_L \cos\lambda_S - (\sin\phi_0 \sin\theta_L \cos\epsilon - \cos\phi_0 \sin\epsilon)\sin\lambda_S}, \quad \text{(E.18a)}$$

$$\sin a_S = \zeta_S = \cos\phi_0 \cos\theta_L \cos\lambda_S + (\cos\phi_0 \sin\theta_L \cos\epsilon + \sin\phi_0 \sin\epsilon)\sin\lambda_S. \quad \text{(E.18b)}$$

From Eqs. (E.18), A_S and a_S can be derived by applying inverse trigonometric functions. While arcsin uniquely delivers an altitude between $-90°$ and $90°$, applying arctan leads to ambiguities. For deriving an azimuth between $0°$ and $360°$, we have to look in which quadrant it is lying. Therefore we use ζ and v from Eqs. (E.17a) and (E.17b), respectively. We find

$$\zeta > 0 \wedge v > 0 \Rightarrow A_S = \arctan f(\ldots), \quad \text{(E.19a)}$$

$$\zeta < 0 \quad\quad\quad\quad \Rightarrow A_S = \arctan f(\ldots) + 180°, \quad \text{(E.19b)}$$

$$\zeta > 0 \wedge v < 0 \Rightarrow A_S = \arctan f(\ldots) + 360°. \quad \text{(E.19c)}$$

$f(\ldots)$ denotes the function at the right-hand side of Eq. (E.18a). Note that an altitude $a_S < 0°$ corresponds to the Sun being below the horizon. This means that the Sun is not visible and no solar energy can be harvested.

The approximations presented on the previous pages are accurate within arcminutes for 200 centuries, of the year 2000. Several years ago, NREL presented a much more complicated model, the so-called solar position algorithm (SPA), with uncertainties of only $\pm 0.0003°$ in the period from 2000 BC to AD 6000 [201].

Example

As an example we will calculate the position of the Sun in **Delft on 14 April 2014 at 11:00 local time***.*

To determine the solar position we need the date and time (in UTC), and the latitude and longitude. Since the time zone in Delft on 14 April is CEST, central European summer time, the time difference with UTC is +2 hours, such that 11:00 CEST corresponds to 9:00 UTC. According to `Google Maps`*, the latitude and longitude of the* Markt *in the centre of Delft are given by*

$$\phi_0 = 52.01° \text{ N} = +52.01°,$$
$$\lambda_0 = 4.36° \text{ E} = + 4.36°.$$

For the calculation we first have to express the date and time as the time elapsed since 1 January 2000 noon UTC.

$$D = 4 \times 366 + 10 \times 365 + 2 \times 31 + 28 + 13 - 0.5 + \tfrac{9}{24} = 5216.875.$$

Now we can calculate the mean longitude q and the mean anomaly g of the Sun according to Eqs. (E.2) and (E.3),

$$q = 280.459° + 0.98564736° D = 22.4580712°,$$
$$g = 357.529° + 0.98560028° D = 99.28246073°,$$

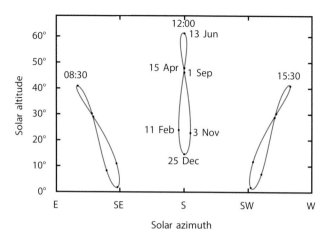

Figure E.4: The analemma, i.e. the apparent curve of the Sun throughout the year when observed at the same mean solar time every day. The analemma is shown for Delft (52° N latitude) at three points in time during the day.

where the values were normalized to $[0°, 360°)$. From Eq. (E.4) we thus obtain for the latitude of the Sun in ecliptic coordinates

$$\lambda_S = q + 1.915° \sin g + 0.020° \sin 2g = 24.34162696°.$$

For the axial tilt ϵ of the Earth we obtain from Eq. (E.7)

$$\epsilon = 23.429° - 0.00000036° D = 23.42712193°.$$

The Greenwich mean sidereal time (GMST) is given by Eq. (E.10),

$$\text{GMST} = 18.697374558\,\text{h} + 24.06570982441908\,\text{h} \times D + 0.000026\,\text{h} \times T^2 = 22.49731535\,\text{h},$$

where we used $T = D/36525$ and normalized to $[0\,h, 24\,h)$. We then find for the local mean sidereal time θ_L

$$\theta_L = \text{GMST}\frac{15°}{\text{hour}} + \lambda_0 = 341.8197303°.$$

Now we have all the variables required to calculate the solar position. From Eqs. (E.18) and (E.19) we thus find

$$\tan A_S = \frac{-\sin\theta_L \cos\lambda_S + \cos\theta_L \cos\epsilon \sin\lambda_S}{-\sin\phi_0 \cos\theta_L \cos\lambda_S - (\sin\phi_0 \sin\theta_L \cos\epsilon - \cos\phi_0 \sin\epsilon)\sin\lambda_S} = -1.318180633,$$

$$\sin a_S = \cos\phi_0 \cos\theta_L \cos\lambda_S + (\cos\phi_0 \sin\theta_L \cos\epsilon + \sin\phi_0 \sin\epsilon)\sin\lambda_S = 0.589415473,$$

which leads to the solar altitude $a_S = \mathbf{36.1°}$ and the solar azimuth $A_S = \mathbf{127.2°}$.

E. Some aspects on location issues

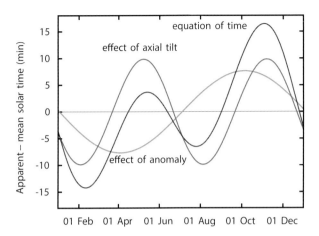

Figure E.5: The effect of the anomaly of the terrestrial orbit and the axial tilt of the Earth rotation axis on the difference between apparent and mean solar time. The equation of time nearly is the sum of these two effects.

E.2 The equation of time

Figure E.4 shows the position of the Sun throughout the year in Delft at 8:30, 12:00 and 15:30 mean solar time (MST), i.e. UTC $+ \lambda_0$, where the longitude λ_0 is expressed in hours. We see that the position of the Sun at these instances not only varies in the altitude but also in azimuth in the course of a year, such that it seems to run along the shape of an *eight*. This closed curve is called the *analemma*.

The difference between the apparent solar time (AST), i.e. the timescale where the Sun really is highest at noon every day, and the mean solar time is described by the so-called equation of time (EoT), which is defined as

$$\text{EoT} = \text{AST} - \text{MST}. \tag{E.20}$$

The equation of time is given as the difference between the mean longitude q, as defined in Eq. (E.2), and the right ascension α_S of the Sun in the equatorial coordinate system,

$$\text{EoT}(D) = [q(D) - \alpha_S(D)] \frac{1}{15} \frac{\text{hour}}{\text{deg}}. \tag{E.21}$$

The right ascension is connected to the ecliptic longitude of the Sun λ_S, as given in Eq. (E.4) via

$$\tan \alpha_s = \cos \epsilon \tan \lambda_S, \tag{E.22}$$

and ϵ is the axial tilt as given in Eq. (E.7).

The equation of time has two major contributors: the anomaly due to the elliptic orbit of the Earth around the Sun and the Axial tilt of the rotational axis of the Earth with respect to the ecliptic. Both effects are shown in Figure E.5. In this figure, the total EoT is also shown, which is nearly the sum of the two contributors (The maximal deviation is less than a minute).

We see that the largest negative shift is on 11 February, where, the apparent noon is about 14 min 12 s prior to the mean solar noon. The largest positive shift is on 3 November, when the apparent solar noon is about 16 min 25 s past the mean solar noon. These points are also marked in Figure E.4. There, the zeros of the EoT are also shown, which are on 15 April, 13 June, 1 September, and 25 December.

E.3 Angle between the Sun and a PV module

As seen in Eq. (18.2), the irradiance on a PV module G_M^{dir} is given by

$$G_M^{\text{dir}} = I_e^{\text{dir}} \cos \gamma, \tag{E.23}$$

where I_e^{dir} is the direct normal irradiance irradiance and $\gamma = \sphericalangle(A_M, a_M)(A_S, a_S)$ is the angle between the surface normal and the incident direction of the sunlight. The different variables are depicted in Figure. 18.6.

For calculating $\cos \gamma$, we utilise that the scalar product of two unit vectors is equal to the cosine of the enclosed angle. Thus, we can write,

$$\cos \gamma = \mathbf{n}_M \times \mathbf{n}_S. \tag{E.24}$$

The normal vectors are given by

$$\mathbf{n}_M = \begin{pmatrix} \xi_M \\ \upsilon_M \\ \zeta_M \end{pmatrix} = \begin{pmatrix} \cos a_M \cos A_M \\ \cos a_M \sin A_M \\ \sin a_M \end{pmatrix}, \tag{E.25}$$

$$\mathbf{n}_S = \begin{pmatrix} \xi_S \\ \upsilon_S \\ \zeta_S \end{pmatrix} = \begin{pmatrix} \cos a_S \cos A_S \\ \cos a_S \sin A_S \\ \sin a_S, \end{pmatrix}, \tag{E.26}$$

where we used the relationship between Cartesian and spherical horizontal coordinates given in Eq. (18.1). Hence, we find

$$\begin{aligned} \cos \gamma &= \mathbf{n}_M \cdot \mathbf{n}_S \\ &= \cos a_M \cos A_M \cos a_S \cos A_S \\ &\quad + \cos a_M \sin A_M \cos a_S \sin A_S + \sin a_M \sin a_S \\ &= \cos a_M \cos a_S \left(\cos A_M \cos A_S + \sin A_M \sin A_S \right) \\ &\quad + \sin a_M \sin a_S \\ &= \cos a_M \cos a_S \cos (A_M - A_S) + \sin a_M \sin a_S. \end{aligned} \tag{E.27}$$

E.4 Modules mounted on a tilted roof

When a module is mounted on a horizontal plane, it is easy to determine its normal $\mathbf{n}_M$. However, when a module is to be mounted on an arbitrarily tilted roof things become more complicated. We thus will calculate the normal of the module in horizontal coordinates when the coordinates with respect to the roof are given. In fact, we have to transform the module normal from the *roof coordinate system* to the horizontal coordinate system.

E. Some aspects on location issues

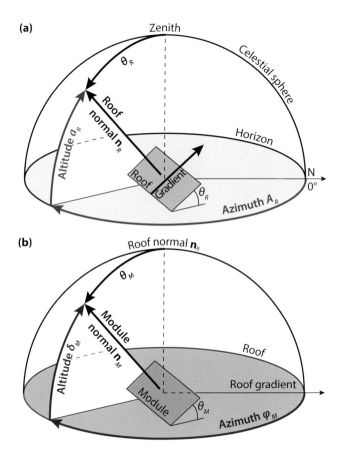

Figure E.6: The angles used to describe (a) the orientation of a roof on a horizontal plane and (b) the orientation of a module mounted on a roof.

As illustrated in Figure E.6 (a), the orientation of the roof in horizontal coordinates is characterized by the azimuth A_R and the altitude a_R of its normal $\mathbf{n}_R$. The module is installed on the roof, and its orientation with respect to the roof is best described in the roof coordinate system, where the fundamental plane is parallel to the roof and the principal direction is along the gradient of the roof, as illustrated in Figure E.6 (b). In this system, the module normal is given by the azimuth ϕ_M and the altitude is given by δ_M. The coordinate transform itself is transformed by combining two rotations.

First, we rotate with the angle $90° - a_R$ around the axis that is perpendicular to both $\mathbf{n}_R$ and the gradient direction of the roof. Secondly, we rotate with the angle $A_R + 180°$ along the zenith. We thus obtain

$$\begin{pmatrix} \xi_M \\ \upsilon_M \\ \zeta_M \end{pmatrix} = \begin{pmatrix} \cos a_M \cos A_M \\ \cos a_M \sin A_M \\ \sin a_M \end{pmatrix}$$

$$= \begin{pmatrix} -\cos A_R & \sin A_R & 0 \\ -\sin A_R & -\cos A_R & 0 \\ 0 & 0 & 1 \end{pmatrix} \begin{pmatrix} \sin a_R & 0 & -\cos a_R \\ 0 & 1 & 0 \\ \cos a_R & 0 & \sin a_R \end{pmatrix} \begin{pmatrix} \cos \delta_M \cos \phi_M \\ \cos \delta_M \sin \phi_M \\ \sin \delta_M \end{pmatrix}. \quad \text{(E.28)}$$

The coordinates of the module in the horizontal coordinate system are then given by

$$\xi_M = \cos a_M \cos A_M \quad \text{(E.29a)}$$
$$= -\cos A_R \sin a_R \cos \delta_M \cos \phi_M + \sin A_R \cos \delta_M \sin \phi_M + \cos A_R \cos a_R \sin \delta_M, \quad \text{(E.29b)}$$
$$\upsilon_M = \cos a_M \sin A_M \quad \text{(E.29c)}$$
$$= -\sin A_R \sin a_R \cos \delta_M \cos \phi_M - \cos A_R \cos \delta_M \sin \phi_M + \sin A_R \cos a_R \sin \delta_M, \quad \text{(E.29d)}$$
$$\zeta_M = \sin a_M \quad = \cos a_R \cos \delta_M \cos \phi_M + \sin a_R \sin \delta_M. \quad \text{(E.29e)}$$

Dividing Eq. (E.29c) by Eq. (E.29a) and leaving Eq. (E.29e) unchanged leads to the final expressions for the module orientation in horizontal coordinates,

$$\tan A_M = \frac{-\sin A_R \sin a_R \cos \delta_M \cos \phi_M - \cos A_R \cos \delta_M \sin \phi_M + \sin A_R \cos a_R \sin \delta_M}{-\cos A_R \sin a_R \cos \delta_M \cos \phi_M + \sin A_R \cos \delta_M \sin \phi_M + \cos A_R \cos a_R \sin \delta_M}, \quad \text{(E.30a)}$$

$$\sin a_M = \cos a_R \cos \delta_M \cos \phi_M + \sin a_R \sin \delta_M. \quad \text{(E.30b)}$$

Finally, the cosine of the angle between the module orientation and the solar position is given by

$$\cos \gamma = \mathbf{n}_M \times \mathbf{n}_S = \cos a_S \cos(A_R - A_S)(\cos a_R \sin \delta_M - \sin a_R \cos \delta_M \cos \phi_M)$$
$$+ \cos a_S \sin(A_R - A_S) \cos \delta_M \sin \phi_M \quad \text{(E.31)}$$
$$+ \sin a_S (\cos a_R \cos \delta_M \cos \phi_M + \sin a_R \sin \delta_M).$$

We will try to understand these results by discussing easy examples.

First, we look at a roof that faces eastward and has a tilt angle θ_R. Then, $a_R = 90° − \theta_R$ and $A_R = 90°$. On this roof a solar module is installed at a tilting angle θ_M with respect to the roof. The modules are mounted parallel to the gradient of the roof. We thus have $\delta_M = 90° − \theta_M$ and $\phi_M = 270°$. From Eqs. (E.30) we thus obtain

$$\sin a_M = \cos a_R \cos \delta_M \times 0 + \sin a_R \sin \delta_M. \tag{E.32a}$$
$$= \sin a_R \sin \delta_M$$

$$\tan A_M = \frac{-0 - 0 + 1 \times \cos a_R \sin \delta_M}{0 + \sin A_R \cos \delta_M \sin \times 1 + 0} \tag{E.32b}$$
$$= \cos a_R \tan \delta_M.$$

In the second example, the roof is facing southward and tilted at an angle θ_R. We thus have $a_R = 90° − \theta_R$ and $A_R = 180°$. Now the module is tilted at an angle θ_M with respect to the roof and mounted perpendicular to the gradient of the roof. Hence, $\delta_M = 90° − \theta_M$ and $\phi_M = 180°$. Using Eqs. (E.30) we find

$$\sin a_M = \cos a_R \cos \delta_M \times (-1) + \sin a_R \sin \delta_M. \tag{E.33a}$$
$$= \cos(a_R + \delta_M) = \sin(a_R - \theta_M),$$

$$\tan A_M = \frac{-0 - 0 + 0}{-\sin a_R \cos \delta_M + 0 + \cos A_R \sin \delta_M} = 0. \tag{E.33b}$$

E.5 Length of the shadow behind a PV module

In this section we will determine how far behind the module the shadow reaches dependent on the solar position, the module orientation and the length l of the module. Figure E.7 shows the important parameters that we need to derive the length of the shadow behind a PV module. For the determination we look at a module that is tilted at an angle θ_M. Its normal vector has an azimuth A_M. This module touches the ground at two corner points that we call E and F. Without loss of generality, we may assume that E is at the origin of our horizontal coordinate system, $E = (0, 0, 0)$. Further, P is the top corner point of the module lying above E. The length of the module, i.e. the distance between E and P is l, $\overline{EP} = l$. The shadow of this point on the horizontal plane, we denote by P'. Then, the *length of the shadow d* is defined as the shortest distance between P' and the line g, which connects E with F.

For the determination of d we first must derive P, and then P'. The normal of the module $\mathbf{n}_M$ is given by

$$\mathbf{n}_M = \begin{pmatrix} \cos a_M \cos A_M \\ \cos a_M \sin A_M \\ \sin a_M \end{pmatrix} = \begin{pmatrix} \sin \theta_M \cos A_M \\ \sin \theta_M \sin A_M \\ \cos \theta_M \end{pmatrix}. \tag{E.34}$$

The direction vector $\mathbf{r}$ of the line g connecting E with F is then given as

$$\mathbf{r} = \begin{pmatrix} \sin A_M \\ -\cos A_M \\ 0 \end{pmatrix}. \tag{E.35}$$

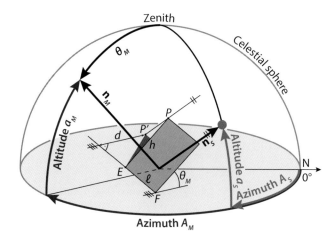

Figure E.7: Derivation of the length d of a shadow behind a PV module with ground length l and height h. The module has the normal $\mathbf{n}_M$ while the position of the Sun is given by the direction $\mathbf{n}_S$. The length of the shadow d is given as the distance between the line connecting the lower module corners E and F and the projection of the upper corner P on the ground (P').

Subsequently, we can calculate the direction vector $\mathbf{h}$ of the line that connects E with P with the vector product

$$\mathbf{h} = -\mathbf{n}_M \times \mathbf{r} = -\begin{pmatrix}\sin\theta_M \cos A_M\\ \sin\theta_M \sin A_M\\ \cos\theta_M\end{pmatrix} \times \begin{pmatrix}\sin A_M\\ -\cos A_M\\ 0\end{pmatrix} = \begin{pmatrix}-\cos\theta_M \cos A_M\\ -\cos\theta_M \sin A_M\\ \sin\theta_M\end{pmatrix}, \quad (E.36)$$

where the $-$ sign is used because of the fact that the horizontal coordinate system is left-handed. Since $\mathbf{h}$ has length 1, i.e. it is a unit vector, we can easily derive the position of the point P with

$$P = E + l \times \mathbf{h} = l\begin{pmatrix}-\cos\theta_M \cos A_M\\ -\cos\theta_M \sin A_M\\ \sin\theta_M\end{pmatrix}. \quad (E.37)$$

To calculate the position of P' we define the line s, which goes through P and points towards the Sun,

$$s(t) = P + t \times \mathbf{n}_S. \quad (E.38)$$

When the position of the Sun is described by its altitude a_S and azimuth A_S, we find

$$s(t) = l\begin{pmatrix}-\cos\theta_M \cos A_M\\ -\cos\theta_M \sin A_M\\ \sin\theta_M\end{pmatrix} + t\begin{pmatrix}\cos a_S \cos A_S\\ \cos a_S \sin A_S\\ \sin a_S,\end{pmatrix}. \quad (E.39)$$

We find the shadow P' of the point P as the *intersection* of the line s with the horizontal plane, $z = 0$,

$$l\sin\theta_M + t\sin a_S = 0. \quad (E.40)$$

E. Some aspects on location issues

Hence,
$$t = -l\frac{\sin\theta_M}{\sin a_S}. \tag{E.41}$$

The coordinates of $P' = (P'_x, P'_y, 0)$ are then given as
$$P'_x = -l\left(\cos\theta_M \cos A_M + \sin\theta_M \cot a_S \cos A_S\right), \tag{E.42a}$$
$$P'_y = -l\left(\cos\theta_M \sin A_M + \sin\theta_M \cot a_S \sin A_S\right). \tag{E.42b}$$

As stated earlier, the length of the shadow d is given as the shortest distance between P' and the line g connecting E and F. Let g' be the line through P' that is perpendicular to g, $g \perp g'$. Since $E = (0, 0, 0)$ we find for g and g'
$$g(u) = \quad u \times \mathbf{r}, \tag{E.43a}$$
$$g'(v) = P' + v \times \mathbf{r}', \tag{E.43b}$$

where the direction vector $\mathbf{r}'$ is given as
$$\mathbf{r}' = \begin{pmatrix} \cos A_M \\ \sin A_M \\ 0 \end{pmatrix}. \tag{E.44}$$

d is the distance between P' and the intersection of g with g'. At this intersection we have
$$g(u) = g'(v). \tag{E.45}$$

Solving this equation, we obtain
$$u\begin{pmatrix} \sin A_M \\ -\cos A_M \\ 0 \end{pmatrix} = P' + v\begin{pmatrix} \cos A_M \\ \sin A_M \\ 0 \end{pmatrix}$$

$$u\sin A_M = P'_x + v\cos A_M$$
$$-u\cos A_M = P'_y + v\sin A_M$$

$$u\sin A_M \cos A_M = P'_x \cos A_M + v\cos^2 A_M$$
$$-u\cos A_M \sin A_M = P'_y \sin A_M + v\sin^2 A_M$$

By adding the last two equations we find
$$P'_x \cos A_M + v\cos^2 A_M + P'_y \sin A_M + v\sin^2 A_M = 0, \tag{E.46}$$

and hence
$$v = -P'_x \cos A_M - P'_y \sin A_M. \tag{E.47}$$

Using Eqs. (E.42), we derive
$$\begin{aligned} v &= l[\,(\cos\theta_M \cos A_M + \sin\theta_M \cot a_S \cos A_S)\cos A_M \\ &\quad + (\cos\theta_M \sin A_M + \sin\theta_M \cot a_S \sin A_S)\sin A_M] \\ &= l(\cos\theta_M \cos^2 A_M + \sin\theta_M \cot a_S \cos A_S \cos A_M \\ &\quad + \cos\theta_M \sin^2 A_M + \sin\theta_M \cot a_S \sin A_S \sin A_M). \end{aligned} \tag{E.48}$$

Because the direction vector **r**′ of the line g' is a unit vector, the length of the shadow d is equal to v. Therefore we obtain from Eq. (E.48), with some trigonometric operations,

$$d = l \left[\cos \theta_M + \sin \theta_M \cot a_S \cos(A_M - A_S)\right]. \tag{E.49}$$

F
Derivations for DC-DC converters

As we have discussed in Section 19.2.2, a DC-DC converter is an electronic circuit which converts a source of direct current from one voltage level to another. An ideal PV converter should draw the maximum power from the PV panel and supply at the load side. We discussed three types of DC-DC converters: *buck* converters that convert voltage from a higher to a lower level, *boost* converter that convert voltage from a lower to a higher level, and *buck-boost* converters, which can perform both operations. In this Appendix, we will derive important relations for these three types in more detail.

F.1 Buck converter

For a buck converter [see Figure 19.9 (a)] in the *on* mode, the voltage across the inductor can be expressed as:

$$V_L = V_d - V_o, \tag{F.1}$$

where V_d is the voltage level to be converted and V_o is the output voltage. The inductor regulates the flow of current during the switching operation. This component functions according to Lenz's law for which the time derivative of current flowing through the coil is linked to the voltage across the inductor via the proportionality factor L, known as the inductance of the coil [see Eq. 19.19]. Moving from the derivative to the integral form of Lenz's law the increase of current in the inductor in the *on* mode can be written as

$$\Delta i_{L-\text{on}} = \frac{1}{L} \int_0^{t_{\text{on}}} V_L dt = \frac{V_d - V_o}{L} t_{\text{on}}, \tag{F.2}$$

where t_{on} is the duration of the *on* mode. Similarly, during the *off* mode, the voltage across the inductor can be expressed as

$$V_L = -V_o. \tag{F.3}$$

The decrease of current in the inductor can be written as

$$\Delta i_{L-\text{off}} = \frac{1}{L}\int_{t_{\text{on}}}^{t_S} V_L dt = -\frac{V_o}{L} t_{\text{off}}, \qquad (F.4)$$

where t_S is the period of the signal and t_{off} is the duration of the *off* mode [see Figure 19.8 (b)]. In *steady-state* operation the flow of current across the inductor has to be zero,

$$\Delta i_{L-\text{on}} + \Delta i_{L-\text{off}} = 0. \qquad (F.5)$$

Alternatively,

$$\frac{V_d - V_o}{L} t_{\text{on}} = \frac{V_o}{L} t_{\text{off}}, \qquad (F.6)$$

which can be arranged as

$$V_o = D \cdot V_d, \qquad (F.7)$$

which is Eq. (19.11).

F.2 Boost converter

For a boost converter (see Figure 19.9 (b)) in the *on* mode the voltage across the inductor can be expressed as

$$V_L = V_d, \qquad (F.8)$$

where V_d is the voltage level to be converted. In the *on* mode, the increase of current in the inductor can be written as

$$\Delta i_{L-\text{on}} = \frac{1}{L}\int_0^{t_{\text{on}}} V_L dt = \frac{V_d}{L} t_{\text{on}}. \qquad (F.9)$$

Similarly, during the *off* mode, the voltage across the inductor can be expressed as

$$V_L = V_d - V_o, \qquad (F.10)$$

while the decrease of current in the inductor can be written as

$$\Delta i_{L-\text{off}} = \frac{1}{L}\int_{t_{\text{on}}}^{t_S} V_L dt = \frac{V_d - V_o}{L} t_{\text{off}}. \qquad (F.11)$$

In *steady-state* operation the flow of current across the inductor has to be zero,

$$\Delta i_{L-\text{on}} + \Delta i_{L-\text{off}} = 0, \qquad (F.12)$$

or, alternatively,

$$\frac{V_d - V_o}{L} t_{\text{off}} = -\frac{V_d}{L} t_{\text{on}}. \qquad (F.13)$$

This equation can be rearranged as

$$V_o = \frac{V_d}{1-D},$$
$$V_d = (1-D)V_o, \qquad (F.14)$$

which is the same expression as Eq. (19.15)

F. Derivations for DC-DC converters

F.3 Buck-boost converter

For a boost converter [see Figure 19.10] in the *on* mode the voltage across the inductor can be expressed as

$$V_L = V_d, \tag{F.15}$$

where V_d is the voltage level to be converted. In the *on* mode, the increase of current in the inductor can be written as

$$\Delta i_{L-\text{on}} = \frac{1}{L} \int_0^{t_\text{on}} V_L dt = \frac{V_d}{L} t_\text{on}. \tag{F.16}$$

Similarly, during the *off* mode, the voltage across the inductor can be expressed as

$$V_L = -V_o, \tag{F.17}$$

while the decrease of current in the inductor can be written as

$$\Delta i_{L-\text{off}} = \frac{1}{L} \int_{t_\text{on}}^{t_s} V_L dt = -\frac{V_o}{L} t_\text{off}. \tag{F.18}$$

In *steady-state* operation the flow of current across the inductor has to be zero,

$$\Delta i_{L-\text{on}} + \Delta i_{L-\text{off}} = 0, \tag{F.19}$$

or, alternatively:

$$\frac{V_o}{L} t_\text{off} = \frac{V_d}{L} t_\text{on}. \tag{F.20}$$

This equation can be rearranged as

$$V_o = \frac{D}{1-D} V_d,$$
$$V_d = \frac{1-D}{D} V_o, \tag{F.21}$$

which is the same as Eq. 19.15.

G
Fluid-dynamic model

As we have seen in Section 20.3, it is very important to develop an accurate thermal model of the solar module, which evaluates the effects of the major meteorological parameters on the final temperature of a solar module.

In this appendix, we will develop an accurate fluid-dynamic (FD) model for estimating the PV module working temperature T_M as a function of various meteorological parameters. The model is based on a detailed energy balance between the module itself and the surrounding environment. Both the installed configuration of the array together with external parameters such as direct incident solar irradiance on the panels, wind speed and cloud cover will be taken into account. The model presented here is mostly based on the work by Fuentes [164].

G.1 Framework of the fluid-dynamic model

We consider three types of heat transfer, which we already discussed in Section 22.1, between the PV module and the surroundings: *conduction*, *convection* and *radiation*. They are illustrated in Figure G.1. In the FD model, the following contributions are considered:

- Heat received from the Sun in the form of irradiance αG_M, where α is the absorptivity of the module.
- Convective heat exchange with the surrounding air from the front and rear sides of the module
$$h_c(T_M - T_a),$$
where h_c denotes the overall convective heat transfer coefficient of the module.
- Radiative heat exchange between the upper module surface and the sky
$$\epsilon_{\text{top}} \sigma \left(T_M^4 - T_{\text{sky}}^4 \right),$$

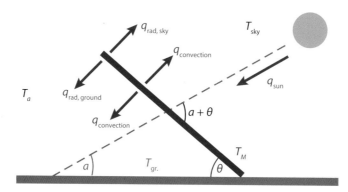

Figure G.1: Representation of heat exchange between a tilted module surface and the surroundings.

where $\epsilon_{top} = 0.84$ is the emissivity of the module front glass and σ is the Stefan–Boltzmann constant as defined in Eq. (5.20), and between the rear surface and the ground

$$\epsilon_{back}\sigma\left(T_M^4 - T_{gr.}^4\right),$$

where the emissivity of the back is assumed to be $\epsilon_{back} = 0.89$.

- Conductive heat transfer between the module and the mounting structure. Because of the small area of the contact points, we may neglect this contribution [202].

By considering each of the contributions mentioned above separately, we can write down the heat transfer balance [164],

$$mc\frac{dT_M}{dt} = \alpha G_M - h_c(T_M - T_a) - \epsilon_{back}\sigma\left(T_M^4 - T_{gr.}^4\right) - \epsilon_{top}\sigma\left(T_M^4 - T_{sky}^4\right). \quad (G.1)$$

Before providing a solution for this differential equation, we want to point out that we consider the entire module to be a single mass with a uniform temperature T_M. This is different to Section 20.3.1, where we presented simple thermal models that were used to estimate the temperature of the *solar cell* within a PV module. The assumption of the PV module being at a uniform temperature T_M is not entirely realistic, because PV modules are made of various layers of different materials that surround the actual solar cells. It is the purpose of this section to evaluate the temperature of the solar cell itself, which is where the solar radiation is effectively absorbed. This temperature will be higher than the surface module temperature T_M, because of the heat produced in the cell due to light absorption. However, our uniform-temperature approximation is justified, because the solar cell has a relatively low thickness and its heat capacity is low w.r.t. that of the other layers [202]. This results in a very low thermal resistance of the cell to heat flow.

Further, we consider a *steady-state* situation, meaning that the module temperature will not change over time for each of the 10-minute time steps used when evaluating the model. In reality, the temperature follows an exponential decay lagging behind variations in irradiation level. The *time constant* of a PV module is the 'time it takes for the module to reach 63% of the total change in temperature resulting from a step change in irradiance' [202].

G. Fluid-dynamic model

Time constants for PV modules are generally in the order of about seven minutes. For time steps exceeding the time constant, as it is in our case, the module can be approximated to be in a steady-state condition. Under this assumption, the term on the left-hand side of Eq. (G.1) vanishes.

It is now possible to proceed with the solution of the thermal energy balance equation. The formula can be linearized by noticing that

$$\left(a^4 - b^4\right) = \left(a^2 + b^2\right)(a+b)(a-b). \tag{G.2}$$

Since

$$\left(T_M^2 + T_{sky}^2\right)(T_M + T_{sky})$$

changes less than 5%, when T_M varies 10 °C, we can consider this term to be constant [164]. Therefore the energy balance can be simplified, becoming linear with respect to T_M. By defining

$$h_{r,\,sky} = \epsilon_{top}\sigma \left(T_M^2 + T_{sky}^2\right)(T_M + T_{sky}), \tag{G.3a}$$

$$h_{r,\,gr.} = \epsilon_{back}\sigma \left(T_M^2 + T_{gr.}^2\right)(T_M + T_{gr.}), \tag{G.3b}$$

we can rewrite Eq. (G.1) and find

$$\alpha G_M - h_c(T_M - T_a) - h_{r,\,sky}(T_M - T_{sky}) - h_{r,\,gr.}(T_M - T_{gr.}) = 0. \tag{G.4}$$

By rearranging the terms, the formula can be explicitly expressed as a function of T_M,

$$T_M = \frac{\alpha G + h_c T_a + h_{r,\,sky} T_{sky} + h_{r,\,gr.} T_{gr.}}{h_c + h_{r,\,sky} + h_{r,\,gr.}}. \tag{G.5}$$

However, since $h_{r,\,gr.}$ and $h_{r,\,sky}$ are also functions of T_M, the equation must be solved iteratively: an initial module temperature is assigned and $h_{r,\,gr.}$ and $h_{r,\,sky}$ are updated after each iteration. A nearly exact solution can be obtained after five iterations.

Before we can solve Eq. (G.5) iteratively, there are still many unknown variables that must be determined. This will be done in the following sections.

G.2 Convective heat transfer coefficients

Convection is a form of energy transfer from one place to another caused by the movement of a fluid. Convective heat transfer can be either *free* or *forced* depending on the cause of the fluid motion.

In *free convection*, heat transfer is caused by temperature differences which affect the density of the fluid itself. Air starts circulating due to a difference in buoyancy between hot (less dense) fluid and cold (denser) fluid. A circular motion is therefore initiated with rising hot fluid and sinking cold fluid. Free convection only takes place in a gravitational field [203].

Forced convection, on the other hand, is caused by a fluid flow caused by external forces, which therefore enhance the convective heat exchange. The heat transfer depends very much on whether the induced flow over a solid surface is laminar or turbulent. In the case of turbulent flow, an increased heat transfer is expected with respect to the laminar situation. This is due to an increased heat transport across the main direction of the flow. On the contrary, in laminar flow regime, only conduction is responsible for transport in the cross direction. For this reason, forced convection is always studied separately in the laminar and turbulent regimes [203].

The overall convection transfer is made up of the two relative contributions from free and forced components. The mixed convective coefficient can be obtained by taking the cubic root of the cubes of the forced and convective coefficients [163],

$$h_{mixed}^3 = h_{forced}^3 + h_{free}^3. \tag{G.6}$$

Since the convective heat transfer coefficients will be different on the top and rear surfaces of the module, the total heat transfer coefficient has to be decoupled between the top h_c^T and rear h_c^B surfaces. The overall convective heat transfer will eventually be determined by the sum of the two components.

G.2.1 Convective heat transfer on the top surface

Convective heat transfer has to be distinguished in free and forced components. For the forced component we further have to distinguish between laminar and turbulent flow. We obtain for the laminar and turbulent convective heat transfer coefficients

$$h_{forced}^{lam.} = \frac{0.86\,\text{Re}^{-0.5}}{\text{Pr}^{0.67}} \rho c_{air} w, \tag{G.7a}$$

$$h_{forced}^{turb.} = \frac{0.028\,\text{Re}^{-0.2}}{\text{Pr}^{0.4}} \rho c_{air} w. \tag{G.7b}$$

Re is the *Reynolds number* that expresses the ratio of the inertial forces to viscous forces,

$$\text{Re} = \frac{w D_h}{\nu}, \tag{G.8}$$

where w is the *wind speed* at the height of the PV array, D_h is the *hydraulic diameter* of the module, which is used as relevant length scale, and ν is the *kinematic viscosity* of air. Pr is the *Prandtl number* which is the ratio between the momentum diffusivity and the thermal diffusivity. It is considered to be 0.71 for air. Finally, ρ and c_{air} are the density and heat capacity of air, respectively. The hydraulic diameter of a rectangle of length L and width W, and thus of the PV module, is given as

$$D_h = \frac{2LW}{L+W}. \tag{G.9}$$

Figure G.2 shows the forced heat transfer coefficient as a function of wind speed. We notice an overall increase in the heat transfer with increasing wind speeds due to the increased force transfer component. There are two different regions in the graph which represent the laminar and turbulent regimes. The laminar flow extends until around 3 m/s and is characterized by a lower convective heat exchange compared to the turbulent regime.

G. Fluid-dynamic model

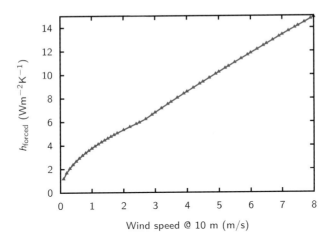

Figure G.2: Forced convective heat transfer coefficient with increasing wind speeds.

In a good approximation, h_{forced} and w are proportional to each other and we may write

$$h_{\text{forced}}^{\text{lam.}} \simeq w^{0.5}, \tag{G.10a}$$

$$h_{\text{forced}}^{\text{turb.}} \simeq w^{0.8}. \tag{G.10b}$$

To determine the free heat transfer coefficient we only need to utilize the dimensionless *Nusselt number* Nu, which expresses the ratio between the convective and conductive heat transfer [164]

$$\text{Nu} = \frac{h_{\text{free}} D_h}{k} = 0.21 (\text{Gr} \times \text{Pr})^{0.32}, \tag{G.11}$$

where k is the heat conductivity of air and Gr is the *Grashof number*, which is the ratio between the buoyancy and viscous forces,

$$\text{Gr} = \frac{g \beta (T - T_a) D_h^3}{\nu^2}. \tag{G.12}$$

Here, g is the *acceleration due to gravity on Earth* and β is the volumetric thermal expansion coefficient of air, which can be approximated to be $\beta = 1/T$.

With the value of both the free and forced coefficients we can calculate the total mixed heat convective mass transfer coefficient using Eq. (G.6):

$$h_{\text{mixed}} = h_c^T = \sqrt[3]{h_{\text{forced}}^3 + h_{\text{free}}^3}. \tag{G.13}$$

G.2.2 Convective heat transfer on the rear surface

Convection on the back side of the module will be lower than on the top because of the mounting structure and the relative vicinity to the ground. For example, a rack mount configuration, which is approximately installed at 1 m height, will achieve a larger heat exchange rate than a standoff mounted array that is mounted only 20 cm above ground. We model the effect of the different mounting configurations by scaling the convection coefficient obtained for the top of the module. We determine the scaling factor by performing an energy balance at the INOCT conditions [164],

$$\alpha G_M - h_c^T(T_{\text{INOCT}} - T_a) - h_{r,\,\text{sky}}(T_{\text{INOCT}} - T_{\text{sky}}) = h_c^B(T_{\text{INOCT}} - T_a) + h_{r,\,\text{gr.}}(T_{\text{INOCT}} - T_{\text{gr.}}). \tag{G.14}$$

We define R as the ratio of the actual to the ideal heat loss from the back side,

$$R = \frac{h_c^B(T_{\text{INOCT}} - T_a) + \epsilon_{\text{back}}\sigma(T_{\text{INOCT}}^4 - T_{\text{gr.}}^4)}{h_c^T(T_{\text{INOCT}} - T_a) + \epsilon_{\text{top}}\sigma(T_{\text{INOCT}}^4 - T_{\text{sky}}^4)}. \tag{G.15}$$

Substituting this into Eq. (G.14) at INOCT conditions yields

$$R = \frac{\alpha G_M - h_c^T(T_{\text{INOCT}} - T_a) - \epsilon_{\text{top}}\sigma(T_{\text{INOCT}}^4 - T_{\text{sky}}^4)}{h_c^T(T_{\text{INOCT}} - T_a) + \epsilon_{\text{top}}\sigma(T_{\text{INOCT}}^4 - T_{\text{sky}}^4)}. \tag{G.16}$$

The back side convection is therefore given by

$$h_c^B = R \times h_c^T. \tag{G.17}$$

We therefore find the overall convective heat transfer coefficient to be

$$h_c = h_c^T + h_c^B. \tag{G.18}$$

G.3 Other parameters

G.3.1 Sky temperature evaluation

The sky temperature can be expressed as a function of the measured ambient temperature, humidity, cloud cover and cloud elevation [164]. On a cloudy day, usually when the cloud cover is above 6 okta[1] and the sky temperature will approach the ambient temperature, $T_{\text{sky}} \approx T_a$ [202]. However, on a clear day the sky temperature can drop below T_a and can be estimated by

$$T_{\text{sky}} = 0.0552 \times T_a^{3/2}. \tag{G.19}$$

G.3.2 Wind speed at module height

Wind speed is measured with an *anemometer*. As an anemometer usually is installed at a higher height than the PV array, the wind speed experienced by the module has to be scaled down with

$$w = w_r \left(\frac{y_M}{y_r}\right)^{\frac{1}{5}}, \tag{G.20}$$

---

[1] Okta is a measure for the cloud cover, where 0 is clear sky and 8 is a completely cloudy sky.

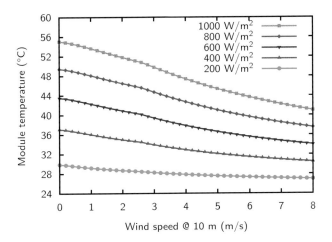

Figure G.3: Influence of the wind speed on the temperature of a solar module at different irradiance values between 200 and 1,000 W/m² and ambient temperature of 25 °C on a clear day.

where y_M and y_r denote the module and anemometer heights, respectively [164]. The elevation factor of 1/5 is determined by the landscape surrounding the installation, which is supposed to be open country.

G.3.3 PV module absorptivity and emissivity

In the thermal model discussed in this appendix, absorptivity α is defined as the fraction of the incident radiation that is converted into thermal energy in the module. This value is linked to the reflectivity R and efficiency η of the module by the equation [164]

$$\alpha = (1-R)(1-\eta). \tag{G.21}$$

A typical values for reflectivity of a solar module is 0.1 [164]. For the emissivity, we use a value of $\epsilon_{\text{top}} = 0.84$ for the front glass surface and $\epsilon_{\text{back}} = 0.89$ for the back surface [204].

Now that all the unknown variables have been determined, we now can perform the iterations in order to calculate the final PV module temperature as a function of the irradiance, wind speed and ambient temperature.

G.4 Evaluation of the thermal model

In the previous section, all the variables that are required to solve the thermal model have been derived. Therefore, the thermal model can now be solved iteratively.

Figure G.3 shows the effect of the wind speed on the module temperature at a fixed level of irradiance for irradiance values between 200 and 1,000 W/m², and an ambient temperature of 25 °C on a clear day. For the higher irradiance values the temperature is significantly above the value of 25 °C for all wind speeds. Thus, STC are not representative of real operating cell conditions. Two different regions are visible in the graph, which correspond to laminar and turbulent flow.

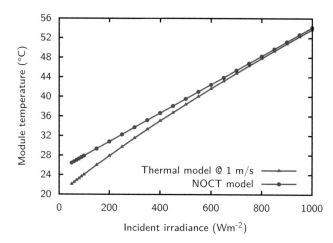

Figure G.4: The module temperature calculated with the NOCT and the thermal model with respect to the incident irradiance for 1 m/s wind speed.

In Figure G.4 the thermal model and the NOCT model presented in Eq. (20.2) are compared to each other for a situation of varying incident radiation and a constant wind speed of 1 m/s. We see that for 1,000 W/m² the NOCT model predicts a module temperature of 54.3 °C. The NOCT value is independent of the wind speed, which is above the values obtained with the thermal model (see also Figure G.3). The contribution of free convective and radiative heat exchange lowers the module temperature with respect to the NOCT model. The difference between the two models is more pronounced at low levels of radiation, but for irradiance values exceeding 1,000 W/m² the two curves start diverging again (not shown). This can be understood when we take into account that the NOCT model was developed as a linearization of the temperature/irradiation dependency around the NOCT conditions [162]. The accuracy of the model therefore decreases with the increasing distance from the NOCT.

Bibliography

[1] R. Feynman, R. Leighton, and M. Sands, *The Feynman Lectures on Physics*, Vol. 1 (Addison-Wesley, Menlo Park, California, 1963) Chap. 4.

[2] A. Kydes, (2011), http://www.eoearth.org/view/article/155350.

[3] OECD/IEA, *2013 Key World Energy Statistics* (Paris, France, 2013).

[4] TomTheHand, (2013), commons.wikimedia.org/wiki/File:Brent_Spot_monthly.svg and commons.wikimedia.org/wiki/File:Oil_Prices_1861_2007.svg.

[5] BP p.l.c., "BP Energy Outlook 2035," London, United Kingdom (2015), http://www.bp.com/content/dam/bp/pdf/Energy-economics/energy-outlook-2015/Energy_Outlook_2035_booklet.pdf.

[6] OECD/IEA, *2014 Key World Energy Statistics* (Paris, France, 2014).

[7] D. L. Turcotte, E. M. Moores, and J. B. Rundle, Physics Today **67**, 34 (2014).

[8] J. R. Petit, J. Jouzel, D. Raynaud, N. I. Barkov, J.-M. Barnola, I. Basile, M. Bender, J. Chappellaz, M. Davis, G. Delaygue, M. Delmotte, V. M. Kotlyakov, M. Legrand, V. Y. Lipenkov, C. Lorius, L. PÉpin, C. Ritz, E. Saltzman, and M. Stievenard, Nature **399**, 429 (1999).

[9] C. MacFarling Meure, D. Etheridge, C. Trudinger, P. Steele, R. Langenfelds, T. van Ommen, A. Smith, and J. Elkins, Geophys. Res. Lett. **33**, L14810 (2006).

[10] C. D. Keeling, S. C. Piper, R. B. Bacastow, M. Wahlen, T. P. Whorf, M. Heimann, and H. A. Meijer, in *A History of Atmospheric CO_2 and Its Effects on Plants, Animals, and Ecosystems*, Ecological Studies, Vol. 177, edited by I. Baldwin, M. Caldwell, G. Heldmaier, R. Jackson, O. Lange, H. Mooney, E.-D. Schulze, U. Sommer, J. Ehleringer, M. Denise Dearing, and T. Cerling (Springer, New York, NY, USA, 2005) pp. 83–113.

[11] T. F. Stocker, D. Qin, G.-K. Plattner, M. M. B. Tignor, S. K. Allen, J. Boschung, A. Nauels, Y. Xia, V. Bex, and P. M. Midgley, eds., *Climate Change 2013 – The Physical Science Basis: Working Group I Contribution to the Fifth Assessment Report of the Intergovernmental Panel on Climate Change* (Cambridge University Press, Cambridge, United Kingdom, 2014).

[12] L. Freris and D. Infield, *Renewable Energy in Power Systems* (John Wiley & Sons Inc, Chichester, United Kingdom, 2008).

[13] M. Beerepoot and A. Marmion, *Policies for renewable heat* (OECD/IEA, Paris, France, 2012).

[14] The Worldbank, (2014), wdi.worldbank.org/table/3.7#.

[15] Centraal Bureau voor de Statistiek, (2014), statline.cbs.nl/StatWeb/publication/?VW=T&DM=SLNL&PA=80030NED&D1=0,3&D2=0&D3=a&D4=14-15&HD=140818-2139&HDR=T,G1,G3&STB=G2.

[16] Ministry of Energy and Mines, "Decennial plan of Expansion of Energy 2007/2016," Brazilia, Distrito Federal, Brazil (2007).

[17] G. Masson, M. Latour, M. Rekinger, I.-T. Theologitis, and M. Papoutsi, "Global Market Outlook for Photovoltaics 2013–2017," European Photovoltaic Industry Association, Brussels, Belgium (2013).

[18] G. Masson, S. Orlandi, and M. Rekinger, "Global Market Outlook for Photovoltaics 2014–2018," European Photovoltaic Industry Association, Brussels, Belgium (2014).

[19] S. Chowdhury, U. Sumita, A. Islam, and I. Bedja, Energy Policy **68**, 285 (2014).

[20] K. Ardani and R. Margolis, *2010 Solar Technologies Market Report*, Tech. Rep. DOE/GO-102011-3318 (U.S. Department of Energy, Energy Efficiency & Renewable Energy, Washington, DC, 2011).

[21] R. P. Raffaele, *Current Trends in Photovoltaics*, Santa Barbara Summit on Energy Efficiency (The Institute for Energy Efficiency, Santa Barbara, California, 2011).

[22] V. Shah, J. Booream-Phelps, and S. Mie, "2014 Outlook: Let the Second Gold Rush Begin," Deutsche Bank, New York, New York (2014).

[23] A. Einstein, Ann. Phys. (4) **322**, 132 (1905).

[24] D. Neamen, *Semiconductor Device Physics: Basic Principles*, 4th ed. (McGraw-Hill, New York, NY, USA, 2012).

[25] P. Würfel, *Physics of Solar Cells* (WILEY-VCH Verlag, Weinheim, Germany, 2005).

[26] G. Conibeer, Mater. Today **10**, 42 (2007).

[27] W. Shockley and H. J. Queisser, J. Appl. Phys. **32**, 510 (1961).

[28] NASA/Goddard, (2012), www.nasa.gov/mission_pages/sunearth/multimedia/Sunlayers.html.

[29] "IEC 60904-3, Photovoltaic devices – Part 3: Measurement principles for terrestrial photovoltaic (PV) solar devices with reference spectral irradiance data." (2008).

[30] B. Mills, (2007), commons.wikimedia.org/wiki/File:Silicon-unit-cell-3D-balls.png.

[31] S. M. Sze, *Semiconductor Devices Physics and Technology*, 2nd ed. (John Wiley & Sons Inc, New York, NY, USA, 2002).

[32] W. Shockley and W. T. Read, Phys. Rev. **87**, 835 (1952).

[33] R. N. Hall, Phys. Rev. **87**, 387 (1952).

[34] A. Richter, S. W. Glunz, F. Werner, J. Schmidt, and A. Cuevas, Phys. Rev. B **86**, 165202 (2012).

[35] M. A. Green, *Solar cells: Operating principles, technology, and system applications* (Prentice-Hall, Inc., Englewood Cliffs, NJ, USA, 1982).

[36] T. Markvart, Appl. Phys. Lett. **91**, 064102 (2007).

[37] T. Markvart, Phys. Status Solidi A **205**, 2752 (2008).

[38] T. Markvart and G. H. Bauer, Appl. Phys. Lett. **101**, 193901 (2012).

[39] R. J. van Overstraeten and R. P. Mertens, *Physics, Technology and Use of Photovoltaics* (A. Hilger, Bristol, United Kingdom, 1986).

[40] Chetvorno, (2012), commons.wikimedia.org/wiki/File:Optical_flat_interference.svg.

[41] P. Nadar, (circa 1900), commons.wikimedia.org/wiki/File:Edmond_Becquerel_by_Nadar.jpg.

[42] AT&T Archives, (1954).

[43] D. M. Chapin, C. S. Fuller, and G. L. Pearson, J. Appl. Phys. **25**, 676 (1954).

[44] D. C. Reynolds, G. Leies, L. L. Antes, and R. E. Marburger, Phys. Rev. **96**, 533 (1954).

[45] NASA, (2008), ISS017-E-012652 spaceflight.nasa.gov/gallery/images/station/crew-17/html/iss017e012652.html.

[46] R. Bergmann and J. Werner, Thin Solid Films **403 - 404**, 162 (2002), proceedings of Symposium P on Thin Film Materials for Photovoltaics.

[47] M. A. Green, K. Emery, Y. Hishikawa, W. Warta, and E. D. Dunlop, Prog. Photovolt: Res. Appl. **22**, 1 (2014).

[48] M. A. Green, K. Emery, Y. Hishikawa, W. Warta, and E. D. Dunlop, Prog. Photovolt: Res. Appl. **23**, 1 (2015).

[49] United States Geological Survey, (2003), http://www.usgs.gov.

[50] K. L. Chopra, P. D. Paulson, and V. Dutta, Prog. Photovolt: Res. Appl. **12**, 69 (2004).

[51] "PHOTON module price index," (2012), 25 May, www.photon-international.com/newsletter/document/65647.pdf.

[52] J. Poortmans and V. Arkhipov, eds., *Thin Film Solar Cells: Fabrication, Characterization and Applications* (John Wiley & Sons, Inc., Hoboken, NJ, USA, 2006).

[53] K. Bädeker, Ann. Phys. (4) **327**, 749 (1907).

[54] T. Koida, H. Fujiwara, and M. Kondo, Jpn. J. Appl. Phys. **46**, L685 (2007).

[55] O. Kluth, *Texturierte Zinkoxidschichten für Silizium-Dünnschichtsolarzellen*, Ph.D. thesis, RWTH Aachen, Forschungszentrum Jülich, Germany (2001).

[56] P. Drude, Ann. Phys. (4) **306**, 566 (1900).

[57] N. W. Ashcroft and N. D. Mermin, *Solid State Physics, HRW International Editions* (Saunders College, Philadelphia, PA, USA, 1976).

[58] D. Mergel and Z. Qiao, J. Phys. D Appl. Phys. **35**, 794 (2002).

[59] F. Ruske, A. Pflug, V. Sittinger, B. Szyszka, D. Greiner, and B. Rech, Thin Solid Films **518**, 1289 (2009).

[60] J. Sap, O. Isabella, K. Jäger, and M. Zeman, Thin Solid Films **520**, 1096 (2011).

[61] E. Fortunato, D. Ginley, H. Hosono, and D. C. Paine, MRS Bull. **32**, 242 (2007).

[62] K. H. Wedepohl, Geochim. Cosmochim. Ac. **59**, 1217 (1995).

[63] B. Mills, (2007), commons.wikimedia.org/wiki/File:Gallium-arsenide-unit-cell-3D-balls.png.

[64] H. Binder, *Lexikon der chemischen Elemente* (Hirzel Verlag, Stuttgart, Germany, 1999).

[65] A. Tanaka, Toxicol. Appl. Pharm. **198**, 405 (2004).

[66] R. A. Street, *Hydrogenated amorphous silicon* (Cambridge University Press, Cambridge, UK, 1991).

[67] M. Zeman, "Advanced Amorphous Silicon Solar Cell Technologies," in *Thin Film Solar Cells: Fabrication, Characterization and Applications*, edited by J. Poortmans and V. Arkhipov (John Wiley & Sons, Inc., Hoboken, NJ, USA, 2006) Chap. 5.

[68] M. A. Wank, *Manipulating the Hydrogenated Amorphous Silicon Growing Surface*, Ph.D. thesis, Delft University of Technology, Delft, the Netherlands (2011).

[69] A. H. M. Smets and M. C. M. van de Sanden, Phys. Rev. B **76**, 073202 (2007).

[70] L. Houben, *Plasmaabscheidung von mikrokristallinem Silizium: Merkmale der Mikrostruktur und deren Deutung im Sinne von Wachstumsvorgängen*, Berichte des Forschungszentrums Jülich, Vol. 3735 (2000).

[71] O. Vetterl, F. Finger, R. Carius, P. Hapke, L. Houben, O. Kluth, A. Lambertz, A. Mück, B. Rech, and H. Wagner, Solar Energy Materials and Solar Cells **62**, 97 (2000).

[72] D. E. Carlson and C. R. Wronski, Appl. Phys. Lett. **28**, 671 (1976).

[73] D. L. Staebler and C. R. Wronski, Appl. Phys. Lett. **31**, 292 (1977).

[74] F. Dross, K. Baert, T. Bearda, J. Deckers, V. Depauw, O. El Daif, I. Gordon, A. Gougam, J. Govaerts, S. Granata, R. Labie, X. Loozen, R. Martini, A. Masolin, B. O'Sullivan, Y. Qiu, J. Vaes, D. Van Gestel, J. Van Hoeymissen, A. Vanleenhove, K. Van Nieuwenhuysen, S. Venkatachalam, M. Meuris, and J. Poortmans, Prog. Photovolt: Res. Appl. **20**, 770 (2012).

[75] J. Haschke, D. Amkreutz, L. Korte, F. Ruske, and B. Rech, Sol. Energ. Mat. Sol. C. **128**, 190 (2014).

[76] D. Amkreutz, J. Haschke, S. Kuhnapfel, P. Sonntag, and B. Rech, IEEE J. Photovoltaics **4**, 1496 (2014).

[77] M. A. Green, K. Emery, Y. Hishikawa, W. Warta, and E. D. Dunlop, Prog. Photovolt: Res. Appl. **22**, 701 (2014).

[78] B. Mills, (2007), commons.wikimedia.org/wiki/File:Chalcopyrite-unit-cell-3D-balls.png.

[79] R. Klenk and M. C. Lux-Steiner, "Chalcopyrite Based Solar Cells," in *Thin Film Solar Cells: Fabrication, Characterization and Applications*, edited by J. Poortmans and V. Arkhipov (John Wiley & Sons, Inc., Hoboken, NJ, USA, 2006) Chap. 6.

[80] M. Contreras, L. Mansfield, B. Egaas, J. Li, M. Romero, R. Noufi, E. Rudiger-Voigt, and W. Mannstadt, in *37th IEEE Photovoltaic Specialists Conference (PVSC)* (2011) pp. 000026–000031.

[81] H. Zhu, J. Hüpkes, E. Bunte, and S. Huang, Appl. Surf. Sci. **256**, 4601 (2010).

[82] A. Chirilă, P. Reinhard, F. Pianezzi, P. Bloesch, A. R. Uhl, C. Fella, L. Kranz, D. Keller, C. Gretener, H. Hagendorfer, D. Jaeger, R. Erni, S. Nishiwaki, S. Buecheler, and A. N. Tiwari, Nat. Mater. **12**, 1107 (2013).

[83] J. Fritsche, D. Kraft, A. Thißen, T. Mayer, A. Klein, and W. Jaegermann, Thin Solid Films **403 - 404**, 252 (2002), proceedings of Symposium P on Thin Film Materials for Photovoltaics.

[84] The Economist, "Solar power: A painful eclipse," (2011), 15 October.

[85] PV Tech, "First solar analyst day: Next major production capacity expansion in 2015," (2014), 19 March.

[86] R. U. Ayres and L. Ayres, *A Handbook of Industrial Ecology* (Edward Elgar Publishing, Cheltenham, UK, 2002).

[87] M. A. Green, A. Ho-Baillie, and H. J. Snaith, Nat. Photon. **8**, 506 (2014).

[88] E. L. Unger, E. T. Hoke, C. D. Bailie, W. H. Nguyen, A. R. Bowring, T. Heumuller, M. G. Christoforo, and M. D. McGehee, Energy Environ. Sci. **7**, 3690 (2014).

[89] G. Papakonstaniou, *Investigation and Optimization of the Front Metal Contact of Silicon Heterojunction Solar Cells*, Master's thesis, Delft University of Technology (2014).

[90] P. Babal, *Doped nanocrystalline silicon oxide for use as (intermediate) reflecting layers in thin-film silicon solar cells*, Ph.D. thesis, Delft University of Technology, The Netherlands (2014).

[91] A. Rockett, in *The Materials Science of Semiconductors* (Springer, New York, NY, 2008) Chap. 11, pp. 505–572.

[92] M. M. Hilali, *Understanding and Development of Manufacturable Screenprinted Contacts on High Sheet-resistance Emitters for Low-cost Silicon Solar Cells*, Ph.D. thesis, Georgia Institute of Technology, Atlanta, Georgia (2005).

[93] A. R. Burgers, *New Metallisation Patterns and Analysis of Light Trapping For Silicon Solar Cells*, Ph.D. thesis, Utrecht University, The Netherlands (2005).

[94] L. J. Caballero, in *Solar Energy*, edited by D. R. Rugescu (In Tech, 2010) Chap. 16, pp. 375–398, http://www.intechopen.com/books/solar-energy.

[95] A. Khanna, Z. P. Ling, V. Shanmugam, M. B. Boreland, I. Hayashi, D. Kirk, H. Akimoto, A. G. Aberle, and T. Mueller, in *28th European Photovoltaic Solar Energy Conference and Exhibition* (2013) p. 1336.

[96] A. International, "ASTM Standard B374, Standard Terminology Relating to Electroplating," (2011).

[97] N. Kanani, in *Electroplating: Basic Principles, Processes and Practice* (Elsevier, Oxford, UK, 2004) Chap. 4, pp. 87–140.

[98] Y. Huang and H. H. Lou, "Electroplating," in *Encyclopedia of Chemical Processing*, edited by S. Lee (Taylor & Francis Group, New York, NY, 2006) Chap. 81, pp. 839–848.

[99] J. Bartsch, A. Mondon, C. Schetter, M. Hörteis, and S. Glunz, in *Photovoltaic Specialists Conference (PVSC), 2010 35th IEEE* (2010) p. 1299.

[100] J. L. Hernández, D. Adachi, K. Yoshikawa, D. Schroos, E. van Assche, A. Feltrin, N. Valckx, N. Menou, J. Poortmans, M. Yoshimi, T. Uto, H. Uzu, M. Hino, H. Kawasaki, M. Kanematsu, K. Nakano, T. Mishima, R.and Kuchiyama, G. Koizumi, C. Allebé, T. Terashita, M. Hiraishi, N. Nakanishi, and K. Yamamoto, in *27th European Photovoltaic Solar Energy Conference and Exhibition* (2012) p. 655.

[101] H. El-Sayed, M. Greiner, and P. Kruse, Appl. Surf. Sci. **253**, 8962 (2007).

[102] A. A. Istratov and E. R. Weber, J. Electrochem. Soc. **149**, G21 (2002).

[103] J. You, J. Kang, D. Kim, J. J. Pak, and C. S. Kang, Sol. Energ. Mat. Sol. C. **79**, 339 (2003).

[104] J. Kang, J. You, C. Kang, J. J. Pak, and D. Kim, Solar Energy Materials and Solar Cells **74**, 91 (2002).

[105] I. Tobías, C. del Cañizo, and J. Alonso, "Crystalline silicon solar cells and modules," in *Handbook of Photovoltaic Science and Engineering*, edited by A. Luque and S. Hegedus (John Wiley & Sons Ltd, Chichester, England, 2003) Chap. 7, pp. 255–306.

[106] École polytechnique fédérale de Lausanne, PV-Lab, (2012), pvlab.epfl.ch/pv_module_design/module_reliability/efficient_encapsulation.

[107] "IEC 61215, Crystalline silicon terrestrial photovoltaic (PV) modules – Design qualification and type approval," (2005).

[108] "IEC 61646, Thin-film terrestrial photovoltaic (PV) modules – Design qualification and type approval," (2008).

[109] J. Wohlgemut, *IEC 61215: What it is and isn't*, Tech. Rep. NREL/PR-5200-54714 (National Renewable Energy Laboratory, 2012).

[110] "IEC 61730-1, Photovoltaic (PV) module safety qualification – Part 1: Requirements for construction," (2013).

[111] "IEC 61730-2, Photovoltaic (PV) module safety qualification – Part 2: Requirements for testing," (2012).

[112] "IEC 61701, Salt mist corrosion testing of photovoltaic (PV) modules," (2011).

[113] "IEC 62716, Photovoltaic (PV) modules – Ammonia corrosion testing," (2013).

[114] K. Jäger, J. Lenssen, P. Veltman, and E. Hamers, in *28th European Photovoltaic Solar Energy Conference and Exhibition* (2013) p. 2164.

[115] S. P. Philipps, A. W. Bett, K. Horowitz, and S. Kurtz, *Current Status of Concentrator Photovoltaic (CPV) Technology*, Tech. Rep. (Fraunhofer Institute for Solar Energy Systems ISE and National Renewable Energy Laboratory NREL, Freiburg, Germany and Golden, Colorado, 2015).

[116] T. Markvart, J. Opt. A-Pure Appl. Op. **10**, 015008 (2008).

[117] R. M. Swanson, "Crystalline silicon solar cells and modules," in *Handbook of Photovoltaic Science and Engineering*, edited by A. Luque and S. Hegedus (John Wiley & Sons Ltd, Chichester, England, 2003) Chap. 11, pp. 449–504.

[118] M. Schmid and P. Manley, J. Photon. Energy **5**, 057003 (2014).

[119] W. van Sark, J. de Wild, J. Rath, A. Meijerink, and R. E. Schropp, Nanoscale Res. Lett. **8**, 81 (2013).

[120] T. F. Schulze and T. W. Schmidt, Energy Environ. Sci. , (2014).

[121] J. C. Goldschmidt and S. Fischer, Adv. Opt. Mater. **3**, 510 (2015).

[122] Y. Y. Cheng, B. Fuckel, R. W. MacQueen, T. Khoury, R. G. C. R. Clady, T. F. Schulze, N. J. Ekins-Daukes, M. J. Crossley, B. Stannowski, K. Lips, and T. W. Schmidt, Energy Environ. Sci. **5**, 6953 (2012).

[123] A. Nattestad, Y. Y. Cheng, R. W. MacQueen, T. F. Schulze, F. W. Thompson, A. J. Mozer, B. Fückel, T. Khoury, M. J. Crossley, K. Lips, G. G. Wallace, and T. W. Schmidt, J. Phys. Chem. Lett. **4**, 2073 (2013).

[124] J. de Wild, T. Duindam, J. Rath, A. Meijerink, W. van Sark, and R. Schropp, Photovoltaics, IEEE Journal of **3**, 17 (2013).

[125] W. G. J. H. M. van Sark, A. Meijerink, and R. E. I. Schropp, "Third generation photovoltaics," (InTech Europe, Rijeka, Croatia, 2012) Chap. 1, p. 1.

[126] D. Jurbergs, E. Rogojina, L. Mangolini, and U. Kortshagen, Appl. Phys. Lett. **88**, 233116 (2006).

[127] N. Gupta, G. F. Alapatt, R. Podila, R. Singh, and K. F. Poole, Int. J. Photoenergy **2009**, 154059 (2009).

[128] O. E. Semonin, J. M. Luther, S. Choi, H.-Y. Chen, J. Gao, A. J. Nozik, and M. C. Beard, Science **334**, 1530 (2011).

[129] K. Tanabe, Electron. Lett. **43**, 998 (2007).

[130] M. B. Smith and J. Michl, Chem. Rev. **110**, 6891 (2010), pMID: 21053979.

[131] M. J. Y. Tayebjee, A. A. Gray-Weale, and T. W. Schmidt, J. Phys. Chem. Lett. **3**, 2749 (2012).

[132] A. Luque, A. Marti, and C. Stanley, Nat. Photon. **6**, 146 (2012).

[133] D. König, K. Casalenuovo, Y. Takeda, G. Conibeer, J. Guillemoles, R. Patterson, L. Huang, and M. Green, Physica E **42**, 2862 (2010), 14th International Conference on Modulated Semiconductor Structures.

[134] R. Clady, M. J. Y. Tayebjee, P. Aliberti, D. König, N. J. Ekins-Daukes, G. J. Conibeer, T. W. Schmidt, and M. A. Green, Prog. Photovolt: Res. Appl. **20**, 82 (2012).

[135] O. Schaefer, S. Willborn, S. Goeke, J. A. Toledo, and V. Cassagne, "Self Consumption of PV Electricity," European Photovoltaic Industry Association, Brussels, Belgium (2013).

[136] Gehrlicher Solar, (2010), commons.wikimedia.org/wiki/File:Energiepark_Lauingen_Gehrlicher_Solar_AG.JPG.

[137] Univ. of Oregon Solar Radiation Monitoring Laboratory, (2008), solardat.uoregon.edu/PolarSunChartProgram.php.

[138] F. Kasten and A. T. Young, Appl. Opt. **28**, 4735 (1989).

[139] E. Laue, Solar Energy **13**, 43 (1970).

[140] C. Honsberg and S. Bowden, (2014), pveducation.org/pvcdrom/properties-of-sunlight/air-mass.

[141] C. Rigollier, O. Bauer, and L. Wald, Solar Energy **68**, 33 (2000).

[142] H. C. Hottel and B. B. Woertz, Trans. ASME **64**, 91 (1942).

[143] J. E. Hay, Renew. Energ. **3**, 373 (1993).

[144] R. Perez, P. Ineichen, R. Seals, J. Michalsky, and R. Stewart, Solar Energy **44**, 271 (1990).

[145] P. Loutzenhiser, H. Manz, C. Felsmann, P. Strachan, T. Frank, and G. Maxwell, Solar Energy **81**, 254 (2007).

[146] PVPerformance Modeling Collaborative, (2015), https://pvpmc.sandia.gov/modeling-steps/1-weather-design-inputs/plane-of-array-poa-irradiance/calculating-poa-irradiance/poa-sky-diffuse.

[147] W. G. Rees, *Physical Principles of Remote Sensing*, 3rd ed. (Cambridge University Press, Cambridge, United Kingdom, 2013).

[148] (2014), http://files.pvsyst.com/help/albedo.htm.

[149] A. Falk, M. Meinhardt, and V. Wachenfeld, in *Proceedings of the European Conference on Power Conversion and Intelligent Motion (PCIM)*, Nuremberg, Germany (2009) pp. 14–20.

[150] S. Araujo, P. Zacharias, and R. Mallwitz, IEEE T. Ind. Electron. **57**, 3118 (2010).

[151] C. Rodriguez and J. Bishop, IEEE T. Ind. Electron. **56**, 4332 (2009).

[152] N. Mohan, T. M. Undeland, and W. P. Robbins, *Power Electronics: Converters, Applications, and Design*, 3rd ed. (John Wiley & Sons Inc, Hoboken, NJ, USA, 2003).

[153] K. Mertens, *Photovoltaics: Fundamentals, Technology and Practice* (John Wiley & Sons Ltd, Chichester, United Kingdom, 2014) texbook-pv.org.

[154] State of California, (2014), www.gosolarcalifornia.ca.gov/equipment/inverter_tests/summaries/.

[155] C. Honsberg and S. Bowden, (2014), pvcdrom.pveducation.org/BATTERY/charlead.htm.

[156] W. Böhnstedt, C. Radel, and F. Scholten, J. Power Sources **19**, 307 (1987).

[157] NASA Surface meteorology and Solar Energy, (2014), eosweb.larc.nasa.gov/cgi-bin/sse/sse.cgi?skip@larc.nasa.gov.

[158] PVGIS, (2014), re.jrc.ec.europa.eu/pvgis/.

[159] M. Šúri, T. A. Huld, E. D. Dunlop, and H. A. Ossenbrink, Solar Energy **81**, 1295 (2007).

[160] T. Huld, R. Müller, and A. Gambardella, Solar Energy **86**, 1803 (2012).

[161] J. W. Stultz, *Journal of Energy*, J. Energy **3**, 363 (1979).

[162] R. J. Ross and M. Smokler, *Flat-Plate Solar Array Project Final Report*, Tech. Rep. JPL-PUB-86-31-VOL-6 (Jet Propulsion Lab., California Inst. of Tech, Pasadena, CA, United States, 1986).

[163] E. A. de la Breteque, Solar Energy **83**, 1425 (2009).

[164] M. K. Fuentes, *A Simplified Thermal Model for Flat-Plate Photovoltaic Arrays*, Tech. Rep. (Sandia National Laboratories, 1987).

[165] J. A. Duffie and W. A. Beckman, *Solar Engineering of Thermal Processes*, 4th ed. (John Wiley & Sons Inc, Hoboken, NJ, USA, 2013).

[166] J. Randall and J. Jacot, Renewable Energy **28**, 1851 (2003).

[167] E. Lorenz, T. Scheidsteger, J. Hurka, D. Heinemann, and C. Kurz, Prog. Photovolt: Res. Appl. **19**, 757 (2011).

[168] Delft University of Technology, "Dutch PV Portal," (2014), http://dutchpvportal.tudelft.nl.

[169] W. Bower, M. Behnke, W. Erdman, and C. Whitaker, "Performance test protocol for evaluating inverters used in grid-connected photovoltaic systems," (2004), www.gosolarcalifornia.org/equipment/documents/2004-11-22_Test_Protocol.pdf.

[170] D. L. King, S. Gonzalez, G. M. Galbraith, and W. E. Boyson, *Performance Model for Grid-Connected Photovoltaic Inverters*, Tech. Rep. SAND2007-5036 (Sandia National Laboratories, 2007).

[171] Sandia National Laboratories, "Inverter library," (2014), sam.nrel.gov/sites/sam.nrel.gov/files/sam-library-sandia-inverters-2014-1-14.csv.

[172] A. Driesse, P. Jain, and S. Harrison, in *33rd IEEE Photovoltaic Specialists Conference (PVSC)* (2008) pp. 1–6.

[173] E. Sjerps-Koomen, E. Alsema, and W. Turkenburg, Solar Energy **57**, 421 (1996).

[174] J. Schmid, "Die energetische bewertung von photovoltaikanlagen," in *Themen 92/93* (Forschungsverbund Sonnenenergie, Köln, Germany, 1993).

[175] C. Baltus, J. Eikelboom, and R. van Zolingen, in *14th European Photovoltaic Solar Energy Conference and* (1997) p. 1547.

[176] S. Silvestre, "Review of System Design and Sizing Tools," in *Practical Handbook of Photovoltaics*, edited by A. McEvoy, T. Markvart, and L. Castañer (Academic Press, Boston, Massachusetts, 2012) Chap. IIA-4, 2nd ed.

[177] K. Bullis, "Solar City and Tesla Hatch a Plan to Lower the Cost of Solar Power," (2014).

[178] *Annual Energy Outlook 2014 With Projections to 2040* (U.S. Energy Information Administration, Washington, D.C., 2014).

[179] P. R. Wolfe, "Solar Parks and Solar Farms," in *Practical Handbook of Photovoltaics*, edited by A. McEvoy, T. Markvart, and L. Castañer (Academic Press, Boston, Massachusetts, 2012) Chap. IIE-2, 2nd ed.

[180] M. Bazilian, I. Onyeji, M. Liebreich, I. MacGill, J. Chase, J. Shah, D. Gielen, D. Arent, D. Landfear, and S. Zhengrong, Renew. Energ. **53**, 329 (2013).

[181] Deutsche Gesellschaft für Sonnenenergie, *Planning and Installing Photovoltaic Systems: A Guide for Installers, Architects and Engineers*, Planning and Installing Series (Earthscan, London, United Kingdom, 2008).

[182] E. Alsema, "Energy Payback Time and CO_2 Emissions of PV Systems," in *Practical Handbook of Photovoltaics*, edited by A. McEvoy, T. Markvart, and L. Castañer (Academic Press, Boston, Massachusetts, 2012) Chap. IV-2, 2nd ed.

[183] H. Vogel, *Gerthsen Physik*, 19th ed. (Springer-Verlag, Berlin, Germany, 1997).

[184] "Residential Energy Consumption Survey (RECS) 2009," (2009).

[185] K. Jäger, (2014), personal communication.

[186] V. Ramanathan and Y. Feng, Atmos. Environ. **43**, 37 (2009).

[187] C. Andraka, Energy Procedia **49**, 684 (2014).

[188] KJkolb, (2005), commons.wikimedia.org/wiki/File:Parabolic_trough_solar_thermal_electric_power_plant_1.jpg.

[189] R. Montoya, (2014), share.sandia.gov/news/resources/news_releases/images/2014/areva.jpg.

[190] B. Appel, (2005), commons.wikimedia.org/wiki/File:Dish-stirling-at-odeillo.jpg.

[191] afloresm, (2007), commons.wikimedia.org/wiki/File:PS10_solar_power_tower_2.jpg.

[192] H. Zhang, J. Baeyens, J. Degrève, and G. Cáceres, Renew. Sust. Energ. Rev. **22**, 466 (2013).

[193] F. F. Abdi, L. Han, A. H. M. Smets, M. Zeman, B. Dam, and R. van de Krol, Nat. Commun. **4**, 2195 (2013).

[194] P. Klapetek, *Characterization of randomly rough surfaces in nanometric scale using methods of modern metrology*, Ph.D. thesis, Masaryk University, Brno, Czech Republic (2003).

[195] K. Jäger, *On the Scalar Scattering Theory for Thin-Film Solar Cells*, Ph.D. thesis, Delft University of Technology, Delft, the Netherlands (2012).

[196] K. Sato, Y. Gotoh, Y. Wakayama, Y. Hayashi, K. Adachi, and N. Nishimura, Rep. Res. Lab., Asahi Glass Co. Ltd. **42**, 129 (1992).

[197] D. Showdown — D. Dominé, P. Buehlmann, J. Bailat, A. Billet, A. Feltrin, and C. Ballif, Phys. Status Solidi-R **2**, 163 (2008).

[198] O. Kluth, B. Rech, L. Houben, S. Wieder, G. Schöpe, C. Beneking, H. Wagner, A. Löffl, and H. Schock, Thin Solid Films **351**, 247 (1999).

[199] M. Berginski, J. Hüpkes, W. Reetz, B. Rech, and M. Wuttig, Thin Solid Films **516**, 5836 (2007).

[200] Astronomical Applications Department of the U.S. Naval Observatory, (2015), reproduced from The Astronomical Almanac Online and produced by the U.S. Naval Observatory and H.M. Nautical Almanac Office. http://aa.usno.navy.mil/faq/docs/SunApprox.php.

[201] I. Reda and A. Andreas, *Solar Position Algorithm for Solar Radiation Applications*, Tech. Rep. NREL/TP-560-34302 (National Renewable Energy Laboratory, 2008).

[202] A. Jones and C. Underwood, Solar Energy **70**, 349 (2001).

[203] H. E. A. van den Akker and R. F. Mudde, *Fysische transportverschijnselen*, 3rd ed. (VSSD, 2008) p. 299.

[204] M. Muller, B. Marion, and J. Rodriguez, in *Photovoltaic Specialists Conference (PVSC), 2012 38th IEEE* (2012) pp. 000697–000702.

Index

Absorber, 22, 31, 116, 126, 128
    thickness, 138
Absorption, 21, 112, 211
    coefficient, 22, 32, 67, 140, 155, 181, 190
    free carrier, 179
    parasitic, 116, 134, 184, 241, 385
    profile, 67, 68
Abundance of elements, 177, 202, 205
Acceptor, 47, 50, 56, 83, 87, 385
Adams, William G., 149
Aerosol, 43, 272
Air mass, 43, 271
    relative optical, 272
Albedo, 44, 274
Alferov, Zhores, 151
Aluminium, 168, 180, 193, 203, 230, 311
Ammonia, 232
Ampere meter, 117
Anemometer, 442
Annealing, 202
Anti-reflective coating, 141–144, 165, 168, 169
Antimony, 305, 306
Antimony telluride, 203
Arc furnace, 158
Archimedes, 377
Arctic Circle, 264
Argon, 159
Arsenic, 181
Arsine, 187
Asia Pacific, 13
Atmosphere, 7, 37, 43–44, 231, 263, 271–273, 356
Atomic force microscopy, 413
Atomic number, 48
Australia, 264

Balance of system, 16, 236
Band bending, 87
Band diagram, 52, 56, 79, 83, 90
    CdTe cell, 203
    CIGS cell, 201
    equilibrium, 85
    forward bias, 92
    hetero junction, 99
    III-V triple junction, 184
    illuminated, 95
    metal, 100
    metal-semiconductor junction, 101
    photoelectrochemical cell, 387
    reverse bias, 92
Bandgap, 22, 47, 52, 128, 131, 140, 178, 246
    direct, 22, 66–67, 138, 155, 181, 200, 203
    indirect, 22, 66–67, 155
    states, 56
Battery, 256, 261, 297–306, 323–324, 343–344, 354
    ageing, 299, 305–306
    C-rate, 303, 308
    capacity, 256, 303, 305, 323, 345
    charging, 300, 306
    corrosion, 309
    current-voltage characteristic, 343
    cycle lifetime, 304–307
    depths of discharge, 304, 345
    discharging, 300
    efficiency, 303–304
        coulombic, 304, 308
        voltaic, 304
    equivalent circuit, 302, 343
    lead-acid, 298, 300, 306, 309, 324
        flat plate, 306
        gassing, 306, 307
        rod plate, 306
        tubular plate, 306
    lithium ion, 299, 300
    lithium-ion polymer, 299
    nickel cadmium, 298
    nickel-metal hydride, 298
    overheating, 305
    primary, 298
    redox-flow, 299–300
    secondary, 298
    self discharge, 306
    state of charge, 300, 302, 304, 346
    temperature effects, 305, 308
    voltage, 303, 307
Becquerel, Alexandre-Edmond, 149
Bell Laboratories, 150
Biomass, 5, 8
Bismuth vanadate, 385, 389
Blackbody, 39, 126, 367
    radiation, 39–41, 129
Boltzmann approximation, 93, 113
Boltzmann constant, 40
Boltzmann, Ludwig, 41
Bonding model, 49–50
Boron, 50, 180, 189, 193
Bromine, 203, 210

Buoy, 344
Buoyancy, 439, 441

Cadmium, 203, 298
    toxicity, 204
Cadmium stannate, 203
Cadmium sulphide, 151, 200, 203
Calcium, 306
Calcium carbinate, 372
Calcium chloride, 372
Canada, 310
Capacitor, 289, 298
Capacity factor, 18
Carbon, 158
Carbon dioxide, 7, 43, 198, 217, 356, 357, 382
Carbon monoxide, 382
Carlson, Dave E., 151, 191
Carnot engine, 126
Carrier
    collection, 23
    concentration, 51
        intrinsic, 49, 74
    density
        free charge, 179
    majority, 51, 87
    minority, 51, 112, 401, 406, 409
        injection, 93
Catalysis, 385
    efficiency, 387
Celestial sphere, 263, 417
Chapin, Daryl M., 150
Charge, 395
    density, 32
    elementary, 4
    neutrality, 50
    equation, 55
Chemical vapour deposition, 158, 188
    atmospheric-pressure, 178, 415
    low-pressure, 178, 415
    metal-organic, 178, 187
    plasma enhanced, 162, 168, 174, 198, 217–218
China, 13, 15, 152
Chlorine, 203
Chlorofluorocarbons, 374
Climate, 7, 320
Closed-space sublimation, 204
Coal, 5, 149
Cobalt phosphate, 387
Colombia, 264
Combustion, 382
Concentrated solar power, 151, 236, 376–378
    dish Sterling, 377
    power tower, 377
Concentrator photovoltaics, 77, 127, 137, 151, 180, 186, 235–237, 240
Conductance, 279, 311
Conduction band, 21, 52

Continuity equation, 33, 61, 409, 410
Continuous random network, 189
Coordinate system
    altitude, 263
    azimuth, 37, 264
    Cartesian, 264
    declination, 419
    ecliptic, 417
    equatorial, 419
    fundamental plane, 263, 417, 419
    horizontal, 263, 419, 422
    meridian, 264
    polar angle, 37
    principal direction, 264, 417
    right ascension, 419, 425
    spherical, 264
Coordination number, 48
Copper, 201, 203, 311
Copper indium gallium diselenide, 200
Copper iron disulphide, 199
Copper sulphide, 151
Copper zinc tin sulphide, 203
Correlation length, 413
Critical angle, 31
Crucible, 159
Crystal
    lattice, 153
        cubic, 210
        defect, 72
        diamond cubic, 49, 153
        matching, 185
        mismatch, 100, 156, 187
        tetragonal, 199
        vibration, 22, 66
        zincblende, 181, 203
    momentum, 66, 153
    orientation, 153, 159
    seed, 159
Current density, 33, 79, 395
    dark, *see* Current density, saturation
    generation
        photo, 95, 112
        thermal, 93, 94
    illuminated, 407
    matching, 187, 195, 197, 240
    recombination, 93, 130, 131
    saturation, 94, 112, 118, 131, 404–405, 410, 412
    short circuit, 111–112, 117, 130, 135, 182
Current source, 117
Current-density voltage characteristic, 118
    dark, 94, 401–405
    hysteresis, 211
    illuminated, 97, 111, 405–407
    photoelectrochemical cell, 387
Current-voltage characteristic, 226, 275, 328
Czochralski process, 159
Czochralski, Jan, 150, 159

Da Vinci, Leonardo, 377
Degree of freedom, 73
Density of states, 52
    effective, 54, 74, 99
Detailed balance, 128
Diborane, 159, 198
Diesel, 259
Discharge lamp, 117
Distillation, 158
Donor, 47, 50, 56, 83, 87, 385
Drude model, 179
Drude, Paul, 179
Duty cycle, 288

Ecliptic, 417
Ecology, 357–359
    carbon footprint, 357
    energy payback time, 357–359
    energy yield ratio, 357
    life cycle assessment, 357
    pollution, 359
Economy, 351–356
    direct consumption, 354
    feed-in tariff, 152, 353
    grid parity, 355–356
    installed PV, 13–16
    learning curve, 16
    levelized cost of electricity, 354–355
    net metering, 352
    payback time, 351–352
    self consumption, 353–354
    socket parity, 355–356
Efficacy
    recombination, 73
Efficiency
    battery, *see* Battery, efficiency
    inverter, *see* Power electronics, efficiency
    PV module, 330
    round-trip, 303
    solar cell, *see* Solar cell, efficiency
    solar-to-hydrogen, 384, 387
Einstein, Albert, 21, 27, 150
Electric displacement, 395
Electricity, 5, 10
    hydro, 10, 17
    meter
        analogue, 352
        smart, 352
    wind, 18
Electrochemistry, 298, 382
Electrolysis, 9, 221, 383–384
    activation energy, 384
    loss, 384
Electrolyte, 149, 151, 210, 299
Electromagnetic wave, 28–29
    derivation, 396–397
    properties, 397–399

Electromagnetism, 27
Electron
    affinity, 85, 99, 100, 206
        rule, 99
    bond
        covalent, 48, 78
        dangling, 78, 162, 189
        hybrid, 206
    dispersion diagram, 66
        gallium arsenide, 181
        silicon, 153
    free, 49
    momentum, 22, 190
    orbital, 48, 206
    valence, 52, 78
Electron-hole pair, 22, 126
Electroplating, 221–222
    copper, 222
Emissivity, 438
Emittance, 126
    radiant, 39, 367
Energy, 3–4
    chemical, 9, 22, 126, 298
    consumption, 5, 6, 340
    density
        gravimetric, 299
        volumetric, 299
    electrical, 298
    electrochemical, 79
    excess, 128, 307
    Fermi, 53, 54, 56, 73, 85, 90
        quasi, 74, 79, 92, 95, 137
    final, 5
    Gibbs free, 382, 384
    ionization, 206
    mechanical, 9
    nuclear, 9, 17
    primary, 5, 358
    problem, 6
    radiative, 22
    renewable, 9
    secondary, 5
    solar, 10
    specific, 297
    storage, 297, 371–372, 381
    thermal, *see* Heat
    unit, 4
Energy band, 52
Entropy, 126
Epitaxy, 181, 186–187, 199
Equation of Time, 425–426
Equator, 264, 324, 381
Equivalent circuit
    two diode model, 118
Equivalent sun hours (ESH), 317, 320–321
Erbium, 241
Etching, 197

anisotropic, 168
Étendue, 235
Ethene, 206
Europe, 13, 335
European Solar Radiation Atlas, 272
European Union, 310
EVA, 229
Excess carrier, 47, 65, 410, 411
Excitation, 65
Exciton, 207
External quantum efficiency, 115–117, 135, 163

Faraday constant, 384
Fermi-Dirac distribution, 52, 53, 73
Feynman, Richard, 3
Fick's law, 168
Field
    electric, 27, 192, 395, 409
    magnetic, 27, 395
Firing, 168
Fischer-Tropsch process, 382
Float zone process, 159, 169, 170, 172, 173
Fluid, 367
Fluid bed reactor, 158
Fluorine, 180, 203
Flywheel, 298
Force, 4
Fossil fuels, 7, 368, 381
    hydraulic fracturing, 7
    refined, 5
    tar sands, 7
Fourier transform, 292
Francia, Giovanni, 151
Fresnel equations, 30, 141, 144
Fresnel lense, 236
Fritts, Charles, 149
Fuel cell, 9, 299
Fuller, Calvin S., 150

Gallium, 180, 181, 203
Gallium arsenide, 67, 69, 133, 138, 140, 151, 155, 181
Gallium phosphide, 181
Galvanic isolation, 284, 294, 296
Gas, 259
Generation, 47, 61, 137
    photo-, 67, 68, 94, 405, 406
    radiative, 67–68
    thermal, 69, 404
Germanium, 140, 155, 182, 186
    amorphous hydrogenated, 188
    nanocrystalline hydrogenated, 188
Germany, 13, 152
Glass, 192, 229
Gold, 203
Grätzel, Michael, 151
Grasshoff number, 441
Gravitation, 439, 441

Greenhouse gases, 7
Grey body, 368

Hall, Robert N., 72
Heat, 5, 365
    capacity, 365
    engine, 9, 125
    exchange coefficient, 372
    latent, 366, 372, 374, 375
    process, 368
    pump, 374
    sensible, 365
    sink, 125
    storage, 368, 371–372
        collector storage wall, 372
        packed bed, 372
        seasonal, 372
    transfer, 437–443
        conductive, 366–367
        convective, 367, 439–442
        radiative, 367–368
    trap, 149
Heating, 5
    solar thermal, 10, 368–373
Heliostat, 377
Hematite, 385
Hertz, Heinrich, 28, 150
Hofmann, August Wilhelm von, 383
Hole, 49
    transporting material, 211
Hour angle, 420
Hydrofluoric acid, 168
Hydrogen, 9, 198, 217, 298, 306, 381
Hydrogen chloride, 158, 415

IEC 61215, 231
IEC 61370, 232
IEC 61646, 231
IEC 61701, 232
IEC 62716, 232
IGBT, 295
Impedance of free space, 398
Impurity, 49, 72, 170
Index, 244
India, 310
Indium, 177, 180, 202, 203
Indium arsenide, 181
Indium oxide, 178
Indium phosphide, 140, 155, 181
Indium tin oxide, 174, 178, 208
Inductor, 289, 296, 433–435
Ingot
    monocrystalline, 159
    multicrystalline, 159–160
Insulation, 372
Intensity, 30
Interference, 141–142, 165–166, 179, 193

Internal quantum efficiency, 134
International Panel on Climate Change, 7
Inverter, 226
Iodine, 203, 210
Irradiance, 38, 114, 137, 231, 235, 276, 344
    diffuse, 271
    diffuse horizontal, 273
    direct, 269
    direct normal, 269, 271–273
    global horizontal, 274, 320
    spectral, 38, 68
Isopropanol, 168

Japan, 13, 15
Jeans, James, 40
Joule, James Prescott, 4
Julian date, 419

Kepler's second law, 417
Kepler, Johannes, 417
Kerf, 160
Kesterite, 203

Lambert cosine law, 38
Lambert–Beer law, 32, 67, 140, 155, 160
Laminar flow, 440
Lamination, 230
Lanthanides, 240
Laser scribing, 169, 232–234
Lead, 210, 298, 300, 309
Lead dioxide, 300, 309
Lead selenide, 244
Lead sulphate, 309
Lenz's law, 293, 433
Lenz, Heinrich, 293
Lifetime, 47, 156, 162, 247
    minority carrier, 70–71, 76, 137
Light induced degradation, 195–197
Light management, 116, 139–144, 155, 193, 195, 230, 370, 413
    photoelectrochemical cell, 389
    scattering, 180
Lithium, 299
Lord Rayleigh, 40

Magnesium chloride, 372
Magnesium fluoride, 169
Magnetic induction, 395
Mask, 198
Mass
    effective, 52, 73, 104, 179
Maxwell equations, 395–397
Maxwell, James Clerk, 27
Mercury, 180
Metallurgical junction, 84, 90
Methane, 198, 206, 217, 382
Midnight sun, 264

Mobility, 47, 51, 59
Molybdenum, 200, 203, 233
Monochromator, 117
Mouchot, Augustin, 149

Nameplate capacity, 205
NASA, 36, 151, 326
Netherlands, 264, 321, 326, 333, 340
Neutrino, 36
Newton's law, 367
Newton, Isaac, 4
Nickel, 298, 384
Nitrogen, 159
Nominal operating cell temperature (NOCT), 326, 444
    installed (INOCT), 327, 442
Norway, 264
Nuclear
    fission, 9
    fusion, 36
Nusselt number, 441

Oil
    crisis, 151
    crude, 5
Optical thickness, 116, 140
    Rayleigh, 273
Optics
    absorptive media, 31–32
    flat interfaces, 29–31
    geometric limit, 144
Organic materials, 205–206, 240
    HOMO and LUMO levels, 206
Overpotential, 384, 385
Oxidation, 383
Oxygen, 43, 199, 306
Ozone, 43
    layer, 374

Pakistan, 310
Passivation, 78, 164, 168, 174, 188, 189
    chemical, 162
    thermal oxide, 170
Pearson, Gerald L., 151
Penetration depth, 32, 140, 182
Permeability, 396
Permittivity, 179, 395
    complex, 31
Perovski, Lev A., 210
Perovskite, 210
Phase change, 366, 374
    material, 371–372
Phonon, 66, 76
Phosphine, 159, 187, 198
Phosphorus, 50, 168, 180, 189, 193
Phosphosilicate glass, 168
Photo diode, 115
Photoanode, 151, 385, 387

Photoelectric effect, 21, 150
Photoelectrochemical cell, 208, 387
  band diagram, 387
  current-density voltage characteristic, 387
Photon, 21, 42, 150
  flow, 39, 115
  flux, 39, 68, 111, 128
Photosphere, 36
Photosynthesis, 382
Photovoltaic effect, 21, 149
Photovoltaics, 10
  terawatt scale, 177, 202
Physical vapour deposition
  coevaporation, 201–202
  evaporation, 198, 219–220
  sputtering, 178, 197, 218–219
Planck constant, 40
Planck law, 40, 368
Planck, Max, 40
Plasma frequency, 179, 370
Platinum, 149, 210, 384, 389
Poisson equation, 32–33
Polar night, 264
Polarisation, 30
Polonium, 199
Polymer, 202
  conductive, 205
  conjugated, 206
  thermoplastic, 229
Polyvinyl fluoride, 230
Potassium, 201
Potassium hydroxide, 168
Potential
  electrochemical, 53, 85, 95
  electrostatic, 87, 89, 401
  redox, 210
Power, 4
  density, 297
  peak, 13, 111, 259
  rating, 319
  spectral, 38
Power electronics, 281–297
  continuous conduction mode, 289
  DC-AC converter, 291–295
    full-bridge, see H-bridge
    H-bridge, 291–295
    half-bridge, 294–295
  DC-DC converter, 261, 275, 281, 285–291, 433–435
    buck-boost, 289, 435
    step-down (buck), 288–289, 433–434
    step-up (boost), 289, 295, 434
  efficiency, 296–297, 320, 334–339
  harmonic content, 288, 292, 295
  inverter, 256, 261, 275, 281–282, see also DC-AC converter, 300, 334–339, 341, 345
    California Energy Commission efficiency, 335
    European efficiency, 335
    requirements, 282–283
    Sandia National Laboratories model, 335–338
    single-phase, 283
    three-phase, 283, 294
    transfomer-less, 295
  low-pass filter, 288, 292, 295
  pulse-width modulation, 288, 292, 295, 341
  steady-state operation, 289, 434
  switch, 295–296, 337
  transformer, 284, 294
Prandtl number, 440
Precursor, 186, 188, 198
Production yield, 202
Prosumer, 352
PV array, 223, 259, 283, 307, 341, 442
PV module, 223–234, 255, 259, 283, 317
  absorptivity, 443
  active area, 226, 234
  ageing, 284
  angle of incidence, 269, 426–429
  bypass diode, 227–229, 284
  efficiency, 226, 328, 331, 443
  encapsulation, 229
  fabrication, 229–230
  flexible, 234
  interconnection, 202, 323
  irradiance, 269–271, 273–274, 317, 326, 330–332, 437
  junction box, 230
  lifetime, 229
    testing, 231–232
  meteorological effects, 325–332
  operating point, see Solar cell, operating point
  parallel connection, 226
  potential-induced degradation, 296
  series connection, 223–226
  shading, 227, 270–271, 280, 285, 429–432
  temperature coefficient, 328
  temperature effects, 437–444
  thin-film, 232–234, 296
  time constant, 438
PV system
  architecture, 282–286
    central inverter, 283–284
    central inverter w. optimizers, 285
    micro inverter, 284–285
    string inverter, 285
    team concept, 286
  back-up generator, 255
  balance of system, 261
  cable, 309–311
  charge controller, 256, 261, 275, 286, 306–309, 341
    series, 308
    shunt, 308
  components, 259–261, 275–311
    battery, see Battery

cable, 261, 284, 285
charge controller, see PV systems, charge controller
DC-DC converter, see Power electronics, DC-DC converter
inverter, see Power electronics, inverter
mounting structure, 261
power electronics, see Power electronics
design, 317–347
  economic, 317
  energy balance, 317, 333–334, 339–340, 344–347
efficiency, 339
energy storage, 261, 341
energy yield, 332, 335
grid connected, 256–259, 282, 333–340
hybrid, 259
islanding, 296
load, 255, 256, 261, 333, 341, 344, 345
  profile, 325
location, 263–274
loss, 320
loss of load probability, 341, 347
maximum power point tracking, 112, 275–281, 283, 286, 291, 295, 307, 334, 339, 341, 343
  direct, 278–280
  fixed voltage, 277
  fractional open-circuit voltage, 277–278
  incremental conductance, 279–280
  indirect, 277–278
  perturb and observe algorithm, 278–279
off-grid, see stand alone
operating point, see Solar cell, operating point
performance ratio, 339
self consumption, 259
shading, 284
stand alone, 256, 282, 286, 319–325, 341–347
  days of autonomy, 323, 344
  nominal voltage, 319, 345
Pyranometer, 274, 326
Pyrheliometer, 274

Quantum confinement, 242
Quantum dot, 240, 242, 246
Quantum efficiency, 68
Quantum mechanics, 42
Quartzite, 158
Queisser, Hans-Joachim, 128

Radiance, 37–38, 235
Radiation
  blackbody, 39–41, 131
  solar, 79
  thermal, 367
Radio frequency (RF), 159, 198
Radiometry, 37–39
Ragone chart, 297, 299, 381

Rankine cycle, 374, 376
Rankine, William J. M., 376
Rare earth element, 180
Read, William T., 72
Recombination, 23, 47, 61, 112, 113, 118, 237, 404
  Auger, 76–77, 133, 138, 156, 162, 235
  CIGS cell, 201
  direct, 69–71, 131, 138, 181
  non-radiative, 72, 211, 242
  radiative, see Recombination, direct
  Shockley-Read-Hall, 72–76, 138, 156, 162, 191, 193, 201
    capture rate, 73
    rate, 75
  surface, 77–78, 407, 409–413
    velocity, 78, 162
  thermal, 70
Redox reaction, 383
Reduction, 383
Reflectivity, 30, 112, 134, 165
Refractive index, 28, 396
  complex, 31, 47, 134, 179
  grading, 141, 165
Refrigerant, 375
Relaxation time, 179
Resistance, 310
Reversible hydrogen electrode, 387
Reynolds number, 440
Reynolds, D. C., 151
Root-mean-square roughness, 413

Sandia National Laboratories, 335
Satellite, 151
Saussure, Horace-Bénédict de, 149
Sawing, 160
Schottky barrier, 83, 103–104
Schottky, Walter, 103
Screen printing, 168, 220–221
Second law of thermodynamics, 125, 235
Selenium, 177, 199
Selenization, 202
Semiconductor, 22
  n-type, 55
  p-type, 56
  capture cross section, 73
  defect density, 201
  depletion approximation, 87
  depletion region, 84, 87, 95, 97, 99, 403, 406
    width, 91, 100, 103
  diffusion, 47, 59, 84, 93, 403
    coefficient, 47
    device, 192
    length, 47, 71, 94, 97, 112, 138, 191, 403
  doping, 50–51, 55–57, 189, 198
    gradient, 387
  drift, 47, 58–59, 93, 173
    device, 192

emission coefficient, 73
equilibrium, 52, 55
II-IV, 203
intrinsic, 49, 54
junction, 47, 83–105, 295
    breakdown, 229
    emitter, 160
    forward bias, 92, 112, 401
    hetero-, 83, 97–100, 172, 201, 204, 207
    homo-, 83–97, 401–407
    localized, 172
    metal-semiconductor, 83, 100–105, 163, 164
    ohmic, 100, 104–105
    p-i-n, 191–192
    reverse bias, 92
    step, 87
    tunnel, 184–185
low-level injection, 70, 75, 77, 401, 402, 407
MOSFET, 295
non-equilibrium, 79–80
nondegenerate, 54
pentary, 203
physics, 47
quarternary, 202
quasi-neutral region, 84, 403, 409, 411
space-charge region, 47, 84, 401
steady state, 70, 406, 409, 411
ternary, 199
thermionic emission, 100, 103
transport, 58–61
trap density, 73, 135
trap state, 72, 73, 135
tunneling, 100, 105, 173
Shockley equation, 94, 405
Shockley, William B., 72, 128
Siemens process, 158, 159
Silane, 198, 217
Silica, 375
Silicon, 140
    amorphous hydrogenated, 172, 188–189
    crystalline, 47, 67, 112, 116, 138, 150, 153–156
    doping, 159
        solid state diffusion, 168
    electronic grade, 158
    ground state, 48
    metallurgical, 158
    microcrystalline, *see* Silicon, nanocrystalline hydrogenated
    monocrystalline, 156
    multicrystalline, *see* polycrystalline
    nanocrystalline hydrogenated, 188–190
    polycrystalline, 156
    ribbon, 160
    wafer, 160
        cleaning, 166
Silicon carbide, 193, 198
    amorphous, 188, 385

Silicon nitride, 168
Silicon oxide, 193, 198
    nanocrystalline hydrogenated, 189, 190
Silver, 168, 180, 193, 203
Sintering, 202
Sky
    clear, 272, 442
    model, 274
    temperature, 442
    view factor, 273
Snell's law, 30, 144
Sodium, 201, 296
Sodium sulphate, 372
Solar
    constant, 37, 272
    irradiance, 271–273
    radiation, 10
    spectrum, 42–44
        AM0, 43, 129, 185
        AM1.5, 44, 111, 114, 117, 129, 239, 263, 317, 320
        diffuse, 44
        direct, 44
        global, 44
        transmitted, 387
    tracking, 236, 371, 377
        dual axis, 377
Solar cell
    3rd generation
        hot carrier, 246–247
        intermediate band, 246
        multi-exciton generation, 244
        multi-junction, 240
        spectral downconversion, 241–242
        spectral upconversion, 241
    active area, 134
    back contact, 200
    back reflector, 193
    back surface field, 164–165, 169, 170
    buried junction, 201
    busbar, 163, 223, 232
    crystalline silicon, 153–174, 330
        design, 160–166
        fabrication, 166–169
        light induced degradation, 170
    design rule
        bandgap utilization, 131, 137–139, 156, 162, 193, 201
        light management, *see* Light management
        spectral utilization, 139, 155
    efficiency, 17, 111, 114, 136
        ultimate, 128, 129
    efficiency limit
        Shockley-Queisser, 24, 128–133, 139, 182, 239
        silicon, 133
        spectral utilization, 185, 195
        thermodynamic, 22, 125–127, 376

equivalent circuit, 117–120
fill factor, 111–114, 118, 131, 163, 413
flexible, 177
front surface field, 172
hetero junction, 172–174
history, 149–152
ideality factor, 113, 118
interdigitated back contact, 170–172
loss, 23
    collection, 135
    optical, 134, 141, 165–166, 172
    shading, 134, 163, 169
    spectral mismatch, 128–129
maximum power point, 112, 276, 284, 285, 307, 328, 331
multi-junction, 139, 151, 182–186, 193–197
    intermediate reflector, 197
    metamorphic, 186
    micromorph, 193–195
operating point, 120, 275
PERL concept, 169–170
resistance
    series, 118, 135, 163
    shunt, 118, 135
shunting, 197
temperature effects, *see* PV module, meteorological effects
thin-film
    cadmium telluride, 203–205
    chalcogenide, 199–205
    chalcopyrite, 199–203
    CIGS, 200–202
    dye-sensitized, 151, 208–211
    hybrid, 208–211
    III-V, 180–187
    kesterite, 202–203
    organic, 205–208
    silicon, 151, 188–199, 389
    substrate, 193
    superstrate, 192, 232
Solar collector, 149, 368–371
    concentrating, 371
    covered, 370
    flat plate, 371
    Fresnel concentrator, 377
    loss, 369
    parabolic trough, 149, 377
    uncovered, 370
    vacuum, 370
Solar cooling, 373–375
    absorption cooling cycle, 374–375
    desiccant, 375
    vapour compression cycle, 373–374
Solar fuels, 10, 381–389
Solar panel, 223
Solar Position Algorithm, 423
Solid angle, 37

Solvent, 375
Special relativity, 27
Spectral response, 116
Speed of light, 28, 396
Staebler-Wronski effect, 197
Standard test conditions (STC), 43, 111, 114, 317, 325, 334
Stefan, Jožef, 41
Stefan–Boltzmann constant, 41, 368
Stefan–Boltzmann law, 367
Sterling engine, 377
String, 223, 283, 285
    diode, 284
Sugar, 382
Sulphation, 305
Sulphur, 199
Sulphuric acid, 300, 384
Sun, 5, 35–37, 231, 263, 317, 437
    analemma, 425
    anomaly, 425
    mean anomaly, 419
    position, 263–264, 417–424
Superposition approximation, 407
Susceptibility
    dielectric, 179
Synthesis gas, 382

Telecommunications, 323
Tellurium, 177, 199, 203, 205
Texturing, 142–144, 166, 168, 180, 193, 197
    inverted pyramid, 169
Thermal equilibrium, 52, 131, 410, 411
Thermal velocity, 73, 78
Thermalization, 79, 128, 246
Thermodynamics, 365
Thyristor, 295
    gate turn-off, 295
Time
    apparent solar, 264, 425
    Greenwich mean sidereal, 420
    local mean sidereal, 420
    mean solar, 425
Tin, 180, 210
Tin oxide, 178, 203, 211, 415
Titanium dioxide, 208, 211, 385
Total reflection, 31
Transition metals, 240
Transmittance, 30, 134
Transparent conducting oxide, 174, 178–180, 193, 296, 370
Trichlorosilane, 158
Trimethyl
    -aluminium, 187
    -gallium, 187
    -indium, 187
Trivich, Dan, 150
Tungsten oxide, 385

Turbidity, 272
    Linke factor, 272
Turbine, 376
Turbulent flow, 440

Unit cell, 48
United States, 310, 335, 368
Uranium, 5

Vacuum
    level, 206
    ultra high, 186
Valence band, 21, 52, 173
Vernal equinox, 417, 419
Very-high frequency (VHF), 198
Viscosity, 440
Volta, Alessandro, 10
Voltage
    built-in, 90, 92, 101
    open circuit, 111–112, 130, 131, 137, 156, 170, 182, 201, 277, 285, 323, 328, 331, 334, 412–413
Voltameter, 383

Wafer bonding, 202
Water, 43, 272, 371
        splitting, 306, 382–389
            efficiency, 387
            photoelectrochemical, 384–389
Water-gas shift reaction, 382
Watt, James, 4
Wave equation, 28, 396–397
Wave vector, 66
Wave-particle duality, 42
Wavelength, 28, 68
Wavenumber, 31
Weather, 210, 230, 256, 261, 279, 297, 317, 323
Wien displacement law, 41
Wien, Wilhelm, 40
Wind, 5, 10, 259, 326, 332, 440, 442–443
Work, 125
Work function, 99, 100
Wronski, Chris R., 151, 191

Xenon, 117

Ytterbium, 241

Zenith, 264, 271
Zinc, 180
Zinc oxide, 178, 200, 415
Zinc sulphide, 169

Arno Smets studied physics at Eindhoven University of Technology, where he pursued a PhD in plasma processing of silicon. He has worked several years as PV researcher at AIST in Japan. Since November 2015 Arno is full professor within the group Photovoltaic Materials and Devices (PVMD) at Delft University of Technology. His research interests range from fundamental silicon material research, solar fuels to PV system design. Arno lectures on photovoltaics at the university and in the Massive Open Online Course (MOOC) on edX and Edraak.

Klaus Jäger studied physics at the University of Innsbruck, and obtained his diploma in physics from the Swiss Federal Institute of Technology Zurich in 2008. In 2012 he obtained his PhD degree from Delft University of Technology. After working in industry and a research position in Delft, he became a researcher at the Helmholtz-Zentrum Berlin in 2015. His main research interest is nanophotonics for solar energy, where he currently focuses on theoretical and numerical work for light-management in solar cells.

Olindo Isabella studied electronic engineering at University of Naples Federico II. He obtained his PhD degree from Delft University of Technology in the Netherlands in January 2013. During his PhD work he spent one year at AIST in Japan. Olindo now works as assistant professor in the group Photovoltaic Materials and Devices of the same university, with his research focusing on opto-electrical modelling, implementation of advanced light management techniques, and development of high efficiency wafer-based silicon solar cells. He is responsible for the PV laboratory and gives lectures on PV systems.

René van Swaaij studied physics at Utrecht University and did his PhD on amorphous silicon photoconductors at the same university. After working in the United Kingdom René came to Delft University of Technology, where he was appointed as associate professor in 2002. From 2009 onwards he works in PVMD group. His research interests lie mainly with silicon based solar-cell technology, in particular the processing of and physics underlying the operation of thin-film silicon solar cells and silicon heterojunction solar cells. René lectures on semiconductor physics and photovoltaic technologies.

Miro Zeman graduated in materials science from Slovak University of Technology in Bratislava in 1981. He received his PhD degree at the same university in 1989 for research on amorphous silicon alloys. Subsequently he joined Delft University of Technology in the Netherlands where he was appointed associate professor in 2001 and started lecturing a course on solar cells. In 2009 he was appointed full professor at Delft University of Technology for the chair of Photovoltaic Materials and Devices in the department of Electrical Sustainable Energy, of which he is the head since 2011.

Photograph of KJ by Robert Largent. Other photographs by Laya Zindel.

Also published by UIT Cambridge

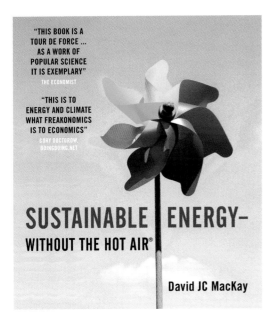

"... may be the best technical book about the environment that I've ever read. This is to energy and climate what Freakonomics is to economics."
Cory Doctorow, Boing Boing

Addressing the sustainable energy crisis in case study format, this enlightening book analyzes the relevant numbers and organizes a plan for change on both a personal level and an international scale

"Researched with long-term co-operation from industry, it emphasizes facts and evidence"
BBC News Magazine

This optimistic and richly-informed book evaluates all the options and explains how we can greatly reduce the amount of material demanded and used in manufacturing, while still meeting everyone's needs.

About UIT Cambridge

Making important topics accesible.

For our full range of titles and to order direct from our website, see www.uit.co.uk

Join our mailing list for new titles, special offers, reviews, author appearances and events: www.uit.co.uk/newsletter-subscribe

For bulk orders (50+ copies) we offer discount terms. Contact sales@uit.co.uk for details.

Send us a book proposal on sustainability, energy, popular science, etc: see www.uit.co.uk/submission-guidelines

 @ uitbooks

 /UITCambridge

Published by Green Books

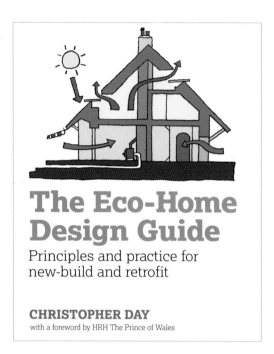

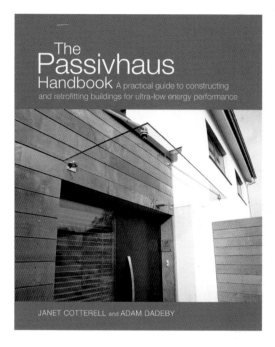

"A unique text in the study of the elements which constitutes effective, sustainable architectural design for healthy living." **Lino Bianco, Senior Lecturer, University of Malta**

This book explains the principles of eco-home design using common sense methods to create a pleasant, comfortable and healthy home.

"As we move towards the 2016 zero carbon target in house building, Passivhaus construction looks like becoming not just popular in the UK, but commonplace." **Kevin McCloud**

This bestselling practical guide takes you through everything you need, focusing on getting the building fabric right, to achieve ultra-low energy consumption.

About Green Books

Environmental publishers for 25 years.

For our full range of titles and to order direct from our website, see www.greenbooks.co.uk

Join our mailing list for new titles, special offers, reviews, author appearances and events: www.greenbooks.co.uk/subscribe

For bulk orders (50+ copies) we offer discount terms. Contact sales@greenbooks.co.uk.

Send us a book proposal on eco-building, science, gardening, etc: see www.greenbooks.co.uk/for-authors

 @Green_Books   /GreenBooks